GLOBAL DERIVATIVES

Products, Theory and Practice

GLOBAL DERIVATIVES

Products, Theory and Practice

Editor

Eric Benhamou

Pricing Partners, France

NEW JERSEY • LONDON • SINGAPORE • BEIJING • SHANGHAI • HONG KONG • TAIPEI • CHENNAI

Published by

World Scientific Publishing Co. Pte. Ltd.
5 Toh Tuck Link, Singapore 596224
USA office: 27 Warren Street, Suite 401-402, Hackensack, NJ 07601
UK office: 57 Shelton Street, Covent Garden, London WC2H 9HE

British Library Cataloguing-in-Publication Data
A catalogue record for this book is available from the British Library.

GLOBAL DERIVATIVES: PRODUCTS, THEORY AND PRACTICE

ISBN-13 978-981-256-689-8
ISBN-10 981-256-689-9

Typeset by Stallion Press
Email: enquiries@stallionpress.com

Printed in Singapore.

PREFACE

The innovation in financial derivatives over the past 20 years is historically unprecedented in its scope and speed. Callable spread options, constant proportion portfolio insurance (CPPI), target redemption notes (TARN), callable snowballs, thunderballs, collaterized debt obligation (CDO) squared, himalaya, podium, galaxy, just to name a few among the more exotic products have now become ordinary. Along the same lines, new accounting standards have urged for a better comprehension of the use and risk management of financial derivatives. And the increase of computing power has also pushed to price more accurately financial derivatives with more advanced mathematical and computational tools.

In *Global Derivatives: Products, Theory and Practice*, the authors aim to provide a comprehensive introduction to the subject from a practitioner point of view with only the minimum necessary quantitative developments. All the authors are experienced financial engineers. Most of them are working for Pricing Partners and have practised cutting edge and innovative technological developments, in order to remain in "best practice". Much has already been written on derivatives in many text books. What differentiates *Global Derivatives* from the other books on the subject is that it provides details on products, models and pricing tricks that are used in banks and often not mentioned in the academic books.

This book is designed as a graduate textbook for financial engineering course. Student should find a practitioner point of view on the subject that could be very useful when considering a carrier in derivatives either as a quant, a trader or a structurer. This book should also appeal to practitioners, to the end users of derivatives, that is, corporate treasurers, portfolio managers, hedge funds concerned by hedging financial risks or looking for investment strategies that fit their risk appetite in a given market environment such as low interest rates and low volatility.

The first two chapters are a gentle introduction to derivatives products and markets. It explains the vanilla products such as forwards, futures, standard call and put options and their use in hedging risk. The basic option strategies are also considered.

The third chapter is about the philosophy of modeling with consideration about pricing and hedging fundamentals for fixed income products, and specific considerations for exotic and hybrid structures. In particular, Chapter 3 tries to give an insight into the models people use to price or hedge and why they make this choice. It emphasizes that the quality of a model should not only be measured in terms of pricing but also hedging. In addition, it is stressed that calibration is the art of pricing and should be closely reviewed in the pricing methodology. Before discussing more advanced pricing models, Chapter 4 provides a presentation to the seminal Black–Scholes model and its underlying hedging strategy.

Then in Chapter 5, the reader will find a description of the foundation of the fixed income derivatives market, with a description on curve bootstrapping techniques, LIBOR-in-arrears swaps, constant maturity swaps (CMS) with replication method, interest rate swaps, cross currency swaps as well as vanilla interest rate options such as caps and floors and European swaptions. Leveraging on these simple products, Chapter 6 discusses the most common models to handle the smile, that is, jump models, local volatility models and stochastic volatility models, in particular the SABR model that we are using extensively in investment banks.

Chapter 7 reviews the classic term structure models with the short-term rate models (Vasicek, Hull and White, CIR and Black and Karasinski) and the market models, especially the LIBOR market model of Brace, Gatarek and Musiela (BGM) which is very popular among practitioners given its ability to reproduce market prices for caps, floors or swaptions when using Black's formula.

Chapter 8 is a detailed introduction to the growing market of inflation derivatives. It gives a description on the main street inflation structures such as year-on-year swaps, inflation bonds and zero-coupon swaps, and the recent market for caps and floors and swaptions. This chapter also gives a detailed account of pricing issues, such as the incorporation of seasonality, which are based on the original work of some of the authors of this book.

Chapter 9 tackles the rising and promising sector of hybrids. In particular, the reader will find a long discussion about power reverse dual currency (PRDC) that have been very popular in Japan. Chapter 10 provides an extensive overview of the latest innovative financial derivatives like callable snowballs and target redemption notes (TARN).

The book finishes with a chapter (Chapter 11) on the latest technologies developed by Pricing Partners. As the key to success in the derivatives business is financial innovation, it has become very relevant to build generic system where the end user can easily develop customized products and the corresponding pricing tools. This chapter describes the type of architecture which allows this flexibility, referred to as generic pricer. More precisely, a generic pricer is a tool that allows the description of any payoff without any new programming development. Once the payoff has been specified, the system assembles the algorithm to solve the pricing model. This is a very powerful architecture that can generate a first mover advantage to the most innovative institutions.

Last but not the least, in the appendix sections, the reader will first find a technical review of stochastic calculus and risk neutral pricing which is central to the pricing and hedging of derivatives transactions. While Appendix A is a rigorous presentation of the fundamentals of financial mathematics, it remains synthetic and does not get lost in unnecessary mathematical details that obscure the understanding of the basic concepts. The copula theory to model multivariate dependence structure between variables is also presented. This approach has become popular in finance to price "correlation products", that is, derivative structures whose underlying is a basket of assets such as first-to-default credit derivatives or single tranche CLOs. A detailed section on model calibration issues and the associated linear and non-linear optimization techniques complete the chapter. Appendix B provides a large review of the Monte Carlo simulation techniques with a detailed presentation of variance reduction techniques, the extension to the pricing of American Monte Carlo options and practical issues related to the derivation of the Greeks. Appendix C describes tree-based and partial differential equations (PDEs) methods, which are of great use for Bermudan and American options.

To the reader: Learn and enjoy.

Bernard Lapeyre
Ecole Nationale des Ponts
et Chaussées

ABOUT THE AUTHORS

Eric Benhamou is the CEO of Pricing Partners. Prior to this, he headed the fixed income quantitative research team at Ixis-CIB, in charge of the modeling for fixed income, inflation, FX hybrids, and funds derivatives for Europe and Asia. Previously he worked for Goldman Sachs on hybrids derivatives, mixing equities, fixed income, funds, FX and commodities. A former alumnus of the Ecole Polytechnique and the French National School of Statistics (ENSAE), he holds a PhD in Financial Mathematics from the London School of Economics and a DEA in Stochastic Calculus from University Paris VI, headed by Nicole El Karoui. He is a regular speaker at Risk conferences and has published various articles on subjects such as advanced Monte Carlo techniques with Malliavin calculus, inflation derivatives, and convexity correction computation.

René Anger is a founding member of Pricing Partners and is currently the Chief Technology Officer. After graduating from Université de Technologie de Compiègne (UTC), he started working for Sungard Infinity, a startup at that time, specialized in Front to Back systems for Interest Rates Derivatives, and worked as an IT consultant in various financial institutions around the world. Prior to joining Pricing Partners, he founded a consulting company specialized in financial markets and worked for various French banks on projects related to pricing, sensitivity calculation, IAS and EAI. René has developed an

IT system expertise on technologies like generic pricing, grid computing and application service providing applied techniques to the financial field.

Nabyl Belgrade is working in Natexis BP as a front office quant. Prior to that, he worked on inflation quantitative modeling for Ixis-CIB. He holds a PhD in Financial Mathematics and a DE in Stochastic Calculus from Paris1 Pantheon–Sorbonne University. His main works are on inflation derivatives: valuation of inflation swap, floors swaptions and hybrid options on forward real yield, seasonality estimation and impact in the pricing of the inflation derivatives. He is also an engineer in Statistics applied in Economy and worked previously on the analysis of the American oil and gas market.

Marian Ciucă is currently a quantitative engineer in Pricing Partners. Marian Ciucă graduated from University of Bucharest (Romania) and holds a PhD in Financial Mathematics from the Aix-Marseille University (France) and a DEA in Applied Mathematics. Prior to joining Pricing Partners, he was a member of the MATHFI research group, from French National Institute for Research in Computer Science and Control (INRIA), where he developed credit derivatives models for their pricer called PREMIA. PREMIA is developed by the MATHFI team in interaction with a consortium of institutions like Société Générale, Calyon, Ixis-CIB, the Crédit Industriel et Commercial (CIC), Summit Systems, EDF and GDF.

Olivier Croissant is currently a senior member of the fixed income quantitative research team at Ixis-CIB. He graduated from Ecole Polytechnique (France) and holds a post-graduate degree in theoretical physics. Prior to joining Ixis-CIB, he was principal mathematician and director of risk projects at Algorithmics, where he developed credit and commodity derivatives pricing functions for risk calculations. Olivier was previously director of financial engineering at Renaissance Software, a company later acquired by Sungard, focusing on fixed income and equity derivatives risk computations. Before starting his financial career in at CIC bank in Paris on an equity derivative desk, he developed expert systems tools in an artificial intelligence company in Monaco and Palo-Alto (California).

Mostafa Ezzine is currently working for Ixis-CIB in Tokyo in trading and structuring. Prior to that Mostafa worked in the Ixis-CIB's fixed income quantitative team in Paris and previously in Calyon.

Noufel Frikha, graduated from the Ecole Nationale Supérieure d' Informatique et de Mathématiques Appliquées (ENSIMAG), is currently a quantitative engineer in Pricing Partners, working on credit derivatives.

Richard Guillemot is currently a member of the fixed income quantitative research team at Ixis-CIB. A former alumni of Ecole des Mines de Nantes, Richard holds two DEAs, one in computing and one in stochastic calculus. Prior to Ixis-CIB, Richard worked for Calyon.

Othman Kabbaj is currently working in the marketing department of Pricing Partners. He has graduated from the Hautes Etudes Commerciales (HEC) school and is expected to join Lehman Brothers in the sales team.

Yosr Khlif is currently a member of the fixed income quantitative research team at Ixis-CIB. Prior to her role at Ixis-CIB, Yosr worked at Calyon in the risk validation group. Yosr holds a PhD in Mathematics.

Etienne Koehler is currently the Global head of Quantitative Risk Analysis at Natexis BP. Prior to this role, he has been the global head of quantitative research at Ixis-CIB, Crédit Agricole, Crédit Lyonnais. Associate professor in the University of La Sorbonne and PhD students' supervisor, Etienne is a former alumnus of Ecole Normale Supérieure. He holds an "Aggregation of Mathematics", a DEA of Mathematics and an MBA of INSEAD. He has published various articles including some in Risk magazine and co-author of several books. He is a regular speaker at quantitative finance conferences.

Leila Korbosli is currently a quantitative engineer in Pricing Partners working on equity derivatives and fixed income equity hybrids. She graduated from the Ecole Nationale Supérieure d' Informatique et de Mathématiques Appliquées (ENSIMAG).

Mohammed Miri is currently a quantitative engineer in Pricing Partners, working on fixed income derivatives. He has graduated from the Ecole Nationale Supérieure d' Informatique et de Mathématiques Appliquées (ENSIMAG).

Jean Marc Prié is currently heading the fixed income quantitative research team at Ixis-CIB. He has an extensive knowledge in developing models and worked for various banks including Crédit Lyonnais, Crédit Agricole. Jean Marc Prié holds a degree from Ecole Centrale de Paris and gives regularly lectures in various DEA and Grandes Ecoles courses.

Arnaud Schauly is currently working for Deutsche Bank in the fixed income quantitative research team. Prior to that, he worked at Ixis-CIB. A former alumnus of Ecole Centrale de Paris, he holds a DEA in Financial Mathematics from Dauphine University.

Amine Triki is currently a member of the FX quantitative team at UBS. Prior to that, he worked in the income quantitative research team at Ixis-CIB. A former alumnus of Ecole Des Mines de Paris, he holds a DEA in Stochastic Calculus.

CONTENTS

Chapter 1

STANDARD PRODUCTS AND MARKETS

This first chapter is a very gentle introduction to the subject of financial derivatives. It describes the organization of the different markets and in particular the organized markets. It also presents a trading room and the various jobs involved in it. It concludes on miscellaneous subjects like technical analysis and some notions on regulation and ethic. People already familiar with derivatives may skip this chapter.

1.1. Introduction to Financial Markets

1.1.1. *Definition of financial markets*

Financial markets are a place where buyers and sellers are ready to exchange various types of financial securities or products. Financial markets enables a variety of corporations and individuals to facilitate short- but also long-term purchase of assets and risk transfer. Financial markets are commonly divided into:

(1) capital markets in turns divided into:
   (a) stock markets, that facilitates equity investment and buying and selling of shares or common stock,
   (b) bond markets, that provides financing through the issue of debt contracts and the buying and selling of bonds and debentures;

(2) money markets, that provides short-term debt financing and investment;

(3) derivatives markets, that provides instruments for handling of financial risks;
(4) futures markets, that provide standardized contracts for trading assets at some forward date (see also forward market);
(5) insurance markets, which facilitates handling of various risks;
(6) foreign exchange markets.

Newly formed (issued) securities are bought or sold in primary markets. Secondary markets allow owners of the security or the product of securities to buy or sell the same.

1.1.2. *The different derivatives markets*

Within the financial markets, one distinguishes traditionally (see Fig. 1.1):

(1) the organized or exchange traded markets for instance the Chicago Board of Trade (www.cbot.com), the Chicago Mercantile Exchange (www.cme.com), the London International Financial Futures Exchange (LIFFE) www.liffe.com which is a part of Euronext group since June 2002.
(2) the non-organized markets called over-the-counter (OTC) markets.

Organized markets (stock exchange with listed products such as stocks, futures, bonds and various options)

Non-organized markets (inter dealer market) referred to as over-the-counter (OTC) market

Figure 1.1: The two markets types.

Cash markets also known as underlying markets (stocks, interest rates, FX, commodity). Fast trading. Very scarce arbitrage.

Derivatives markets
Products that are derived from the cash market and correspond to options and futures markets

Figure 1.2: Cash and derivatives markets.

One also makes the distinction between (see Fig. 1.2).

(1) the cash market where traditional securities, like stocks and indexes are traded;
(2) the derivatives market, which are markets derived from the more traditional ones, as the value of their financial instruments depends on the value of other, more basic and often non-divisible underlying instruments.

Typical derivatives are futures, forward, swaps, call put and warrants. More generally, a derivative security is a financial asset whose value depends on the value of some other assets called underlying of the financial derivatives.

1.1.3. *Importance of the derivatives*

In the recent years, the derivatives business has become increasingly important in the world of finance, being for some underlying securities more liquid, that is to say, having trading volume larger than their underlying markets.

In order to give an idea of the importance of the derivatives market, let us just mention that the total notional value of derivatives

(OTC and exchange based) has been around 400 trillion dollars for the year of 2004 according to the Office of Controller of the Currency (OCC). In comparison, the total market capitalization world wide has only been around 10 trillion dollars in 2004 (*source*: the United Nations) while the GDP of the United States for the same time has only been about 10 trillion of dollars (*source*: CIA). This means that the volume of the derivatives business represents 40 times the total market capitalization and the US GDP. And as shown by Fig. 1.3, the growth of this derivatives market has been impressive over the last years. It has almost double in the last 5 years. When talking about individuals sector, we will show the different components of the derivatives business (Table 1.1).

The fact that the market has been growing very rapidly does not preclude for a bubble burst, hence a greater responsibilities from regulators and controllers to control and preserve the system.

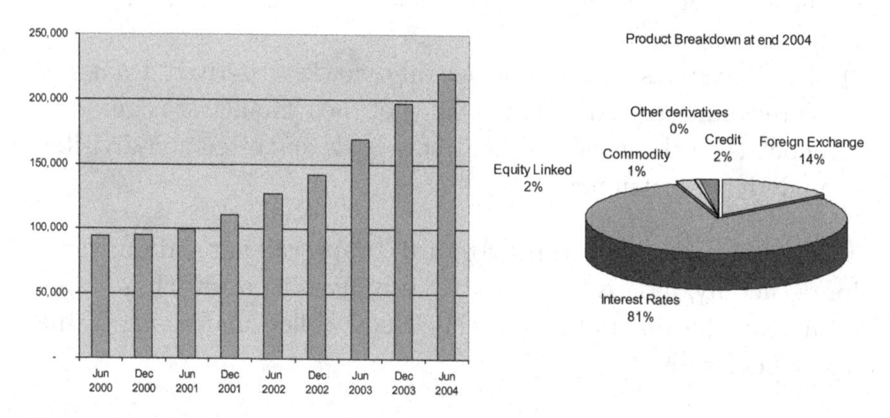

Figure 1.3: Semi-annual total volume of OTC derivatives and corresponding breakdown for end 2004 in billion dollars.

Table 1.1: Semi-annual total volume of OTC derivatives in billions dollars.

Jun 2000	Dec 2000	Jun 2001	Dec 2001	Jun 2002	Dec 2002	Jun 2003	Dec 2003	Jun 2004
94,008	95,199	99,755	111,115	127,564	141,655	169,658	197,167	220,058

1.1.4. *Cash or spot vs. derivatives markets*

A derivative (or financial derivative or financial derivative security) is a financial instrument whose value depends on the value of other more basic underlying variables. The main categories of derivatives are options, futures and forward contracts. Obviously, there exist derivatives of derivatives.

An option is a contract that gives the buyer the right, but not the obligation, to buy (call) or sell (put) some asset(s) at a specified price (the strike) for a specified period of time or point in time. The financial contract of an option contains some rights, hence the terminology. On the contrary, a forward or futures contract is a contract in which the buyer agrees to buy or sell some asset(s) at a specified time for a specified price. What distinguishes options from forwards and futures is precisely the fact that the holder is not forced to buy or sell the underlying asset(s). However, there is a cost in entering an option contract whereas there is no cost in entering a forward or futures one.

Forwards and futures represent a very large part of the derivatives industry. According to the International Swaps and Derivatives Association, Inc (ISDA) survey, forward and futures (including swaps) represent three quarters of the derivatives in notional terms. In practice, many structured deals are a combination of options and other more traditional assets, like swaps or stocks.

There are two main categories of exchange traded options: listed and futures options. Listed options are traded on organized exchange like the Nasdaq (www.nasdaq.com), New York Stock Exchange (www.nyse.com), the London Stock Exchange (LSE, www.londonstockexchange.com) (see Chapter 3 for more details on options). This concerns any type of underlying like equities, equity indexes, currencies, etc. Usually one can find the quotes in the newspaper. The contract can be settled in cash or physical underlying.

Options on futures, traded on a futures exchange, right next to the futures pit, like in the CBOT, LIFFE, etc. concern other type of underlying like commodities (crude oils, electricity, natural gas, aluminum), bond (US Treasury and Bunds), currency futures, etc. The major exchanges are the Chicago Board of Trades (CBOT), the Chicago Mercantile Exchange (CME) and the Philadelphia Stock Exchange (PHLX).

One can find quotes for these options in the newspaper. The big difference is that this type of options is always settled in underlying futures contracts.

Derivatives embody more risk than standard products because of their intrinsic leverage. They are also addressed to specific needs and clients. As a consequence, exchanges have been split between cash markets and derivatives. Moreover, the options and futures markets have specific requirements in terms of minimum margin capital and reserve in order to trade.

1.1.5. *History of financial markets*

Derivatives were originally used to hedge commodities products such as agricultural produce and metals. Indeed, derivatives market can be traced as far as the Antique Greece when farmers may sell in advance the olive crop on a forward basis. Indeed, options concept had already been invented and Thales used them to make a fortune as described in the book of Aristotle "Politics". In the middle age, it was not uncommon for traders in Flander and Champagne to use forward contract, by means of the so called "lettre de faire" where the quality standard and the price were agreed in advance. Later on, in the 17th century, Amsterdam grew an important financial center, which ended up in a large scale financial collapse through the burst of the tulip forward selling activity. Similarly, but with less speculation, Japan developed early derivatives markets, based in Osaka, dealing with agricultural products, especially on rice.

But the real birth of modern futures markets started in the United States, with the creation of the ancestor of the CBOT in 1848. The early expansion went fast. The idea of standardized forward dealing started to appeal to more and more people. Later on, the New York Cotton Exchange was founded in 1870, followed by the Butter and

Cheese Exchange of New York in 1972 and the Chicago Produce Exchange. The period of consolidation came during the First World War when activity was scaled down, followed by the 1929s crisis. The depression took its toll on various exchanges. This led first to more rationalization of exchanges as well as better regulation. Political pressures to regulate futures markets led in 1933 to the commodity exchange act that established more rigorous rules for futures trading exchange.

But the real birth of financial markets can be said to have come really in the 1970s. All the elements were ready for modern financial markets. First of all, the system of fixed exchange rates, which pegged the major currencies to the dollar, which in turn was pegged to gold, went burst. On 15 August 1971, President Nixon decided to abolish the Bretton Woods agreement. With this move, currencies suddenly floated free and were subject to supply and demand like any other product, pushing for more freedom to trade. Second, the exchanges like the CBOT or the CME had decided to standardize their products, making the market more liquid as well as expanding the number of products that they were trading. Options started to become more and more popular although they have been around since the early beginning of the 20th century.

Derivatives markets were ready for their explosion as futures exchanges extended from the United States to Europe and Asia in the 1980s with the opening of various derivatives exchange market such as the LIFFE, the MATTIF (Paris exchange market), which later on merged with the Amsterdam, Bruxelles and Lisbonne exchange markets into the Euronext market, the German market (the Deutsche Börse) which later on merged with the Swiss Exchange into the Eurex.

Undoubtedly, derivatives markets have an important role in the financial markets as they complete underlying markets, enabling us to have more tailor made transactions for hedgers and risk speculators. Because of the complexity of those products, models are needed to try linking and complementing the available information in order to build a consensus of all market players' opinion about the Market's future evolution.

1.1.6. *Listed markets vs. OTC markets*

Exchange traded markets have been created to standardize contract and support the counter party risk through their compensation rooms. On the contrary, in the OTC markets, market players bear the full counterparty risk. The main listed markets are:

(1) EUREX: German's (Frankfurt) future contracts (Bobl, Bund...);
(2) CBOT: American's (Chicago) future contract market;
(3) LIFFE: English's (London) future contract market (Swapnote);
(4) Euronext: previously (Paris) future contract market (Pibor, Notionnel).

Deposit account vs. capital allocation: Deposit account and margin calls process ensure listed markets system to cover for liquidity risk and default risk; mechanism of limit stop quotation prevents the system to derive in a bad spiral.

Other products need to be treated with another designed market player. As the deal engaged the two contractors in future exchange (swap flows, option exercise,...) it can be stopped by the default of one the two before its expiry. To prevent this risk, additional fee, depending mainly of the credit quality of the counterparty and the exposure involved, is charged in these deals.

1.1.7. *Shares and dividends*

A stock, an equity or a share stock is a financial product that provides the ownership of a small piece of a company. When raising capital, company sells the future profits in the form of a stake in the new company. The investor, the share holder, who may be a private investor, institutional, or investment company own the company. There are many types of shares. It is only when a company has satisfy a certain number of criteria, like their capitalization, the number of shares issued and the official publication of information that it can be listed in a given stock market. Typical examples of stocks are Microsoft, Cisco, IBM, General Motors, etc.

Stock markets have often different levels or submarkets to allow for different market capitalization and company types. The advantages of raising capital via the stock market are the low cost and the efficiency. Another constituent of the stock is its dividend. Since the shareholder has a stake in the ownership of the company, he/she is entitled to receive parts of the profits of the firm, which are materialized by the dividends. The value of a portfolio invested in stocks will therefore be not only affected by the change of the price of the stock but also its dividend.

The amount of dividend varies from year to year according to the benefits of the company and its financial strategy. It is decided by the board of directors of the company, usually set a month before being paid. The date at which the amount of dividend is publicly known is called the fixing date. The two others important dates are the settlement date and the pay date. The stock trades-cum-dividend up to the settlement date, meaning that a stock holder will receive the dividend if he/she holds the stock up to the settlement date. He/she will be paid at the pay date. Obviously, after the settlement date, the price of the stock needs to reflect that the dividend will not be paid. As a consequence, after the settlement date, the equity price will drop to offset the disadvantage of receiving not the dividend. The drop will in practice be somehow different from the dividend to reflect the different tax treatment of dividends and stocks. Dividends are considered to be income whereas stocks are considered to be capitals. Stocks trading with dividend are called cum-dividend, whereas when without, ex-dividend. Last but not least, dividends are considered to be a signal about the health of the company. As a rule of thumb, companies try to maintain a constant level of dividend. The price per earning (PER) is the percentage of the dividend compared to the equity price. It indicates whether a company is offering a high dividend or not. We will describe in more details the dividends in the chapter about modeling as it affects substantially the price of equity option. As the price of the stocks evolves, it may reach levels that are very far from its original level. To readjust for this growth, it is not rare to see stock splits. A stock split (like a three-for-one) consists in giving to each stock holder for each share a multiplicative number of shares (three in the example). Stock split should be considered in a derivatives contract.

The total number of stocks is the outstanding number of shares. The outstanding number measures the number of shares potentially in circulation in the financial markets. The open interest is the total number of transactions (buy and sell). It indicates the depth and in a sense the liquidity of the market. The market capitalization of a company is the value of the outstanding number of shares.

Obviously, the behavior of the stock is far from being predictable, leading to major trading successes and losses. We will see that to model a financial derivative, we will assume that the behavior of a stock is somehow predictable (in its drift term) and somehow random, accounting for the fact that no one can really predict perfectly a market.

Last but not the least, buying a stock (and in general a financial asset) is said to assume a long position, while selling it is said to assume short position.

1.1.8. *Typology of the markets participants*

First of all, it is interesting to make the difference between market players and controllers, internal and external, basically regulators. We will see in the following section in more details the market players through an overview of the trading room. But very rapidly, market players (and their staff) are divided into:

- investor
- broker
- trader
- sales
- structurers
- quants and IT
- middle and back office.

1.2. Presentation of the Trading Room and Group Description

(1) *Sales/marketers/originators, traders*: They are the trading room's interface with clients and brokers.
(2) *Support departments*: They work closely with the traders for pricing and hedging: quantitative research, macro economic research, IT platform.
(3) *Middle-Office*: They help to confirm by checking with the counterparty the deal terms and help the trader to plug it in the pipe.
(4) *Back-Office*: They are following the deal until its maturity (repository, cash-flows, ...).

1.2.1. *Different functionality of the trading room*

The trading room is the interface between final client and financial markets. It uses the investment bank's fund to make money out of it. There are three categories of traders: speculators, hedgers and arbitrageurs.

1.2.2. *Job description*

1.2.2.1. *Trader*

The trader is the trading room's interface with brokers on line (and/or with other trading rooms). He has to handle trading room's book positions and its proper hedging. He/she can take view on the market according to certain risk limit agreed in advance with senior management, to deliver a certain objective. The issue at stake for hedgers is to reduce the risk due to the different movements of market variables whereas speculators take short/long positions because they think prices will go up or down. Arbitrageurs make profit by making transactions into several markets without an initial setting. In practice, there are just a few of arbitrage opportunities. Thereafter, in this book, most of arguments are built on the hypothesis that there are no arbitrage opportunities. Traders are closely monitored not only by the internal risk control group through daily VAR limits, agreement on new products and risk committee but also by the trading group's heads. It is his mandate to control the trading group position and to take strategic decision on risky positions.

1.2.2.2. *Sales/originators/structurors*

They are the trading room's interface with client. This can be in the two ways as clients may be interested in both buying and selling. They handle offer and demand between the final client's supply and the traders' book. On the one hand, sales want to help client to access to financial markets, through its trading room. On the other hand, they also take into account trader's will to certain risk and exposition of their book and try to convince clients to have the opposite position. In a sense, they make the market more fluid by intermediating risk takers.

1.2.2.3. *Quant/IT/analyst support*

Quants, IT, analyst and strategist work ahead of the process to provide traders with a variety of tools for a better assessment of their risk and the preferred hedging strategy. They are strategic assets to promote new products both through their pricing and hedging and the measurement of this new product in terms of aggregated risk with the current one of book positions.

Quants and IT department are deeply involved in the trading orientation of the bank by providing models and system tools.

Another strategic department is micro/macro analysts. They provide global micro/macro economic views to forecast trends and market momentum. They use forecasting indicators to help traders to handle specific position.

1.2.2.4. *Middle office/back office support*

They handle the confirmation of the deal, by checking it with the counterparty, ensuring the proper booking of the deal and following the deal's life in the systems till its end.

1.3. Flow Business, Prop Trading and Exotic

1.3.1. *Definition*

Flow business refers to trading done for the client's account that concerns large trading volume and plain vanilla products. Plain vanilla products refer to mainstream product. It alludes to the vanilla flavor that is the most common one for ice cream. By contrast, complex derivatives are referred to as exotic. Proprietary trading refers to any transactions done by a securities firm that affects its accounts and not the one of its client.

1.3.2. *Hedge fund business*

Over the last 10 years, a growing business has become hedge funds business. Those companies are small ones (usually only a tens of people) that collects and invests client's funds as well as their own funds with specified and agreed long-term strategy. They exhibit usually a high risk/reward profile as they are taking huge bets on various markets. They act in various markets and have much more freedom in terms of investing their funds than their mutual funds counterparty.

1.3.3. *Key differences*

Books are mainly driven by clients supply. Bid/ask quotation spreads are tight when flows are great, frequent and two ways sided. P&L is built on small margin based on huge cumulated flows.

On the contrary, when flows are rare or one sided, positions cannot be hedged statically or netted in a book so its hedging process should be done more frequently and/or imperfectly. The bid/ask spread should be larger to take in account transaction costs and/or imperfect hedging.

Books are initiated by the prop trader's opinion of the evolution of the market. Positions can be easily marked to markets. For complex transaction, these may need to be marked to model, as opposed to marked to model to point out that the mark to market given daily is bound to a specific model. In this particular case, there is some risk that the model may be misleading and this is referred to as model risk.

Exotic business needs power models to handle market players about their consensus of what the future should tend to. Some risks cannot be easily sold to the market (and have to be provisioned and scrutinized quite carefully).

Bid/ask quotations should be larger to take account of imperfect day by day Greeks hedging (transaction costs and/or unexplained P&L). Positions are hedged with liquid instruments (case of vanilla products) directly or through its Greeks projection (exotic case).

Generally positions need more capital allocation. This is why prop trading business requires usually more capital than client-oriented business.

1.3.4. *Market making*

Market makers make the commitment to be seller or buyer position on specified range of products. They stand ready to buy and sell a particular stock on a regular and continuous basis at a publicly quoted price. Their engagement is generally expressed as small, bid/offer published quotation. Brokers are good medium of exchange for market makers.

1.3.5. *ECN and electronic trading platform*

Electronic platform, as well as e-trading, is the recent ways to fluidize inter-market players links. The key points are:

(1) insurance that real-time trader's quotation platform has no bug and is correctly fed;
(2) each new deal, once entered, should alert the trader and update its position;
(3) limits (stop loss, cumulated nominal amount, cumulative risk position...) should be present and activated, when needed, on the traders platform.

1.4. Ethic and Deontology

1.4.1. *History*

Bankruptcies history highlights the systemic links risk in the financial industry. We will review some famous derivatives stories.

1.4.1.1. *Barings*

The collapse of Britain's premier bank Barings bank in 1995 is probably one of the most spectacular debacles in the derivatives industry as it took one man to make a solid bank bankrupt. The failure was completely unexpected. This struck the financial community by complete surprise as this happens in a couple of days. Barings was the Britain's oldest merchant bank. It has a strong reputation. And it has been a single man, Nick Leeson, that made this giant collapse.

Lesson had done some good work in the Jakarta office of Barings and was rapidly promoted to be simultaneously head of trading and head of back office for Barings in their Singapore office. Such a position should have rung alarm bells, but the conflict of interest and lack of control went unnoticed within the Barings' senior management.

Leeson and his traders had authority to perform two types of trading:

(1) transacting futures and options orders for clients or for other firms within the Barings organization;
(2) arbitraging price differences between Nikkei futures traded on the SIMEX and Japan's Osaka exchange.

Perhaps it was the inherent lack of risk in such trading that prompted people to not to be concerned about Leeson wearing multiple hats.

Leeson took unauthorized speculative positions primarily in futures linked to the Nikkei 225 and Japanese government bonds (JGB) as well as options on the Nikkei. He hid his trading in an unused BSS error account, number 88888. Exactly why Leeson was speculating is unclear. He claims that he originally used the 88888 account to hide some embarrassing losses resulting from mistakes made by his traders. However, Leeson started actively trading in the

88888 account almost as soon as he arrived in Singapore. The sheer volume of his trading suggests a simple desire to speculate. He lost money from the beginning. Increasing his bets only made him lose more money. By the end of 1992, the 88888 account was under water by about GBP 2MM. A year later, this had mushroomed to GBP 23MM. By the end of 1994, Leeson's 88888 account had lost a total of GBP 208MM. Barings management remained blithely unaware. On February 23, 1995, Leeson hopped on a plane to Kuala Lumpur leaving behind a GBP 827MM hole in the Barings balance sheet.

As a trader, Leeson had extremely bad luck. By mid February 1995, he had accumulated an enormous position — half the open interest in the Nikkei future and 85% of the open interest in the JGB future. The market was aware of this and probably traded against him. Prior to 1995, however, he just made consistently bad bets. The fact that he was so unlucky should not be too much of a surprise. If he had not been so misfortunate, we probably would not have ever heard of him. What really grabbed the world's attention was the fact that the failure was caused by only one man, a trader based in Singapore.

1.4.1.2. *ENRON*

The history of ENRON lies in wrong corporate statement rather and misuse of marked to model that was reported as marked to market. It is more a story of false accounting report than derivatives misuse.

The bankruptcy came in 2001 when the whole Enron empire collapsed. It had been the largest one in US history just 7 months before WorldCom's spectacular bankruptcy.

Bankruptcy came because of various problems:

(1) First, the buying of new power plant and companies consisted in huge investment whose profitability or losses would only be realized 10 or 20 years afterwards.
(2) Second, the large trading activity plus the launch of Enron Online fostered by the internet booming environment made the Enron share price skyrocket giving the company even more cash to invest into dubious and very expensive projects.

(3) Third, complex trading position were marked to model but these were reporting to be marked to market, with very little questioning by the internal risk control group.
(4) Fourth, ENRON, leveraging on its client power with regards to its external auditor Arthur Andersen, forced it to write off his accounting report although the accounting firm has been aware of some wrong reporting.
(5) Fifth, ENRON accumulated large trading bullish position that led to huge losses when the market turned off from the technology bubble to the bear market, bursting the Internet euphoria bubble.

The firm's bankruptcy was the largest in US history, it was only surpassed 7 months later by the WorldCom's bankruptcy.

1.4.1.3. *LTCM*

The long-term capital management (LTCM) is the story of the burst of a star hedge fund. Its founder, a star trader, John Meriwether, had made a strong reputation as one of the best Salomon Brothers bond's trader. Encouraged by its previous trading success, he started his own hedge fund called LTCM. The fund was taking very high leverage position, using sophisticated product as well as sharp quantitative models. But when the crisis in Russia blew up, their highly leveraged position wiped up all the cash of the fund and making it bankrupt.

Their main trading strategy, convergence trades involved buying cheap security and selling the expensive one, betting on the convergence between the two assets. These trades were of four types:

(1) convergence among the US, Japan, and European sovereign bonds;
(2) convergence among European sovereign bonds;
(3) convergence between on-the-run and off-the-run US government bonds;
(4) long positions in emerging markets sovereigns, hedged back to dollars.

Because of tiny differences in prices, the fund was forced to take large and highly leveraged positions as well as to bet on the convergence through complex model. But the fund had underestimated various risk such as:

(1) accounting for liquidity as a real risk and have cash in case of large market moves;
(2) making the difference between marked to model and marked to market;
(3) accounting for large market movement through stress test of their model;
(4) accounting for systemic risk and market collapse by aggregating exposure to common risk factors.

Finally, LTCM's bankruptcy was caused by massive leverage and lack of liquidity. At the end of September 1998, the fund had lost most of investor equity capital and to avoid the threat of a major systemic crisis of the financial market, the Federal Reserve Bank bundled a $3.5 billion rescue package from various investment and commercial banks in exchange for 90% of LTCM equity. *When Genius failed*, by Roger Lowenstein, is the story of LTCM's failure. The Genius mentioned in the title refers to Myron Scholes and Robert Merton (Nobel Prize in economics in 1997).

1.4.2. *Insider trading*

Key indicators calendar announcements: source of bet. There is great temptation to try having in advance indications on the market evolution: history can help detecting an abnormal situation or trend change (charting activity).

Some are tempting to get directly information from those who are building it! If this happened it gives asymmetric information around players and that is no good for the long-term market equilibrium.

"Insider trading" is usually associated with illegal conducts by most investors. But this is not exactly true. Corporate insiders, officers, directors and employees, can buy and sell stock in their own companies as long as they report their trades to the SEC.

Illegal insider trading refers generally to buying or selling a security, in breach of a fiduciary duty or other relationship of trust and confidence, while in possession of material, nonpublic information about the security. Insider trading violations may also include "tipping" such information, securities trading by the person "tipped", and securities trading by those who misappropriate such information.

1.4.3. *Money laundering*

With the recent emphasis on terrorism, money laundering has become a major focus of financial regulators. The sources of money laundering come many illegal activities such as drugs, terrorism financing and other illegal business.

Deontology information and rules are provided to each market player to prevent the proliferation of this plague. Now compliance has been identified by regulatory instance to be full part of the information system beside risk control.

1.5. Summary

Financial markets have been growing for the last 25 years. They are commonly divided into capital markets, derivatives markets, money markets, futures markets. They provide a wide range of instruments from the standardized contracts (forwards, options, . . .) to more exotic ones.

In this chapter, we have seen briefly the organization of those markets and the different participants of a trading room: traders, brokers, sales, structurers, quants and IT, middle and back office. Bankruptcies history shows that the mechanism of financial markets is not so easy and can lead to the crash or the collapse of a hedge fund or a bank. The development of market institutions would diminish the extent of market manipulation and systemic risk.

References

[1] Chance D (1997). *An Introduction to Derivatives*, 4th edn. Dryden Press.
[2] Hull J (2002). *Options, Futures, and other Derivative Securities*, 5th edn. New York: Prentice Hall.
[3] Lowenstein R (2001). *When Genius Failed: The Rise and Fall of Long-Term Capital Management.* Random House Trade Paperbacks.
[4] Wilmott P (2000). *Paul Wilmott Introduces Quantitative Finance.* New York: Wiley.

Chapter 2

THE VANILLA PRODUCTS

In the 1970s, the overall level of volatility of the common assets, that is to say, stocks, interest rates and exchange rates increased considerably. At the same time, financial markets started trading new instruments referred to as derivatives[1] that were adapted to manage in a more efficient way the risk of fluctuation in price of these assets. Arbitrageurs and speculators started using the instruments to get extra leverage, either because they were betting on future movements in asset prices or they wanted to take advantage of markets inconsistency.

Futures, forwards and options contracts are the most common derivatives. Futures contracts are traded in organized markets like CBOT, CME, or MATIF. On the contrary, forwards and options are traded in the OTC markets.

In Section 2.1, we describe forward futures and options contracts. We then explain the interest and usage of these derivatives, both in terms of hedging, speculation and/or arbitrage. In Sections 2.2 and 2.3, we give more details on these derivatives, namely their different types, the way they are priced and traded.

[1]As described in the workmanship of J. De La Vega, the market of futures and options existed in the Amsterdam exchange, since the 17th century. The first options on stocks were first traded on the organized exchange in 1973.

2.1. Interest and Usage of Products

2.1.1. *Interest and usage of products*

Forward, futures and options contracts all share one thing in common. They all have a delivery date, at which the price of the underlying is crucial to determine the profit and loss of the contract. Hedgers, speculators and arbitrageurs use these contracts to take views on the direction of the underlying asset. Their risk on the future price of the asset can be highly rewarded if their prediction is realized.

2.1.2. *Derivatives description*

A forward contract is an agreement between two counter parties to buy or sell at a specified future time an asset for a certain price (the delivery price). The party that agrees to buy the underlying asset assumes a long position on the contract; the other party assumes a short position and consents to sell the asset on the same future date for the same delivery price. At the time the contract is entered to, the delivery price is chosen so that the forward contract costs zero for both parties.

A future contract is a form of a forward contract to buy or sell any kind at a pre-agreed future point in time. This has been standardized for a wide range of usages. Future contracts are different from forward contracts in terms of margin and delivery requirements. They are traded on futures exchange whereas forward contracts are traded over the counter.

An option is a contract whereby the holder has the right not the obligation to buy (in a case of a call) or sell (in a case of a put) the underlying asset of the contract on a future date (the exercise date) at an agreed price (referred to as the strike price or the exercise price). Unlike the buyer, the seller of the option has the obligation to honor the specified feature of the contract.

There is obviously no free lunch. Since an option gives to the buyer an additional right and to the seller an additional obligation, the buyer has to pay to the seller an amount for the option referred to as the option premium. The option offers to its purchaser the right to buy or to sell a specific quantity of a given underlying (e.g. 100 shares for an equity option) at an agreed price either at any time up to maturity date (for an American option) or on a fixed maturity date (for a European option).

2.1.3. *The hedging purpose*

Hedging is a strategy, usually some form of transaction, designed to minimize exposure to an unwanted business risk. Some form of risk taking is inherent to any business activity. Some of the risk is "natural" to a business, whose competitive advantage is to manage this risk well. Other forms of risk are not wanted, but cannot, as things stand, be avoided. Hedging consists in selling off the unwanted risk to those who have the ability or desire to take it.

Now, for these categories of risks, there exist well-developed markets in which the risk is commoditized or securitized, or put into some OTC contracts to be transferred between buyers and sellers.

Futures and forward contracts are a means of hedging against the risk of adverse market movements. These originally started on commodity markets in the 19th century, but over the last 50 years, the market has become much more global and there exists nowadays a huge market to hedge financial market risk, on various instruments.

2.1.3.1. *Hedging with forward contract*

Forward contracts are mostly used to hedge against underlying price movements. Suppose that at January 2005 an American investor needs Euros on July 2005 and wants to hedge against exchange rate (€/$) increase movements. He buys from his bank a forward contract at a delivery price 1.3%.

The difference between the current forward price and the delivery price represents a profit for the investor (see Figure 2.1).

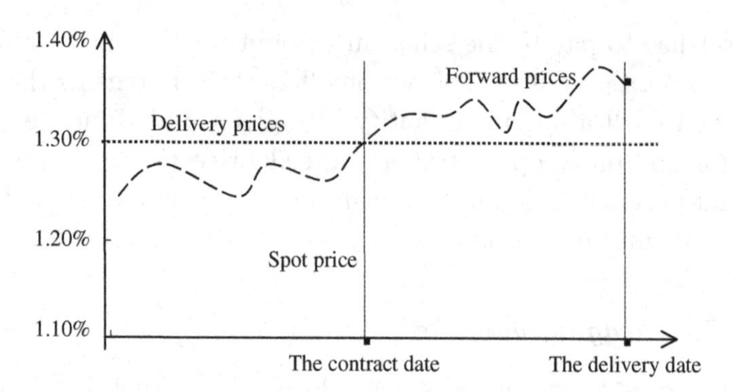

Figure 2.1: Forward contract.

2.1.3.2. *Hedging with options*

By using options one party transfers (buys or sells) risk to or from another. The option is often an insurance instrument; the option holder reduces the risk he bears by paying the option seller a premium to assume it. Following to the sale of the French government for a party of his Air France shares, an investor buys 1000 Air France stocks on January 2005 at 14€ the share. He wants protection for the next 6 months against the adverse price movement, while still allowing the shares to benefit from favorable market movements. So he buys July 2005 put option contracts with strike 13€ at 0.5€ the option price. The investor will limit his loss to 0.5*1000 = 500€ (see Figure 2.2).

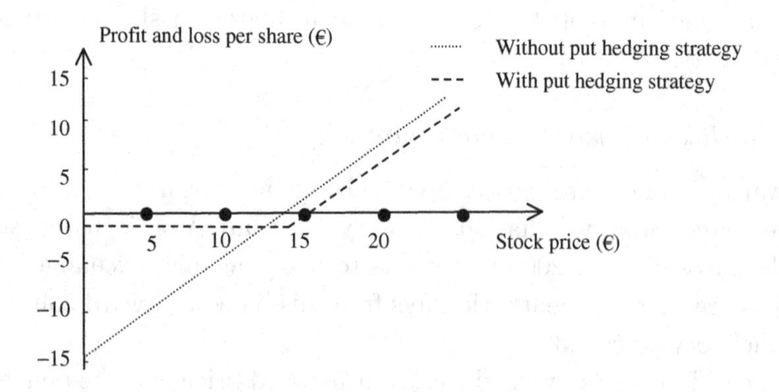

Figure 2.2: Profit and loss of a put.

2.1.4. *The speculation*

The speculators can be considered as the counterparts of hedgers. They accept the risks that hedgers are not willing to take. Speculators are very active in futures markets and other derivatives which involve leverage that can transform a small market movement into a colossal gain or loss. Speculators think that they are good at anticipating future price fluctuations. Thus, they take the risk for their views. In a sense, they allow the market to be efficient by taking the risks, which, taken individually, would seem unreasonably high.

Speculators contribute to increase the market's liquidity. In fact with few players in the market, there would be a larger spread between the bid and ask prices. They are usually well informed on market movements. Thus the market efficiency and transparency are improved. Usually, it is hard to differentiate a speculator from an investor. However, the size of the leverage, the weight of the investment, the frequency of the operations might be indicative of a speculation activity.

Tables 2.1 and 2.2 display how using options and futures, speculators will reach their objective: with the same initial investment, you can make much higher profits than on the spot market. In this example if a speculator expects an increase for the Air France stock price, he/she will choose from the three following strategies, the one that allows him/her the higher leverage.

If the expectation of the speculator is worth, thus the prices of Air France stocks falls down to 11€. The strategy that allows him the higher leverage, leads him to a big loss.

Table 2.1: Strategies for speculators.

Strategies	Initial position (€)	Current position (€)	Profit & loss (€)	Leverage (%)
Long on a stock	14	16	2	14
Long on future contract initial margin 12%	1.68	2	2	119
Long on call option strike at 13€	1.25	3	1.75	140

Table 2.2: Strategies for speculators.

Strategies	Initial position (€)	Current position (€)	Profit & loss (€)	Leverage (%)
Long on a stock	14	11	−3	−21
Long on future contract initial margin 12%	1.68	−3	−3	−178
Long on call option strike at 13€	1.25	0	−1.25	−100

When speculators are very active, the market may move rapidly as speculator may translate in an unnaturally increasing demand or offer. So they cause a growth in prices to rise or fail above their "true value". Further purchasing speculators hope that the price will continue to rise; they create a positive sphere in which prices rise spectacularly above the "true worth". Thus an economic bubble is created. This is usually followed by speculative selling in which the price goes down: we are living at a crash.

2.1.5. *The arbitrage*

Arbitrage is one of three market roles in financial markets, distinct from hedging and speculation. In practice, arbitrage is the procedure of taking advantage from a position of discrepancy between several markets. This is possible when one of three following conditions is not fulfilled:

(1) the same asset must trade at the same price on all markets ("the law of one price");
(2) two assets with identical cash flows must trade at the same price;
(3) an asset with a known price in the future, must today trade at its future price discounted at the risk free rate.

For example in January 2005 the exchange rates:

In Paris 10€ = 13$ = 1378¥.
In Tokyo 1378¥ = 14.3$ = 10€.

So it is profitable to convert 10€ into 14.3$ in Tokyo than converting the 14.3$ into 11€ in Paris. The imbalance between the dollar costs (more expensive in Paris) in the two Exchanges, allows arbitrageurs a non-risky return. These kinds of arbitrage opportunities do not last for long, as the arbitrageurs buys the dollar in Tokyo and sell it in Paris the law of the supply and demand will cause the equilibrium. Arbitrage has the effect of causing prices in different markets to converge. As a result of arbitrage, the currency exchange rates, the price of commodities, and the price of securities in different markets all tend to converge to a fixed price. In the following sections, we explain how the future, forward and option pricing is based on the assumption that in financial markets there are no arbitrage opportunities.

2.2. Pricing of Future Contract

2.2.1. *Trading futures: Margin call and trading account*

2.2.1.1. *The future contract*

In Section 2.1, we gave a brief definition of the future contract. In this section, we will give more details on the standardized features of the contract such as margin requirement, daily settlement, delivery procedure, the bid-offer spreads and the role of the clearinghouse.

The standardization usually requires specifying:

(1) The amount and units of the underlying asset to be traded. This can be a fixed number of: barrels of oil; units of weight (bushels of wheat, ounces of bullion); units of foreign currency; interest rate points; equity index points; National bonds, etc.
(2) The unit of currency in which the asset is quoted.
(3) The class of the delivery: There are two main methods of delivery, viz. the cash and physical delivery. In the case of physical commodities, this specifies not only the quality of the underlying goods but also the manner and location of delivery. For financial assets the delivery is usually in cash, it is simply marked to market on the last trading day and all positions are declared closed.

(4) The delivery month is the month of a final marking- to- market and will be specified by the exchange in the contract specifications. Financial contracts (such as bonds, short-term interest rates, foreign exchange and stock indexes) tend to expire quarterly, in March, June, September and December.
(5) The position limits are the maximum number of contracts that the traders may hold. Some intervenient are not affected by these limits.
(6) The tick size is the minimum permissible price daily fluctuation specified by the exchange.

2.2.1.2. *Product specification*

Although the value of a contract at time of trading should be zero, its price constantly fluctuates. This makes the owner subject to adverse changes in value, and creates a credit risk to the exchange. To minimize this risk, the exchange demands that contract owners post a form of guarantee, known as margin. The amount of margin changes each day, involving movements of cash process by the exchange's clearing house. The gain and loss on the futures contract is calculated by deleting from the initial futures contract price its price at the end of each day, called the marked-to-market price. The margin account is calculated by deleting this gain and loss from the initial margin. Initial margin is paid by both buyer and seller. It represents the loss on that contract, as determined by historical price changes that is not likely to be exceeded on a usual day's trading.

Because a series of adverse price changes may consume the initial margin, a further margin, usually called maintenance margin, is settled by the exchange, which is lower than the initial margin. If the balance in the margin account falls below the maintenance margin, a margin call is required by the Clearinghouse to top up the margin account to the initial margin. If the buyer investor does not provide the margin call, its position is closed out buy selling the contract.

Margin-equity ratio is a term used by speculators, representing the amount of their trading capital that is being held as margin at any particular time. Traders would rarely (and unadvisedly) hold 100% of their capital as margin. The probability of losing their entire capital at some point would be high. By contrast, if the margin–equity ratio is so low as to make the trader's capital equal to the value of the futures contract itself, then they would not profit from the inherent leverage implicit in futures trading. A conservative trader might hold a margin–equity ratio of 15%, while a more aggressive trader might hold 40%.

If we take again the example of the futures contract on Air France stocks. Suppose that the contract size is 100 shares, the futures price is 14€, the initial margin takes 12% on the futures price and the maintenance margin is about 50% of the initial margin. Following to the Air France stock fluctuation prices, we display the behaviors of the futures margin process (see Figures 2.3 and 2.4).

At the 11 trading day the stock price falls to 13.2€, the margin (61€) falls under the maintenance margin (84€) the Clearinghouse

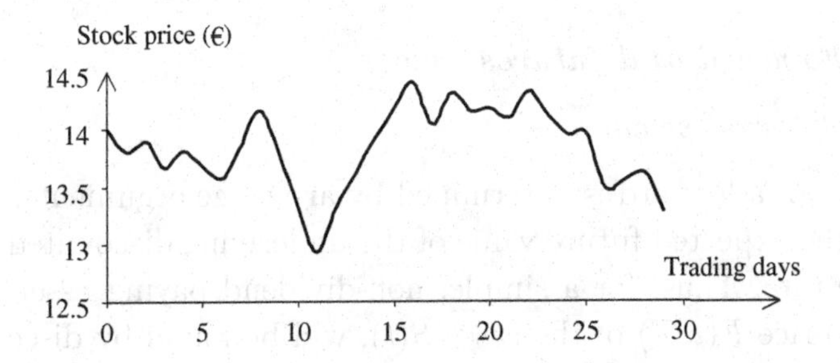

Figure 2.3: The fluctuation of Air France stock price.

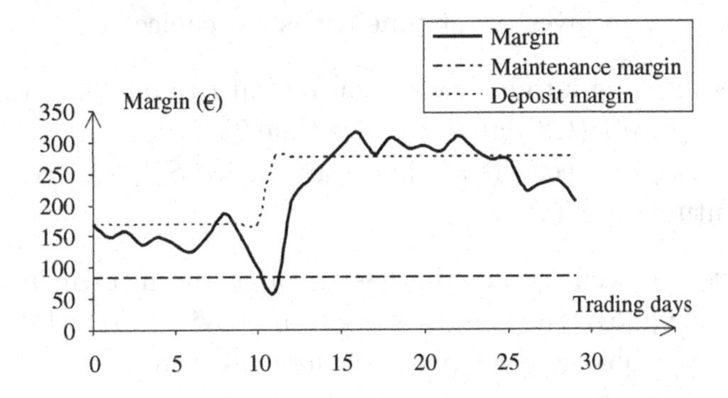

Figure 2.4: The behavior of the future margin process.

requires a margin call from the investor. The margin grows up to 168€, and the deposit margin increase to 275€ (168 + (168 − 61)).

If the investor neutralizes his position at the 16 trading day he makes a profit for about 15% of his investment (275€). Else if he waits until the end of the one month contract, his loss is estimated at −26%.

2.2.2. *Futures vs. forward*

Forward contracts differ from the futures contracts from the non standardization. They are traded on OTC market whereas futures contracts are traded on exchange markets. The total gain or loss is realized at the end of the life of the contract. Because of the settlement procedure, the gain or loss under the futures contract is realized daily. In the following section we will give details on the pricing of the two contracts, and when futures and forward prices converge.

2.2.3. *Forward and futures prices*

2.2.3.1. *Forward pricing*

The price of a forward is determined by arbitrage arguments; it represents the expected future value of the underlying discounted at the risk free rate. Thus, for a simple, non-dividend paying asset S, the forward price $F(t,T)$ of the asset $S(t)$, will be found by discounting the present price $S(t)$ of S at time t to maturity T by the rate of risk-free return r. To own the non-dividend paying security at the future time T, an investor, at time t, has the choice of:

(1) either entering into a forward contract at t to purchase the security at price $F(t,T)$ at the future time T;
(2) or buying the security at time t at a price $S(t)$ and keep it till the future date T.

If he buys the security now he has to carry the investment to the future. Paying $S(t)$ now is equivalent to paying $S(t)^*(1+r)^{(T-t)}$ at T. By no arbitrage arguments we therefore find that:

$$F(t,T) = S(t)^*(1+r)^{(T-t)},$$

else, if the forward price is higher a risk free arbitrage may be possible: investor can buy the security at price $S(t)$ (which costs as borrowing $S(t)$ at a rate r) and at the same time he sells a forward contract at $F(t,T)$. At the future time T the investor delivers the security to the buyer of the forward contract, refunds the loan that costs $S(t)^*(1+r)^{(T-t)}$. Thus he fulfills a riskless profit, since the forward price $F(t,T)$ exceeds the amount $S(t)^*(1+r)^{(T-t)}$). Similarly, by an equivalent argument, if the forward price is lower than the cost of carrying the security a risk free arbitrage is also possible.

With continuous compounding the forward price can be given by the relation

$$F(t,T) = S(t)^* \mathrm{e}^{r(T-t)}.$$

This relationship may be modified for storage costs, dividends, dividend yields and convenience yields. In a perfect market the relationship between forward and spot prices depends only on the above variables.

2.2.3.2. *Futures prices*

When the risk free interest rate is a function of time, in a perfect market the forward price and the futures price are the same. In the case of unpredictably rates (which are the fact in financial market) forward prices and futures prices do not converge. In fact, if we suppose that the underlying asset of the futures and forward contract is strongly correlated with the risk free interest rate. So a long position on a futures contract allows more gain than a long position on a forward contract because of the daily settlement procedure. With an arbitrage argument the futures contract price has to be more expensive than the forward contract price.

In practice, there are various market imperfections (transaction costs, differential borrowing and lending rates, restrictions on short selling) that prevent complete arbitrage. Thus, the forward and futures prices in fact vary within arbitrage boundaries around the theoretical price. The margin treatment and the more liquidity of the futures contracts may accentuate the pricing differences.

Now we will display the proof of the parity of the forward and futures prices in the case of non risky interest rates: suppose that at t_0 an investor takes a long position on α_0 futures contracts at price $\mathrm{Fut}(t_0, t_n)$ maturing at t_n. Every day t_i from t_0 to t_n he increases his position from α_{i-1} to α_i. So at the end of the futures contract the cumulative gain and loss of the investor is given by compounding at the risk free instantaneous rate r_t until t_n

$$\mathrm{LG} = \sum_{i=1}^{n} \alpha_i (\mathrm{Fut}(t_i, t_n) - \mathrm{Fut}(t_{i-1}, t_n)) \mathrm{e}^{\int_{t_i}^{t_n} r_t dt}. \tag{2.1}$$

If the position α_i depends on the time t_i by this way

$$\alpha_i = \mathrm{e}^{\int_{t_0}^{t_i} r_t dt} \tag{2.2}$$

The loss and gain of future structure is given by

$$\begin{aligned} \mathrm{LG} &= \sum_{i=1}^{n} (\mathrm{Fut}(t_i, t_n) - \mathrm{Fut}(t_{i-1}, t_n)) \mathrm{e}^{\int_{t_0}^{t_n} r_t dt} \\ &= (\mathrm{Fut}(t_n, t_n) - \mathrm{Fut}(t_0, t_n)) \mathrm{e}^{\int_{t_0}^{t_n} r_t dt}. \end{aligned} \tag{2.3}$$

As by arbitrage free at any time t the spot future price $\mathrm{Fut}(t, t)$ and the underlying price $S(t)$ are equal, at the end of the contract the investor has exactly cumulated the gain and loss

$$\mathrm{LG} = (S(t_n) - \mathrm{Fut}(t_0, t_n)) \mathrm{e}^{\int_{t_0}^{t_n} r_t dt}. \tag{2.4}$$

Notice that this structure of futures contracts does not require to the investor to invest money at any time. So an investment of $\mathrm{Fut}(t_0, t_n)$ at t_0 combined with this strategy will change his loss and gain to the present value of his position at day t_n

$$PV = S(t_n) \mathrm{e}^{\int_{t_0}^{t_n} r_t dt}. \tag{2.5}$$

This is exactly the present value at t_n of a forward contract settled at t_0 for a delivery date t_n requiring an initial outlay the forward price $\mathrm{Fwd}(t_0, t_n)$. Thus

$$\mathrm{Fwd}(t_0, t_n) = \mathrm{Fut}(t_0, t_n). \tag{2.6}$$

2.3. Options

2.3.1. *Definition and features*

We recall that an option is a contract whereby the buyer has the right to exercise a feature of the contract at a future date and the seller the obligation to honor the specified feature of the contract. Since the option gives the buyer the right and the seller the obligation, the buyer pays the seller an amount for the option called option premium. The option offers for the purchaser the right to buy (for a call option) or to sell (for a put option) a specific quantity of a given underlying at an agreed price (the strike price) either at any time up to maturity date (for an American option) or on a fixed maturity date (for a European option).

The intrinsic value of the option is the value it would have if it were exercised immediately. For example, for a call option the intrinsic value is the maximum between zero and the difference between the exercise price of the option K and the value of the underlying instrument S.

$$\text{IntrinsicVaule}_{\text{Call}}(t) = \max[(S(t) - K), 0]. \tag{2.7}$$

An out-of-the-money option has no intrinsic value, for example, a call option is out-the-money if the strike price is higher than the current underlying price. An in-the-money option conversely does have intrinsic value. The strike price of an in-the-money call option is lower than the current underlying price. An option is at-the-money if the intrinsic value is zero. The strike price for an at-the-money call option is the same as the current price of the underlying security.

Option value or equivalently the premium is the amount that the holder of the option pays to the seller. Intuitively the seller will estimate the worst risk scenario that will happen by selling this option, so he calls the buyer for an amount that reflects the "likelihood" of the option finishing in-the-money. The option value will never be lower than its intrinsic value. The value of an option consists of two components, its intrinsic value and its time value. Time value is simply the difference between option value and intrinsic value. At expiration, where the option value is simply its intrinsic value, time value is zero.

There are six factors affecting the price of the option:

(1) the current underlying price $S(0)$;
(2) the strike price K;
(3) the time to the expiry date T;
(4) the volatility σ of the underlying price: defined such that $\sigma\sqrt{T}$ is the standard deviation of the underlying price at T[2];
(5) the risk free interest rate r;
(6) the eventual dividend or transaction costs.

Suppose that the seller of the option chooses a discrete normal model, to illustrate his anticipation of the fluctuation of the underlying prices at T. Thus:

$$S(T) = \begin{cases} \text{or} \quad S(T)^{\text{Up}} & = \tilde{S}(0) + \sigma\sqrt{T} \quad \text{with probability } p_{\text{up}}, \\ \text{or} \quad S(T)^{\text{down}} & = \tilde{S}(0) - \sigma\sqrt{T} \quad \text{with probability } p_{\text{down}}, \end{cases} \tag{2.8}$$

where $\tilde{S}(0)$ is the expected future value at time T. In the case of non-dividend paying asset the expected value of the underlying price is given by $\tilde{S}(0) = S(0)e^{rT}$.

The option price is then

$$\text{OptionPrice} = [p_{\text{up}}^{*}(S(T)^{\text{up}} - K)_{+} + p_{\text{down}}^{*}(S(T)^{\text{down}} - K)_{+}]^{*}e^{-rT}. \tag{2.9}$$

Notice that this formula uses all the factors given below. In Chapter 4, we will give the Black–Scholes approach to price basic options (call or put options). This approach supposes that the underlying price at time T follows a log-normal variable.

[2]If the underlying price at T is a normal variable and its expected value at 0 is $\tilde{S}(0)$ so the underlying price $S(T)$ at T will fluctuate between $(\tilde{S}(0) - \sigma\sqrt{T})$ and $(\tilde{S}(0) + \sigma\sqrt{T})$.

2.3.2. *Call–put parity*

The call–put parity defines a relationship between the price of a European call option and a European put option on the same non-dividend-paying stock S with the same strike K and expiry date T. Only the assumption of arbitrage-free is used to give this relationship:

$$\mathrm{Call}(t, S_t, K, T) - \mathrm{Put}(t, S_t, K, T) = S_t - K\mathrm{e}^{-r(T-t)}, \qquad (2.10)$$

where r is the risk free interest rate.

Let us consider the following two portfolios:

(1) The first one contains one share S and one European put option contracted at t on the underlying S at strike K and expiry T. So we have to provide at t the amount

$$\mathrm{Potfolio}_1(t) = \mathrm{Put}(t, S_t, K, T) + S_t. \qquad (2.11)$$

A simple calculus explains that at the expiry date T, this portfolio allows his holder the amount

$$\mathrm{Potfolio}_1(T) = \max(S_T, K). \qquad (2.12)$$

(2) The second one contains an amount of cash equal to $K\mathrm{e}^{-r(T-t)}$ and one European call option with same characteristics than the put below. So we have to provide at t the amount

$$\mathrm{Potfolio}_2(t) = \mathrm{Call}(t, S_t, K, T) + K\mathrm{e}^{-r(T-t)}. \qquad (2.13)$$

A simple calculus explains that at the expiry date T, this portfolio allows his holder the same amount than the first portfolio.

By arbitrage-free argument, two quantities of money providing the same amount on a given future date are equals.

2.3.3. *Option strategies: call, put, straddle, strangle, spread and butterfly*

One can combine options and other derivatives to transfer risk, or to create leverage. In this section we will give an overview of the most common trading strategies. They are simply combinations of the basic option positions (short call, long call, short put and long put). Several long or short positions can be combined to form one single position.

In order to control the risk, using options one party transfers (buys or sells) risk to or from another. When using options for insurance, the option holder reduces the risk he endures by paying the option seller a premium for assuming it.

The risk taken on can be anywhere from zero to infinite, depending on the combination of derivatives features used. The payoff from purchasing an option can be much greater than that from purchasing the underlying instrument directly.

2.3.3.1. *Call–put option*

A call option is a financial contract between two parties, the buyer and the seller of the option. The buyer of the option has the right but not the obligation to buy an agreed quantity of a particular commodity or financial instrument (the underlying instrument) from the seller of the option at a certain time for a certain price (the strike price). The seller assumes the corresponding obligations. "Selling" in this context is not the supplying of a physical or financial asset (the underlying instrument), rather it is the granting of the right to buy the underlying, against a fee — the option price or premium.

Exact specifications may differ depending on option style. A European call option allows the holder to exercise, that is to buy, on the delivery date only. An American call option allows exercise at any time during the life of the option. The stock option, the option to buy stock in a particular company, is the most widely known call. However options are traded on many other financial instruments — such as interest rates (see interest rate cap) — as well as on physical assets such as gold or crude oil.

Example of a call option on a stock: I buy a call on Microsoft Corp. strike price $50, exercise on June 1 2005. If the share price is actually $60 on that day (the spot price) then I would exercise my option (i.e. buy the share from the counter-party). I could then sell it in the open market for $60, that is, the option would be worth $10; my profit would be $10 minus the fee I paid for the option.

However, if the share price is only \$40 then I would not exercise the option (if I really wanted to own such a share, I could buy it in the open market for \$40, why waste \$50 on it). The option would expire worthless. Thus, in any future state of the world, I am certain not to lose money on the underlying by owning the option; my loss is limited to the fee I have paid.

From the above, it is clear that a call option has a positive monetary value when the underlying instrument has a spot price (S) above the strike price (K). Since the option will not be exercised unless it is "in-the-money", the payoff for a call option is

$$\mathrm{Max}[(S-K);0] \quad \text{or formally, } (S-K)+$$

Prior to exercise, the option value, and therefore price, varies with the underlying price and with time. The call price must reflect the "likelihood" or chance of the option "finishing in-the-money". The price should thus be higher with a greater maturity, and with a more volatile underlying instrument. The science of determining this value is the central tenet of financial mathematics. The most common method is to use the Black–Scholes formula. Whatever be the formula used, the buyer and the seller must agree to this value initially.

A put option is a financial contract between two parties, the buyer and the seller of the option. The put allows the buyer the *right but not the obligation* to sell a commodity or a financial instrument (the underlying instrument) to the seller of the option at a certain time for a certain price (the strike price). The seller assumes the corresponding obligations.

2.3.3.2. *Straddle*

Volatile market trading strategies are appropriate when the trader believes that the market will move but does not have an opinion on the direction of movement of the market. As long as there is a significant movement upwards or downwards, these strategies offer profit opportunities. A trader need not be bullish or bearish. He must simply be of the opinion that the market is volatile. This market outlook is also referred to as "neutral volatility".

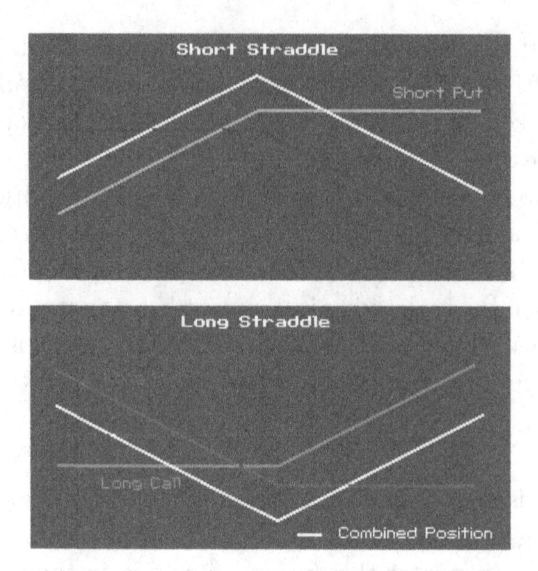

Figure 2.5: Short straddle and long straddle.

A straddle is the simultaneous purchase (or sale) of a call and a put with identical features (the same exercise price and the same expiration date). To "buy a straddle" is to purchase a call and a put with the same exercise price and expiration date (see Figure 2.5).

To "sell a straddle" is the opposite: the trader sells a call and a put with the same exercise price and expiration date (see Figure 2.5).

2.3.3.3. *Strangle*

A strangle is similar to a straddle, except that the call and the put have different exercise prices. Usually, both the call and the put are out-of-the-money. To buy a strangle is to purchase a call and a put with the same expiration date, but different exercise prices. Usually the call strike price is higher than the put strike price. To sell a strangle is to write a call and a put with the same expiration date, but different exercise prices (see Figure 2.6).

A trader, viewing a market as stable, should write strangles. If the market is stable, neither a long call nor a long put will be profitable. The trader would let the options expire worthless, losing the sum of the premiums paid. A "strangle sale" allows the trader to profit from a stable market. Note however, that the strategy was christened

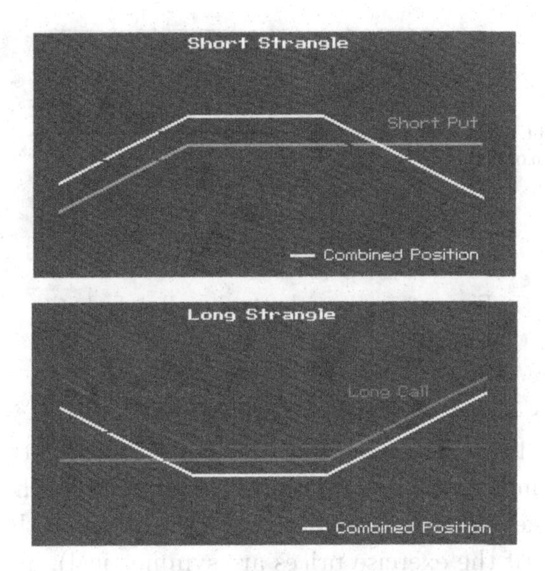

Figure 2.6: Short strangle and long strangle.

"strangle" in memory of a number of good traders who used it when IBM suddenly went volatile in April, 1978. The investor's potential profit is the greatest profit an option writer can make is the amount received for incurring the obligation to deliver — the premium. If the market remains stable, investors having out-of-the-money long put or long call positions will let their options expire worthless. The investor's potential loss is: If the market is volatile, up or down, the trader will have to deliver if exercised. If the price of the underlying interest rises or falls instead of remaining stable as the trader anticipated, he will have to deliver on the call or the put.

2.3.3.4. *Butterfly*

The long butterfly call spread is a combination of a bull spread and a bear spread, utilizing calls and three different exercise prices (see Figure 2.7).

A long butterfly call spread involves:

(1) buying a call with a low exercise price;
(2) writing two calls with a mid-range exercise price;
(3) buying a call with a high exercise price.

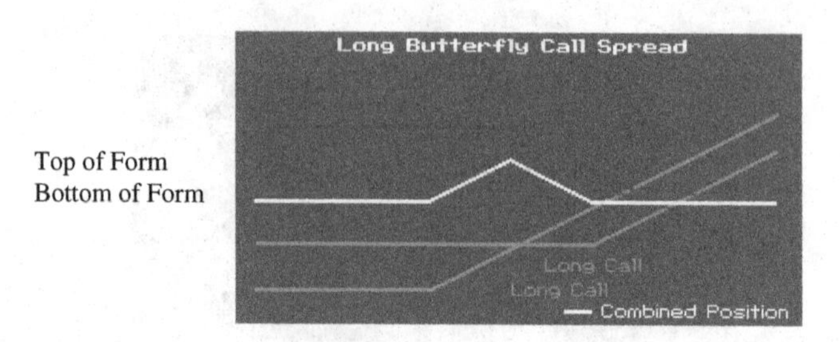

Figure 2.7: Long butterfly call spread.

The investor's profit potential is limited. The trader is short two calls and long two calls. These positions offset one another to a certain degree, limiting the potential profit. Maximum profit is attained when the market price of the underlying interest equals the mid-range exercise price (if the exercise prices are symmetrical). The investor's potential loss is limited. The trader has offsetting positions, but at different exercise prices. The maximum loss is limited to the net premium paid and is realized when the market price of the underlying asset is higher than the high exercise price or lower than the low exercise price.

2.3.4. *Warrants*

A warrant is the right — but not the obligation — to buy or sell a certain quantity of an underlying instrument at an agreed-upon price. The right to buy the underlying instrument is referred to as a call warrant; the right to sell it is known as a put warrant. In this way a warrant is very similar to an option. The difference is primarily that the length of time available to exercise a warrant is much longer than most option contracts. Most warrants have 5–10 years before they must be exercised or expire worthless. In addition, when a warrant is exercised, a new share of stock is created, whereas when an option is exercised, the owner of the option receives an existing share that is delivered by counterparty (except in the case of employee stock options, where new shares are created and issued by the company upon exercise).

There are two types of warrants: "traditional" warrants and the so-called naked warrants.

Traditional warrants are issued in conjunction with a bond (known as a warrant-linked bond) and represent the right to acquire shares in the entity issuing the bond. In other words, the writer of a traditional warrant is also the issuer of the underlying instrument. Warrants are issued in this way as a "sweetener" to make the bond issue more attractive, and to reduce the interest rate that must be offered in order to sell the bond issue.

Naked warrants are issued without an accompanying bond, and like traditional warrants, are traded on the stock exchange. They are typically issued by banks and securities houses. The writer of a naked warrant need not be the issuer of the underlying instrument. A naked warrant is essentially an option with a very long time to expiry. Therefore an employee stock option is also equivalent to a warrant.

Also, when a government agency issues checks which they are unable to pay (due to lack of money) but are redeemable at some point in the future, usually with interest, these are also called warrants. A few years ago, when the State of California had a budget crisis due to a disagreement between the governor and the legislature, the state treasurer was forced to issue warrants paying 18% interest in lieu of being able to pay the state's bills with real money. The state had not had to rely on this practice since before the Depression of the 1930s many places were accepting them at face value because of the interest provision. It was interesting that during this period the Controller of Los Angeles County was buying state warrants to invest the county's surplus funds because the county, on the other hand, actually had money and the interest rate was better than any bank would pay.

2.4. Summary

The main standard products traded on markets are forward, futures and options. Both forward and futures contracts are an agreement to buy or sell at a specified future time an asset for a certain price.

However, there are differences between these two contracts. A forward is a non-standardized contract traded in the OTC markets. Whereas futures are traded on exchanges and they have specified features. An option is the right and not the obligation to buy or sell at a specified time for an agreed price. There are two types of options: calls and puts. In financial markets, we can identify three strategies: hedging, speculation and arbitrage. Forwards and futures are mainly used for hedging. Options are also used for hedging but they are more flexible than forward and futures contracts in the way that they allow traders to profit from favorable market movements. Moreover, options offer various trading strategies ranging from basic positions involving a single option to more complex ones combining several long and short positions.

References

[1] Baxter M and Rennie A (1996). *Financial Calculus — An Introduction to Derivative Pricing.* Cambridge University Press.
[2] Ghassemieh R, Shaw W and Wilson R (1997). Equity-index-linked derivatives, in *The Asian Equity Derivatives Handbook.* Euromoney Publications.
[3] Hull JC (1996). *Options, Futures, and other Derivatives*, 3rd edn. Englewood Cliffs, NJ: Prentice-Hall.
[4] Jarrow RA and Rudd A (1983). *Option Pricing.* Irwin.
[5] Kendall WS (1993). Itovsn3: Doing stochastic calculus with Mathematica, in H.R. Varian (ed.), *Economic and Financial Modeling with Mathematica.* Berlin: Springer-TELOS.
[6] Levy E (1996). Exotic options I, in C. Alexander (ed.) *The Handbook of Risk Management and Analysis.* New York: Wiley.
[7] Reiner E (1992). Quanto mechanics, in *From Black Scholes to Black Holes.* RISK-FINEX Publications.
[8] Schuss Z (1980). *Theory and Applications of Stochastic Differential Equations.* New York: Wiley.
[9] Wilmott P, Dewynne J and Howison S (1993). *Option Pricing — Mathematical Models and Computation.* London: Oxford Financial Press.

Chapter 3

INTRODUCTION TO FINANCIAL MODELING

The aim of this chapter is to give an idea of what kind of modeling people use to propose prices or hedges, why this choice was made and how it happened. The outline of this chapter is the following: first we start by examining why, because of the no arbitrage condition, people show an interest in modeling in Finance, contrary to other domains. Then we show how people now use models from the viewpoint of risk and pricing. In order to see why it was natural to consider some type of random processes, we sketch some mathematical tools that are developed in detail in another chapter. Finally we introduce the notion of risk neutral probability, first in the simple 1-factor case, then in an n factor or multi-currency framework.

3.1. A Bit of History

Financial Modeling — we could also say, modeling using financial mathematics — dramatically increased both in broadness and complexity in the last century and essentially especially in the last 30 years.

We start with a look at the evolution which made this possible, contrary to other fields. Most of the time, for example, when one buys or sells items, no one uses models to estimate the prices. In this example, you can interview people who buy, or sell, vegetables on a market place, the proportion of these people who built a model on potato prices to have an idea of their cost is probably rather small. This situation is now completely different for financial products, where the mathematical complexity involved in modeling is bigger and bigger. Why did this happen and how? The first part of the question has several answers based on the same root. The historical evolution of financial modeling leads to *the* very fundamental idea, the no arbitrage condition: if you are sure — deeply convinced is of course not enough — that a financial strategy will lead to a profit, then it must cost you something to put the strategy in place (the famous motto that makes trading rooms look like restaurants: there is no free lunch!).

This idea, which still looks theoretical, can be described in several practical ways that are appealing to investors or traders and made the current success of the modeling.

(1) The possibility to hedge a position: Assuming that we know the price of basic financial products, we can dynamically (i.e. locally: for small market moves) or sometimes statically (i.e. independently of any market conditions) replicate a financial product by a portfolio of other "basic" financial products. Then the hedged financial product (the portfolio where we are long (i.e. we possess) the financial product and short, (i.e. we sell) the replicating portfolio) bears no financial risk.

(2) The possibility (at least sometimes) to detect mispricings and realize a profit: It may happen that a mispricing exists in the markets. Here "mispricing" means that two products should — relying on financial conditions, not on theory — have the same price but the first one is more expensive than the other one (or reciprocally). The problem is then: how to realize a profit out of this mispricing?

As a (standard) example, it is now well known that an American call option on a non-paying dividends asset has the same price as an European option with same other characteristics (Merton, 1973).

Another example concerns with the pricing of CMS swaps: if the smile of volatility is not properly taken into account, one can show and convince people that CMS swaps are mispriced and see how to realize a profit out of the mispricing.

In both cases, according to the law of supply and demand, it is therefore possible to help the financial community and make the markets more efficient.

These two applications, which have no equivalent on the vegetable market, rely on this consequence of the no arbitrage condition. If we are sure that at a certain moment a product is not cheaper than another one, then the first one is considered to be always not cheaper than the second one.

As a consequence of the above principle: if we are sure that at a certain moment a product has the same price as another one, then we are sure that they always have the same price.

Mathematically speaking, a way to exploit the principle of maximum stated by the first consequence of the no arbitrage condition is to say that the price of an asset is (more or less ...) "an harmonic function". Correctly stated, it will lead to the pricing by partial differential equation (and among others, to the Black–Scholes/Merton option pricing). Looking at the same consequence differently, with a condition a bit more precise, and using an extension by Feyel of the Hahn–Banach theorem to the space of price processes, this could also then lead to pricing by mathematical expectations either in a continuous trading framework or in a discrete trading environment by using Markov chains (in Finance, people talk about "trees").

A look at the history will help us to understand the growing trend for modeling and how people outlined the no arbitrage condition.

This is a rather recent evolution: initially — to make it short, let us say before 1900 — the models were simpler. Many asset comparisons were based on returns, yields to maturity or ratios such as the price earnings ratio (PER). The tools at that time were tables of numbers and linear interpolation. Financial markets were merely regarded as casino games and prices were governed by the law of supply and demand, as for many commodities for this last remark.

The first move towards a modeling of financial products evolution took place in 1900 and it is due to Louis Bachelier. He proposed to see the moves of asset prices as random phenomena.

> In his 1900 dissertation written in Paris, *Théorie de la Spéculation — Random Character of Stock Market Prices* (and in his subsequent work, esp. 1906, 1913), he anticipated much of what was to become standard fare in financial theory: random walk of financial market prices, Brownian motion and martingales (note: all before both Einstein and Wiener!) His innovativeness, however, was not appreciated by his professors or contemporaries. His dissertation received poor marks from his teachers and, consequently blackballed, he quickly dropped into the shadows of the academic underground. After a series of minor posts, he ended up obscurely teaching in Besançon for much of the rest of his life. Virtually nothing else is known of this pioneer — his work being largely ignored until the 1960s. (*Source*: http://cepa.newschool.edu/het/profiles/bachelier.htm)

We see here that there was no perceived need for such a sophisticated modeling in the financial area at that time.

(1) The people who were involved in financial trading were either not aware of this modeling (we can think that it was the vast majority of the traders or investors) or, if by chance had they heard about it, most probably extremely dubious. The great mathematical sophistication made it hard to understand and to believe in. Therefore the investment — intellectual as well as financial — was huge and the profit that could result of it was — or seemed to investors — hugely hypothetical.
(2) Most of the people involved in mathematics did not consider finance as a serious scientific matter and did not look at developments related to finance.

Some of the lessons that can be drawn from this episode are:

(1) The sophistication of a theory is rather a disadvantage than anything else.
(2) A model — even if superior to the existing ones, whatever superior means — is worthless if not accepted by the financial community. This implies, therefore, the question: what to do to make

people believe in the model superiority? Communication is crucial in this matter.

(3) If you cannot imply simply a likely profit of your model, financial or at least in terms of image, you probably have a better usage of your document as wallpaper.

As we all know now, this situation, "no modeling", was to be changed. The next move towards modeling was to take into account arbitrage reasoning. In mathematical finance, people tend to consider that the next step is due to Black, Scholes and Merton in 1973.

This does not mean that nothing — speaking of financial modeling — happened in between. For example, in Economics, Fisher (1906, 1907, 1930) had already considered the risk as part of the process for allocating resources over time. In 1944, von Neumann and Morgenstern published papers on the expected utility theory, that Markowitz (1952, 1959), moving a step further and seeing the notion of expectation on the future as very important, started to use. Tobin (1958) added a chapter to the book, by saying that people should invest in a risk-free asset (cash) and a single portfolio of assets, with weights on each component depending on the investor's aversion, or appetite, for risk. The works of Sharpe (1961, 1964) and Lintner (1965) in their capital asset pricing model (CAPM) allowed practical applications by reducing the calculations to do to some covariance computations (the betas) only.[1]

The door was now opened for practices and empirical studies on the effectiveness of the CAPM lead to further studies and the ideas developed in these papers are the basis of current pricing techniques, such as:

(1) the "intertemporal CAPM" (ICAPM) of Merton (1973);
(2) the "arbitrage pricing theory" (APT) of Ross (1976). In this theory, Ross exploited the notion of pricing by arbitrage, instead

[1]Contrary to what someone once claimed in the famous "pre-Varenne controversy", Sharpe and Lintner are indeed the fathers of the CAPM theory. The mistake comes from a word of Louis XIV of France that was wrongly quoted as "Beta, c'est moi".

of staying on the risk versus return logic of the CAPM. As Ross noted, the non-arbitrage paradigm is the key to finance theories.

The famous theory of option pricing by Black and Scholes (1973) and Merton (1973) derives from arbitrage reasoning: an option and a portfolio that replicates it must have the same price to avoid arbitrage opportunities.

Then they showed that the price of the option was the solution of a partial differential equation (PDE) that they could solve. This technique of pricing by discretization of a PDE is still widely used.

The idea to consider expectation for pricing is also now very widely used, after the works of Harrisson and Kreps (1979) and Harrisson and Pliska (1981). For this last article, the authors show that there is a one-to-one relation between the set of martingale measures and coherent systems of pricing and they explicit the link between these two sets. They also show that there is at least a martingale measure if and only if there is no arbitrage in the market.

This was a starting point for a dramatic increase of probability developments in markets finance. As examples, we can mention the use of local times, stopping time, Malliavin calculus, etc. The readers can find more precisions in http://cepa.newschool.edu/het/schools/finance.htm.

3.2. Usage of Models

We have already mentioned that modeling could lead to pricing and hedging. In this section we start with a review of the more important sources of risk: market, counterparty, operational and model risk.

3.2.1. *Risk categories*

3.2.1.1. *Market risk*

This is the risk to an institution's financial condition resulting from adverse movements in the level or volatility of market prices of interest rate instruments, equities, commodities and currencies. Market risk is usually measured as the potential gain/loss in a position/portfolio that is associated with a price movement of a given probability over a specified time horizon (*Source*: International Financial Risk Institute, http://riskinstitute.ch/00013404.htm).

It is therefore the risk of losses in on- and off-balance-sheet positions arising from undesirable market movements, as the price of a product depends on financial conditions and changes with them. This could lead to extremely high losses that nobody wants.

Let us take as an example the case of a short 5-year call on the asset A. In exchange of an initial amount of money (the premium) we gave to someone (the buyer of the option) the right (if he decides so) to buy 5 years from today (maturity of the option) the stock A (underlying of the option) at the price of 10€ (strike of the option). Even if the spot price (i.e. the price today) of the asset A is very low (let us say: 0.12€), there are a lot of things, for example, bubbles, shortage of materials, political events, etc., that can happen 5 years from now and make the price of the stock take off and reach, for example, 1000€. Even if the option seller (let us call him Mr. S) buys today and keeps for 5 years the stock A, if the option buyer (let us call him Mr. B) decides to exercise the option (to exercise his right to buy) at maturity — he would be non-sensible otherwise — the option seller will have to sell at 10€ something that is worth 1000€.

If Mr. S does not possess the stock at maturity, it can be even worst for him. Depending on the liquidity of the stock, it is not even sure that the option seller can buy the asset in the market at 1000€ if no one wants to sell: In that case, the quoted price will probably be the price of the last transaction, which occurred several days ago. Mr. S will then default. As could conclude Tex Avery: "sad ending, isn't it?"

Though, selling options is not rare: traders can hedge their positions and avoid the above story. In order to do so, they use models which will help them to dynamically neutralize the financial risk of their position.

3.2.1.2. *Counterparty risk*

It is the risk — to each party of a contract — that the counterparty will not completely fulfil his part of the contract. In many financial contracts, counterparty risk is default risk: the failure to make payments because of bankruptcy. In general, this kind of counterparty risk can be reduced by having an organization with extremely good credit act as an intermediary between the two parties. The main purpose of a clearinghouse and of margin requirements is to reduce this counterparty risk.

For other contracts, the risk can come from a downgrade — we can imagine also contracts where it would be an upgrade — of the counterparty or, generally speaking, any credit event in the contract.

In order to have an idea of the amount of money at risk, one must therefore have an idea of the money that could be received at one moment in the future when the counterparty defaults or is downgraded. In order to make this projection, people use models to get ideas of the money at risk.

3.2.1.3. *Operational risk*

The following paragraphs are a quotation of a paper prepared by the Risk Management Group of the Basel Committee on Banking Supervision (reference: http://www.bis.org/publ/bcbs96.pdf)

> The Committee recognizes that the exact approach for operational risk management chosen by an individual bank will depend on a range of factors, including its size and sophistication and the nature and complexity of its activities. However, despite these differences, clear strategies and oversight by the board of directors and senior management, a strong internal control culture (including, among other things, clear lines of responsibility and segregation of duties), effective internal reporting, and contingency planning are all crucial elements of an effective operational risk management framework for banks of any size and scope.

Deregulation and globalization of financial services, together with the growing sophistication of financial technology, are making the activities of banks and thus their risk profiles (i.e. the level of risk across a firm's activities and/or risk categories) more complex.

Developing banking practices suggest that risks other than credit, interest rate and market risk can be substantial. Examples of these new and growing risks faced by banks include:

(1) If not properly controlled, the greater use of more highly automated technology has the potential to transform risks from manual processing errors to system failure risks, as greater reliance is placed on globally integrated systems;
(2) Growth of e-commerce brings with it potential risks (e.g. internal and external fraud and system security issues) that are not yet fully understood;
(3) Large-scale acquisitions, mergers-mergers and consolidations test the viability of new integrated systems;
(4) The emergence of banks acting as large-volume service providers creates the need for continual maintenance of high-grade internal controls and back-up systems;
(5) Banks may engage in risk mitigation techniques (e.g., collateral, credit derivatives, netting arrangements and asset securitizations) to optimize their exposure to market risk and credit risk, but which in turn may produce other forms of risk (e.g. legal risk);
(6) Growing use of outsourcing arrangements and the participation in clearing and settlement systems can mitigate some risks but can also present significant other risks to banks.

The diverse set of risks listed above can be grouped under the heading of "operational risk", which the Committee has defined as "the risk of loss resulting from inadequate or failed internal processes, people and systems or from external events". The definition includes legal risk but excludes strategic and reputational risk.

In the following section we discuss a case of model risk which is indeed an operational risk.

3.2.1.4. *Model risk*

There are not many publications about this risk and we will rather briefly talk about it. One can see at least two different components:

(1) The first one is linked to hedging. Let us consider a single product P, priced with two different models M1 and M2. Even if this product has the same price when one uses the two models, it is not sure at all that the hedge ratios — expressed as quantities of the same references: basic financial product — will be close.
(2) The second one is linked to pricing. Let us consider two products, A and B, each of them being priced with their respective model A and B. It may happen that the market conditions imply that these products are financially "close". Therefore, their prices must also be close under these market conditions.

In order to illustrate this notion, let us look at a portfolio which is the sum of a caplet and a swap and let us then consider an option on this portfolio. Market conditions may imply that the caplet is — and will stay with a high probability — out of the money, that is, of almost null value. Therefore, the option is financially close to an option on a swap (which is also called a swaption) and the option price must be the price of the corresponding swaption.

However, as the models are different, it is not *a priori* sure that the two prices will be almost the same. Depending on the system that is proposed, it may even happen that the very same product has different prices with different models. In this case, we could talk of an operational risk rather than a model risk: the system provides two different prices for the same product! It may then happen that people from different desks but the same — of course not aware of this problem (!) — sell (for the first desk) or buy (for the second desk) two almost (or exactly) identical products and make both profits when one desk should have a negative P&L.

3.2.2. *What are models for?*

There are two broad categories of different uses for a model.

3.2.2.1. *As a predictor*

It is often quicker and easier to get a price with a simple model, with simple assumptions such as normal (i.e. Gaussian) diffusion that allows very often shortcuts or analytical formulae. Most of the time in those cases, the prices that one can get this way are not accurate. However, to propose strategies to a client, to give proposal of capital structure the "accurate" prices are not necessary.

3.2.2.2. *As an "interpolator" or "extrapolator"*

For valuation, analysis and management of exotic securities we need to match standard market "observable" and to use numerical inversion techniques to be sure that for some reference products (the calibration set), prices obtained when using the model are the same as the market observed prices.

This last part is called calibration. It is a very essential part of pricing and hedging with a model. It allows us to numerically determine the model parameters so the model replicates with a good fit to the chosen reference products.

The following two steps are highly sensitive for this procedure:

(1) The choice of the calibration set: This set depends on the product that we want to price and hedge. It should contain the products that we want to use for hedging. One must be aware that changing the calibration set actually means changing the model: usually the model parameters are different with another calibration set. Sometimes changing a calibration set will also imply a volatility function defined by interpolations

on different intervals. This actually means that you work with another volatility function.

(2) The optimization/bootstrap procedure: Apart from some very exceptional cases, we cannot exactly solve an inverse problem. All we can do is to minimize an error. The minimization algorithms, which work well on paper, may have very different efficiency. The naïve use of the Newton algorithm may lead to interesting and surprising results but alas because of wrong answers. Some people say that the good way to make this algorithm work is to start from the solution! This means that it is locally efficient but most of the times find a local optimum when we want a global one.

Another crucial point is that a lot of algorithms use the derivative(s) of the function that we want to minimize. As we usually do not have analytical formulae we are bound to use numerical, therefore inaccurate, and this may have a strong impact on the (in)efficiency of the calibration programs. In other words, Bootstrap method seems to be very useful, precisely when the traditional optimization algorithms fail, it allows us a sequential fitting of the different data, and we arrive to a unique solution whatever the choice of the starting point. This solves the local optimum problem of optimization procedure. But, this method needs the existence of a plausible solution. This may not be the case in general. We also have to keep in mind that we actually want to make optimization under constraints.

All of this may have a lot of impact on the prices of some products. Even if, looking at some stochastic diffusion equations valid in an "ideal" world of continuous trading and infinitely many calibration products, you think that you kept the same model, it is quiet wrong. By the way, it is the same thing with the discretization that you choose: the model changes with the chosen number of steps. What is important is the real model that you use for pricing and hedging: the discretized one that is programmed, not the theoretical limit model.

In this usage the model allows to:

(1) determine the implied market price of risk,
(2) setup hedges,
(3) recognize profits (losses?).

3.2.3. *Simple versus complicated models*

When designing or programming a model it is always tempting, *a priori*, to take into account a lot of possibilities, for example, very general functions for the volatility, with many parameters, many risk factors (to simplify, many Brownian motions), etc.

Sometimes, the techniques that are used are at the same time extremely complex for almost everybody, complicated for specialists and very easy to understand for the person who built the model (at least the few days after he has just finished building it). These techniques may be mathematically elegant, allow a very high level view on the problem and include it in a very general framework.

In order to give an idea of what it can represent, some examples could be: extensions of the Skorohod integral — already an extension of the Ito's integral — to very general sets of stochastic processes — and not only to "simple" continuous semi martingales — and therefore extensions of the Malliavin calculus to some handsome stochastic processes.

The problem is that the notion of elegance is very relative. For example, traders or investors seem to consider as more elegant models that they can use to hedge their positions and make profit. Sometimes, maybe with some exaggeration, we could almost say that the more profit they make, the more elegant they find the model. Therefore, there is often the same question: what is a top model (financially speaking)? Is it better to have a model that will take into account more points and be (maybe) more realistic or to keep it simple?

As far as we know, there is no ready-to-give answer. The following points can help each one to decide on the model:

(1) Does the model incorporate all — or at least, enough — details of the relevant market?
(2) Does it address important risk?
(3) Does it approximate reality?
(4) Can I interpret the model parameters?

(5) Do the model parameters stay rather stable from one day to another?
(6) Do I understand the way the pricing or hedge ratios behave when market conditions move?
(7) Can I — and other people — learn how to use the pricers that are built using the model?
(8) Is my hit ratio (numbers of demands/numbers of done deals) sensible? Here sensible means, not too low (no deals done, no profit) but also not too high (very likely, mispricing).
(9) Will the risk management accept the model? If yes, will the risk managers use it? If not, which consequence will it have on my official P&L?

This point is essential, as it is the one that decides of the official profit of the bank and, also of interest, of the trader's bonus. It is not useful to see a good profit in the system that I use if no one else uses it and that the official P&L is actually an L and not a P.

3.2.4. *Warnings*

After some time, habit becomes a second nature and people tend to use their pricers as black boxes, and not question the results or assumptions made. Here are some warnings that, we think, one always should keep in mind:

(1) The model will be used outside the original intent.
(2) Unreasonable assumptions will be used.
(3) Shortcuts will fail.
(4) Simple models always breakdown. However, they can incorporate many details.
(5) Complicated models are typically more robust about their set of assumptions. It is usually not necessary to change the hypotheses, or at least, not often. However they sometimes miss crucial elements and may be slower to give a price.
(6) The parameters estimations of a complicated model are often not robust: they sometimes change a lot from one day to another.

The reason for this is that the calibration problem is not properly conditioned: there are a lot of unknown parameters and few equations. To the contrary of what one could think, this large degree of freedom is not a good thing: the parameters are free to move a lot and almost not continuously.

(7) Multiple models need independent implementation and frequent checking.

The following points refer to risk detection. They concern with the specific ways or moments to check prices computed by using the model.

(1) Does the portfolio hold its value as trades expire? As there is no more uncertainty at maturity, model life and real life should be the same. For this test, keep track of deals at expiry or legs reset (uncertain cash flows become known).
(2) No future? If we make simulations with the model, which behavior will we see in the future? For example, which forward prices — forward volatilities could be another example — will be implied by the model? Are they coherent with the behavior that we expect or can usually recognize in the market?
(3) Does the model pass the back testing tests? As an example to illustrate this point, suppose that one day you initiate a deal and then daily hedge it following the ratios produced by the model. What will be your P&L at the maturity of the option?

    If your model is not good — there are many reasons that could imply this: wrong assumptions, bugs, instability of calibrated parameters, etc. — you take risks making this test in real life. It is therefore better to do it by back testing in a "what would have happened?" mode.

    For this, it is extremely very important to store your data for model and parameters by back testing.

(4) Do I really see what is going on? A portfolio consisting in a large number of trades can mask the behavior of individual trades; therefore you might miss crucial points.

3.2.5. *Special warnings: calibration issues*

Among the warnings, the category "calibration" deserves a special mention.

(1) Which choice of calibration instruments? The calibration set should always contain at least the products that we want to use for hedging: they are the ones that we will have to buy or sell. Hedging means to build a portfolio that replicates the product that I want to hedge.

If the prices of the products in this portfolio are not the market prices, the portfolio will replicate the product only on paper, not in real life.

A remaining question is: do I consider other products in the calibration set, and why?

As a remark, let us note that a parameterization of model inputs (e.g. choice of interpolation), with a small set of parameters, is simpler.

(2) Exact fit or close fit (e.g. what about bid/ask)? If we want an exact fit between prices computed with the model and prices seen in the market, then we have a problem: there is nothing such as one single price, in comparison to the worst case: great bid/ask difference. Indeed, we know that below a certain amount of money we will not be able to buy a product and above another amount of money we will not be able to sell it. Sometimes, the market is one way (e.g. everybody wants to buy and nobody wants to sell the product) and in these cases, we do not have two bounds but only one (in the given example, it would be an upper bound).

Therefore, under an exact fit calibration, we must also choose which number in the possible range of quotations we want to reproduce. Remember also that depending on the hedge ratio, we will have to buy or sell a certain quantity of each of the products in the replicating portfolio.

(3) Model versus market error: An exact fit calibration may also introduce market errors: one wants to exactly reproduce prices that are not exactly correct, for example, because they concern long maturity products that have not been traded recently. As

the market errors move all the time, they propagate through calibration.

(4) Recalibration frequency? How often do we recalibrate? Here also, it does not seem that there exist ready-to-give answers. This may depend of the frequency at which the market data change.

A constraint which can also intervene is time computation: if a single calibration takes a lot of time, people will be reluctant to often calibrate.

Daily recalibration may also imply that the parameter estimates vary with changes not explicitly modeled but occasional recalibration may imply that the parameter estimates do not reflect mark to market.

(5) Global versus local calibration? Local calibration is as if we had different models. However, if (e.g.) the entire swaption matrix is reproduced, then parameters will change much more often and could lead to violent swings in prices or hedges.

(6) Stability of hedge ratios? For example, the forward curve computed after the yield curve building may vary with no reason.

(7) To recalibrate or not when computing the Greeks? If not carefully thought, calibration is naturally unstable. If the calibrated parameters vary much, the hedge ratios could lead to strange variation in the hedge ratios.

3.3. Reasonable Mathematical Basis for Financial Modeling

3.3.1. *Modeling framework*

When people look at the models that are used in market finance, they all have one point in common: an asset evolution is a "semi martingale" (also called Ito process). We start to show why it was natural to consider them and then develop the notion of risk neutral probability. Before we decide of a modeling framework, let us have a look at a typical graph of a financial variable. In Figure 3.1, we have plotted the S&P 500 PER (price earning ratios) between 1986 and

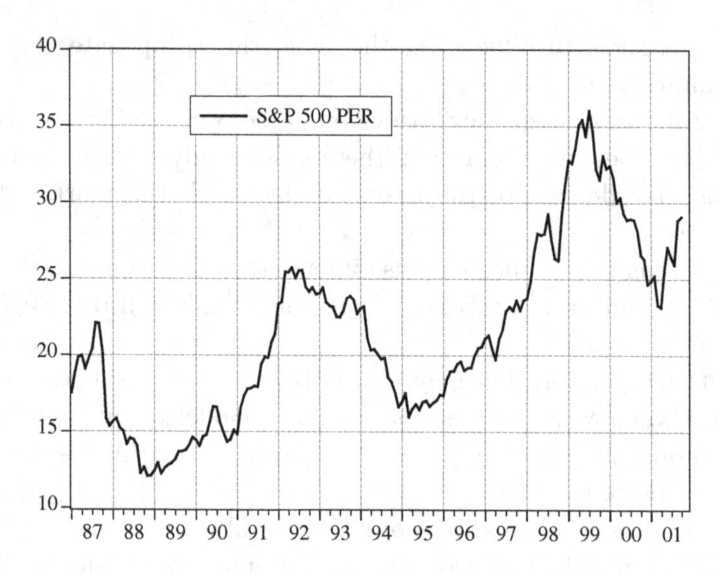

Figure 3.1: S&P 500 Price Earning Ratios.

2001. We could have plotted other indices and the result would have looked the same.

We can see that the asset evolution is the sum of two components:

(1) There is a trend, which is (more or less) linear. This trend varies over time.
(2) There is some additional "noise" which looks random.

Therefore, we will model asset prices (or returns) as stochastic processes that are the sum of a process which is time linear (at least for a short period of time) and of another process, which will add randomness to the first part. For brevity we write S for this set of processes. Then we must look at some properties that we need of this set S in order to make computations.

3.3.1.1. *Words on probability*

As we consider randomness, we must define a probability and a filtration.

(1) The filtration is the set of market data stored everyday (or every hour, or every second) by a provider (or even: all the data

providers), assuming that he never gets rid of any market data. So, we have a set of data that increases over time. Let us note $\Im_t$ this set of data at time t. As we keep all the data as time increases we therefore have $\Im_t \subset \Im_T$ is $t < T$.

(2) We say that a financial quantity is $\Im_t$ measurable if we can compute it with all the data contained in our filtration $\Im_t$.

When we consider credit products, this might lead to some questions: sometimes a credit event (e.g. the default of a firm) is not linked to market conditions. There are recent examples where the default comes from accounting problems, or defalcation. We will not continue is this direction and discuss about needs or not to extend the filtration.

(3) The probability P is defined by historical frequencies: the probability of an event is the frequency at which it happened in the past.

As one can see, these notions do not depend on a specific currency or country but are global.

3.3.1.2. *Types of computations*

Let us recall that we want to be able to propose pricings and hedges to the investors or traders, so we need to produce numbers and prices.

(1) To price portfolios, we need to make additions, if we add products in the portfolio, and subtractions, if we sell, between financial quantities (i.e. elements of S) and still get a financial quantity. S must be stable with "+" and "−".

(2) We want to be able to invest in foreign countries and get prices in our currency. As the price in our currency is the product of the price in the foreign currency by the exchange rate, we must also be able to multiply financial quantities and the result must still be a financial quantity.

Another way to see this is to say that the behavior of a financial quantity is independent of the chosen currency unit. We want the graphs look alike whatever the currency unit that we choose to express prices or financial quantities.

Therefore S must be stable by multiplication — and of course division, due to the symmetry of an exchange rate — between two (or more) financial quantities.

(3) As we want to be able to get the price of any (i.e. sensible) quantity of an asset, S must be also stable by multiplication by a rational number.

(4) There are still at least two operations that are essential and that we want to be able to perform: we want to compute asset returns and, knowing returns of assets, we want to know their prices.

In terms of mathematical operations: we need to compute $\Delta P/P$ for short or long time horizon. As $\Delta P/P$ is linked to $\Delta \log(P)$ we need to be able to make computations with the logarithm of prices. We saw that we need also to do the inverse operation; therefore, we must be able to take exponential of financial quantities and still have a financial quantity.

3.3.1.3. *Ito processes and Ito/Doblin formula*

To sum up we look for a set of processes $X(t)$ such that

(1) $X(t + \Delta t) - X(t)$ can be written under the form $a(t, \ldots)\Delta t + R(t, \ldots)$, where $a(t, \ldots)$ is a function stepwise constant and $R(t, \ldots)$ is a stochastic process.

(2) When we take two such processes and apply to them one of the operations that we listed above, the result must have the same form.

Up to now, we know that a set of processes will do it: the set of Ito processes. The Ito/Doblin formula states that these processes are stable under the action of functions that are continuously twice differentiable (actually, functions that are a bit less smooth will also let this set stable). This includes the operations that we listed.

What is still better: the Ito/Doblin formula is explicit and we know how to compute $f[X(t + \Delta t)] - f[X(t)]$.

Wolfgang Döblin/Vincent Doblin: As Wolfgang Döblin/Vincent Doblin is not well known, let us start with a bit of history. Wolfgang Döblin had fled Germany with his family in March 1933 after the burning of the Reichstag and the vote of full powers to Hitler. His father's notoriety as a writer would have left them no chance to avoid racial persecutions. Together with his parents and his two young brothers, he acquired French citizenship in 1936 and he took the name of Vincent Doblin.

Doblin died in June 1940, at the age of 25. Trying to escape German troops, refusing to give himself up as a prisoner, he committed suicide. At this time, after very few years of research, around 5 years, and the publication of some most remarkable results, he had sent in 1940 during his time in army on the front line in Lorraine, probably in February, a sealed envelop to the Academy of Sciences in Paris.

This sealed envelop could be opened only in May 2000. Among other important results, there was in this document a formula which is another form of what is now called the Ito formula. Actually Ito created Ito's stochastic integral in 1942 and Ito's formula was published in 1951. Readers interested in more information on Döblin's works and life can read "Comments on the life and mathematical legacy of Wolfgang Döblin", *Finance and Stochastics*, Volume 6, Issue 1, January 2002, pp. 3–47, written by Bernard Bru and Marc Yor. They will also find there more explanations on the "Pli cacheté" (i.e. sealed envelop) procedure, which goes back at least to 1701 where one was sent by Johann Bernoulli.

Ito processes: The following sections, more technical even if voluntarily we actually do not state all the technical conditions that would be necessary, explain why it was natural to look at Brownian motions.

It is possible to jump over these sections — explained with the proper details later on — assume the standard usage of Brownian motion and go directly to the risk neutral probability paragraph.

Let us recall that an Ito process is a stochastic process which is the sum of a finite variation process and a martingale.

(1) Among the finite variation processes, we can find all the processes of the form $\int_0^t A(s)ds$ where $A(s)$ is continuous, except maybe on some exceptional points.

If we do not consider "monsters" — functions with some behavioral pathologies that we never have in finance (we talk about the functions) — we could even say that a finite variation process is something of the form $\int_0^t A(s)ds$.

(2) Martingales are processes that keep centered on their current value.

If today (let us say at time t) we look at the future values (let us say at time T) of the process, they are centered on the spot value of time t.

Therefore, we will assume that all financial quantities are of the form $\int_0^t A(s)ds + M(t)$, where $A(s)$ is continuous, except may be on some exceptional points and where $M(t)$ is a martingale for the filtration and probability that we already mentioned.

3.3.1.4. *Usage of Brownian motion*

We therefore, assumed that returns are Ito processes. As the coefficients that we consider (i.e. finite variation process and the martingale parts) are stochastic, it would make no difference to assume that it is the prices which are Ito processes.

The mathematical writing of our assumption is then $\Delta X/X = \mu_X \Delta t + \Delta M_t$ where:

(1) We wrote μ_X as a short writing for $\mu_X(t, (\Im_s)_{0\leq s\leq t})$, i.e. the drift varies over time and may depend on the whole history of the economy. It might therefore be random.
(2) For a time s, M_s is the value at time s of a martingale and $\Delta M_t = M_{t+\Delta t} - M_t$.

Although our modeling framework is getting more precise, it is still too vague to allow computations. For example, we have no precision on the martingale part and it would not be possible to make computations such as Monte-Carlo simulations. We still have to introduce a mathematical tool that simplifies some presentations: the quadratic variation of a stochastic process. Then we will be able to move to Brownian motions and processes of known laws.

Quadratic variation and stochastic integrals: Let us divide an interval $[0, T]$ in pieces $[0, t_1], [t_1, t_2], [t_2, t_3], \ldots, [t_n, T]$.

(1) It can be shown, under reasonable assumptions on the Ito process M, that the sum $[M(0) - M(t_1)]^2 + [M(t_1) - M(t_2)]^2 + \cdots + [M(t_n) - M(T)]^2$ has a limit when we consider smaller and smaller pieces. The limit is called the quadratic variation of the martingale and is noted $\langle M \rangle_T$. It is sometimes more convenient to write $\langle M_T \rangle$, it will mean the same thing. As with a scalar product, let us then define for two Ito processes $\langle M, N \rangle_T = \frac{1}{2}[\langle M + N \rangle_T - \langle M \rangle_T - \langle N \rangle_T] = \langle N, M \rangle_T$, therefore, we also have $\langle M, M \rangle_T = \langle M \rangle_T$. If we look at the classical example of a Brownian motion B, one shows that $\langle B \rangle_T = T$. This is essentially the result that we use in Finance.
(2) Looking at martingales, it can be shown — under reasonable assumptions on the martingale M and the stochastic process $A(s)$ — that $A(t_1)[M(t_2) - M(t_1)] + A(t_2)[M(t_3) - M(t_2)] + A(t_3)[M(t_4) - M(t_3)] + \cdots + A(t_n)[M(t_n) - M(T)]$ has a limit when we consider smaller and smaller pieces. The limit is called an Ito integral and is written $\int_0^T A(s)dM_s$.

Among the classical properties:

(1) $\langle M \rangle_T$ is an increasing function of the time T that is null if $T = 0$.
(2) For "any" Ito process N, if $M_t = \int_0^t K(s)ds$, then at any time T, $\langle M, N \rangle_T = 0$.
(3) For "any" stochastic processes $H(s)$ and $K(s)$:

$$\left\langle \int_0^T H(s)dM_s, \int_0^T K(s)dN_s \right\rangle = \int_0^T H(s)K(s)d\langle M, N \rangle_s.$$

(4) Classical example of a Brownian motion, if $M = N = B$:

$$\left\langle \int_0^T H(s)dB_s, \int_0^T K(s)dB_s \right\rangle = \int_0^T H(s)K(s)ds.$$

(5) For a martingale M that starts at 0 when $t = 0, \mathrm{Var}(M_t) = E(\langle M \rangle_t)$.

Martingales and Brownian motions: We now show that — except cases that we would consider in Finance as "pathological" and not realistic — we can write the martingale part of the Ito process as $\int_0^t \sigma_X(s) \cdot dW_s$ where:

(1) We wrote σ_X — the volatility, which is a positive number — as a short writing for $\sigma_X(t, (\Im_s)_{0 \leq s \leq t})$, that is, a coefficient that varies over time and may depend on the whole history of the economy. It might, therefore, also be random but at time t it is $\Im_t$ measurable so we know its value.
(2) W_s is a Brownian motion (people also talk of Wiener process) which depends on the financial asset that we consider.

The martingale M is given and we want to write $M_t = \int_0^t \sigma_X(s).dW_s$ so we have two unknowns: the volatility and the Brownian motion. Let us assume that they exist then see what they should be. We will then verify that like this we found a solution.

We want $M_t = \int_0^t \sigma_X(s).dW_s$, therefore, we must have $\langle M_t \rangle = \langle \int_0^t \sigma_X(s).dW_s \rangle$ so $\int_0^t \sigma_X^2(s)ds = \langle M \rangle_t$. We suppose that we can differentiate the increasing function $\langle M \rangle_t$ that we will write $f(t)$ for convenience. Then we could, for example, define $\sigma_X(t) = \sqrt{f'(t)}$. In that case, there is no more choice as candidate for a Brownian motion: it would be $\int_0^t \frac{dM_s}{\sqrt{f'(s)}}$, assuming that this integral exists. So now we start from scratch, assume that the martingale M is a nice one, we can differentiate the increasing function $f(t) = \langle M \rangle_t$ and furthermore the integral $W_t = \int_0^t \frac{dM_s}{\sqrt{f'(s)}}$ exists (technical name: the integral will be a local martingale).

Then we have

$$\langle W_t \rangle = \left\langle \int_0^t \frac{dM_s}{\sqrt{f'(s)}} \right\rangle = \int_0^t \frac{d\langle M \rangle_s}{f'(s)} = \int_0^t \frac{d\langle M \rangle_s}{\frac{d\langle M \rangle_s}{ds}} = \int_0^t ds = t$$

and this will show that W_t is a Brownian motion (according to Levy theorem).

Therefore, we can restrain ourselves to integrate Brownian motions and assume that $\frac{\Delta X}{X} = \mu_X \Delta t + \sigma_X \bullet \Delta W_t$ where the Brownian motion may then depend on the asset that we consider. (The attentive reader will have seen that a more exact assumption is $dX/X = \mu_X \cdot dt + \sigma_X \cdot dW_t$.)

3.3.2. *Risk neutral probability*

From now on, we assume that $dX/X = \mu_X.dt + \sigma_X.dW_t$ where the Brownian motion may depend on the financial quantity X.

(1) Here also we wrote σ_X — the volatility, which is a positive number — as a short writing for $\sigma_X(t, (\Im_s)_{0 \leq s \leq t})$, that is, a coefficient that varies over time and may depend on the whole history of the economy. It might, therefore, also be random but at time t it is $\Im_t$ measurable so we know its value.
(2) In the same way, μ_X is a short writing for $\mu_X(t, (\Im_s)_{0 \leq s \leq t})$.

We also use these convenient shortcuts in the n factor case.

One of the problems that we have is that μ_X is linked to the probability that we consider: with some abuse, we could write $E(dX/X) = E(\mu_X)dt$ so, on average the drift would be the expected asset return. But where do we get this value? And if we have it, we only have its average value! The risk-neutral probability framework is useful to avoid this problem.

3.3.2.1. *Example of the one factor case*

Here one factor means that besides all the other assumptions, the Brownian motion that we defined is the same for all the financial quantities.

We consider two assets of respective prices X and Y. Let us look at the variation between two times t and $t + \Delta t$. We have:

$$\frac{\Delta X}{X} = \mu_X \Delta t + \sigma_X \Delta W_t \text{ and } \frac{\Delta Y}{Y} = \mu_Y \Delta t + \sigma_Y \Delta W_t.$$

As mentioned earlier, at time t we know the values of X_t, Y_t, σ_X, σ_Y, μ_X and μ_Y; let us say — just to have values in mind — that $X_t = 500$, $\sigma_X = 17\%$, $Y_t = 300$ and $\sigma_Y = 29\%$. We could also choose values for μ_X and μ_Y (as we also can measure them at time t) but this is not really necessary. We do not know for sure what is the value of ΔX because it depends on the random (by the way, Gaussian) variable $\Delta W_t = W_{t+\Delta t} - W_t$ so for us $X(t + \Delta t)$ is a risky asset. It is the same for ΔY which includes the same random variable.

As we have two equations (one for each asset) but only one unknown (the value of the random variable) it is possible to build a combination of X and Y (i.e. to make a portfolio of Xs and Ys) that eliminates the risky component.

Let us consider a portfolio where we buy 87 (i.e. $Y_t \times \sigma_Y$) assets X and sell 85 (i.e. $X_t \times \sigma_X$) assets Y. The value P_t of the portfolio at time t is

$$\begin{aligned} 87X_t - 85Y_t &= 18{,}000 = Y_t\sigma_Y \times X_t - X_t\sigma_X \times Y_t \\ &= X_tY_t(\sigma_Y - \sigma_X) \end{aligned}$$

The variation of the value of the portfolio is then:

$$\begin{aligned} 87\Delta X_t - 85\Delta Y_t &= 87(500\mu_X\Delta t + 500 \times 0.17\Delta W_t) \\ &\quad - 85(300\mu_Y\Delta t + 300 \times 0.29\Delta W_t) \\ &= (87 \times 500\mu_X - 85 \times 300\mu_Y)\Delta t \\ &= (Y_t\sigma_Y \times X_t\mu_X - X_t\sigma_X \times Y_t\mu_Y)\Delta t \\ &= X_tY_t(\sigma_Y\mu_X - \sigma_X\mu_Y)\Delta t \end{aligned}$$

We have, therefore, the surprise that this portfolio is not risky anymore (of course, we cautiously chose the numbers 87 and 85 for this purpose!).

Therefore, its return must be the risk free rate on the investment period:

$$X_tY_t(\sigma_Y - \sigma_X)r_t\Delta t = X_tY_t(\sigma_Y\mu_X - \sigma_X\mu_Y)\Delta t$$

so

$$(\sigma_Y - \sigma_X)r_t = \sigma_Y\mu_X - \sigma_X\mu_Y$$

that we can write as:

$$\frac{\mu_X - r_t}{\sigma_X} = \frac{\mu_Y - r_t}{\sigma_Y}$$

This is an extremely important result: the quantity $\lambda_t = \frac{\mu_X - r_t}{\sigma_X}$ (known as the Sharpe ratio or the market risk premium) is the same for all the assets that we have in the market. It is an invariant of the market. So we have:

$$\begin{aligned}\frac{dX}{X} &= \mu_X dt + \sigma_X dW_t \\ &= (r_t + \lambda_t \sigma_X)dt + \sigma_X dW_t \\ &= r_t dt + \sigma_X(\lambda_t dt + dW_t).\end{aligned}$$

It seems that we are not better off: we had a problem because we did not know the drift. Now we know it, but as we do not know the Sharpe ratio (otherwise we would know μ_X) it is now the law of $W_t + \int_0^t \lambda_s ds$ that we do not know, so what is the advantage of this new diffusion equation?

Actually, we are really better off because there is one way to make disappear (no violence, of course) the Sharpe ratio: the Girsanov's theorem.

We know that there is another probability Q, which we can express in function of the Sharpe ratio,[2] such that the law of $Z_t = W_t + \int_0^t \lambda_s ds$ under Q is that of a Brownian motion. Q is called the risk neutral probability.

As $dX/X = r_t dt + \sigma_X dZ_t$ we have a better idea of the law (under Q) of the asset return. It will be seen later that we can compute prices by using mathematical expectations under Q: our program will be fulfilled.

3.3.2.2. *The n factor case*

We now suppose that the world economy depends only on a finite number n of stochastic factors. As computers will not take into account an infinite numbers of parameters, this is not in practice a limitation.

For each of these factors we write the martingale part with its Brownian motion $\tilde{W}_i(t)$ under the risk neutral probability Q, as we did before. So,

(1) We start with n assets $X_1(t), X_2(t), \ldots, X_n(t)$ and their n (*a priori*, correlated) martingales.

---

[2]If we note by E^P the expectation under the probability P and E^Q the expectation under the probability Q, for a quantity that we can measure at time t, we have the relation: $E^Q(U) = E^P(U.\mathrm{e}^{-\int_0^t \lambda_s dW_s - \frac{1}{2}\int_0^t \lambda_s^2 ds})$.

(2) Then we deduce the corresponding (*a priori*, correlated) Brownian motions $\tilde{W}_1(t), \tilde{W}_2(t), \ldots, \tilde{W}_n(t)$ and volatilities such that $\frac{dX_i}{X_i} = \mu_i dt + \sigma_i d\tilde{W}_i(t)$.

We will see later on that starting with these n Brownian motions and volatilities, it is possible to deduce other n independent Brownian motions which build the vector $\vec{W}(t) = (W_1(t), W_2(t), \ldots, W_n(t))^t$ and n vectors $\vec{\sigma}_1(t), \vec{\sigma}_2(t), \ldots, \vec{\sigma}_n(t)$ such that $\frac{dX_i}{X_i} = \mu_i dt + \vec{\sigma}_i(t) \bullet d\vec{W}(t)$.

Here, $\vec{\sigma}_i \bullet d\vec{W}(t)$ means the scalar product between the two vectors, that is, $\vec{\sigma}_i \bullet d\vec{W}(t) = \sigma_{i,1} dW_1(t) + \sigma_{i,2} dW_2(t) + \cdots + \sigma_{i,n} dW_n(t)$.

There should be no confusion between the vector $\vec{\sigma}_i(t)$ and the positive number $\sigma_i(t)$. The link between them is the following: we can also impose that the vectors verify: $\|\vec{\sigma}_i\| = \sigma_i$, where $\|\vec{\sigma}_i\|$ means $\sqrt{\sigma_{i,1}^2 + \sigma_{i,2}^2 + \cdots + \sigma_{i,n}^2}$.

The vectorial writing is just a cheap and convenient way to be in an n factor world: write the equations as in a 1-factor case, add an arrow, mentally replace products by scalar products and that's (almost) all.

Concerning this vectorial writing, it may be convenient to note that we still have:

$$\left\langle \int_0^T \vec{H}(s) \bullet d\vec{W}_s, \int_0^T \vec{K}(s) \bullet d\vec{W}_s \right\rangle = \int_0^T \vec{H}(s) \bullet \vec{K}(s) ds.$$

Something important to note is that any asset (even another one that the ones we just considered) has its "volatility" vector but, by construction, the (vectorial) Brownian motion is the same for all the assets in the world. We assume that there is no linear relationship between the vectors $\vec{\sigma}_1(t), \vec{\sigma}_2(t), \ldots, \vec{\sigma}_n(t)$. We now look at another asset X_{n+1} which will then verify $\frac{dX_{n+1}}{X_{n+1}} = \mu_{n+1} dt + \vec{\sigma}_{n+1}(t) \bullet d\vec{W}(t)$.

Each of the $n+1$ vectors $X_i(t)\vec{\sigma}_i(t)$ has only n coordinates and therefore there must exist a relation between these vectors. So we know that there are some real numbers $a_1(t), a_2(t), \ldots, a_{n+1}(t)$ such that:

$$a_1(t)X_1(t)\vec{\sigma}_1(t) + a_2(t)X_2(t)\vec{\sigma}_2(t) + \cdots + a_{n+1}(t)X_{n+1}(t)\vec{\sigma}_{n+1}(t) = \vec{0}.$$

As in the one factor case, we build a portfolio which neutralizes the Brownian motion: let us consider the portfolio where we buy the asset X_i (if $a_i(t) > 0$) or sell the asset (if $a_i(t) < 0$) and let us note P the price of this portfolio.

At any time s, we have therefore: $P(s) = a_1(t)X_1(s) + a_2(t)X_2(s) + \cdots + a_{n+1}(t)X_{n+1}(s)$ (note the mix of time t and time s). To have a lighter writing, we will keep the time t fixed and not continue to write the dependency on it.

We have:

$$\begin{aligned}\Delta P &= a_1\Delta X_1 + a_2\Delta X_2 + \cdots + a_{n+1}\Delta X_{n+1}\\ &= (a_1X_1\mu_1 + a_2X_2\mu_2 + \cdots + a_{n+1}X_{n+1}\mu_{n+1})\Delta t\\ &\quad + (a_1X_1\vec{\sigma}_1 + a_2X_2\vec{\sigma}_2 + \cdots + a_{n+1}X_{n+1}\vec{\sigma}_{n+1})\Delta\vec{W}_t\\ &= (a_1X_1\mu_1 + a_2X_2\mu_2 + \cdots + a_{n+1}X_{n+1}\mu_{n+1})\Delta t.\end{aligned}$$

As this portfolio is not risky we deduce that $\Delta P = r_t dt$ hence:

$$\begin{aligned}a_1X_1\mu_1 &+ a_2X_2\mu_2 + \cdots + a_{n+1}X_{n+1}\mu_{n+1}\\ &= r_t(a_1X_1 + a_2X_2 + \cdots + a_{n+1}X_{n+1})\end{aligned}$$

or

$$a_1X_1(\mu_1 - r_t) + a_2X_2(\mu_2 - r_t) + \cdots + a_{n+1}X_{n+1}(\mu_{n+1} - r_t) = 0.$$

Let us consider now the vector $\vec{v}_i$ of $n+1$ coordinates built this way: the first n coordinates are the coordinates of $\vec{\sigma}_i$ and the last coordinate is $\mu_i - r_t$.

With the last relation, we see that $a_1X_1\vec{v}_1 + a_2X_2\vec{v}_2 + \cdots + a_{n+1}X_{n+1}\vec{v}_{n+1} = \vec{0}$.

The matrix M which has the vectors $\vec{v}_i$ as columns is not invertible, therefore the transposed matrix is neither invertible: there is also a relation between the lines of M.

Writing this relation we see that there exist n real numbers λ_1, $\lambda_2, \ldots, \lambda_n$ such that: $\mu_i - r_t = \lambda_1\sigma_{i,1} + \lambda_2\sigma_{i,2} + \cdots + \lambda_n\sigma_{i,n}$ where $\sigma_{i,1}$ is the first coordinate of the vector $\vec{\sigma}_i$, $\sigma_{i,2}$ its second coordinate, etc.

If we note $\vec{\lambda}$ the vector which has the λ_i as coordinates, the previous relation means that $\mu_i - r_t = \vec{\lambda} \bullet \vec{\sigma}_i$.

As in the one factor case, we can now write the diffusion equation of the assets as:

$$\begin{aligned}\frac{dX_i}{X_i} &= \mu_i dt + \vec{\sigma}_i \bullet d\vec{W}_t \\ &= (r_t + \vec{\lambda} \bullet \vec{\sigma}_i)dt + \vec{\sigma}_i \bullet d\vec{W}_t \\ &= r_t dt + \vec{\sigma}_i \bullet (\vec{\lambda} dt + d\vec{W}_t).\end{aligned}$$

We can again apply Girsanov's theorem, so there is another probability Q, which we can express in the same way in function of the Sharpe ratio,[3] such that the law of $\vec{Z}_t = \vec{W}_t + \int_0^t \vec{\lambda}_s ds$ under Q is that of a Brownian motion.

We also have again $dX/X = r_t dt + \vec{\sigma}_X \bullet d\vec{Z}_t$.

The number of letters is limited and it is better to easily see that we deal with a Brownian motion and to know under which probability we work: a standard notation is to write $\vec{W}_t^Q$ for a process that is a Brownian motion under the probability Q.

In finance we usually do not look at or even mention the underlying filtration. We will always implicitly assume that we can measure a financial quantity as soon as we know the value of the economic shocks. It means that we assume having only one filtration, which is the natural filtration of any of the Brownian motions that we consider.

Remarks on assumptions:

(1) Our reasoning assumes that there are as many "independent" assets, that is, with no linear relation between their volatility vectors — as the random shocks that govern the economy. This had two consequence:

   (a) We obtain a unique vector $\vec{\lambda}$, and then a unique risk neutral probability Q.
   (b) We can replicate any other asset as a portfolio of the independent assets. We therefore assumed that the market is complete.

---

[3]If we note by E^P the expectation under the probability P and E^Q the expectation under the probability Q, for a quantity that we can measure at time t, we have the relation:

$$E^Q(U) = E^P(U.\mathrm{e}^{-\int_0^t \vec{\lambda}_s \bullet d\vec{W}_s - \frac{1}{2}\int_0^t \lambda_s^2 ds}) \quad \text{where}$$
$$\lambda_s^2 = \lambda_1^2(s) + \lambda_2^2(s) + \cdots + \lambda_n^2(s) = \|\vec{\lambda}_s\|^2.$$

(2) We then assumed that whatever the portfolio, a non-risky asset provides the risk free interest rate relevant for the duration of the investment. This is also an assumption: different portfolios can have different tax treatments and therefore, provide different "net" risk free rate.

(3) We assumed to be able to buy or sell any quantity of our basis assets (furthermore with a price that does not depend on the quantity that we buy or sell) is also valid. Of course, the standard remark that this is a strong hypothesis is valid.

3.3.2.3. *The two currency case*

We wanted to have a modeling where the behaviors of the financial quantities look alike whatever the currency, but up to now we did not choose a particular currency.

We now look in more details how to move from one choice of currency to another.

To have an example in mind, we look at the case of an exchange rate. We might also consider another currency unit and decide, in theory, that our money is any particular good instead of banknote.

We, therefore, consider domestic assets (all notations related to them will have the subscript "d") and foreign assets (all notations related to them will have the subscript "f"). A domestic asset is an asset whose cash flows are all denominated in domestic currency, and the equivalent for a foreign asset.

The exchange rate $X(t)$ is defined by: at time t, one unit of the domestic currency is worth $X(t)$ units of the foreign currency (e.g. $1€ = 1.25\$$).

Let us recall that the historical probability P is global. As before, it will be less heavy not to mention all the dependencies.

We sum up what we already know.

$$
\begin{aligned}
\frac{dX_{\mathrm{d}}}{X_{\mathrm{d}}} &= \mu_{X_{\mathrm{d}}} dt + \vec{\sigma}_{X_{\mathrm{d}}} \bullet d\vec{W}_t^P \\
&= r_{\mathrm{d}} dt + \vec{\sigma}_{X_{\mathrm{d}}} \bullet d\vec{W}_t^{Q_{\mathrm{d}}},
\end{aligned}
$$

$$\begin{aligned}\frac{dX_{\mathrm{f}}}{X_{\mathrm{f}}} &= \mu_{X_{\mathrm{f}}}dt + \vec{\sigma}_{X_{\mathrm{f}}} \bullet d\vec{W}_t^P \\ &= r_{\mathrm{f}}dt + \vec{\sigma}_{X_{\mathrm{f}}} \bullet d\vec{W}_t^{Q_{\mathrm{f}}}, \\ \frac{dX}{X} &= \mu_X dt + \vec{\sigma}_X \bullet d\vec{W}_t^P,\end{aligned}$$

We also have:

(1) for any domestic asset $\mu_{X_{\mathrm{d}}} = r_{\mathrm{d}} + \vec{\lambda}_{\mathrm{d}} \bullet \vec{\sigma}_{X_{\mathrm{d}}}$,
(2) for any foreign asset $\mu_{X_{\mathrm{f}}} = r_{\mathrm{f}} + \vec{\lambda}_{\mathrm{f}} \bullet \vec{\sigma}_{X_{\mathrm{f}}}$.

Starting from a foreign price X_{f} of an asset, we now build a domestic one. The asset that proposes the cash-flows of the foreign asset, converted in domestic currency at any time is therefore an asset which provides cash-flows in the domestic currency: it is a domestic asset. Its price Z_{d} is given by: $Z_{\mathrm{d}} = X_{\mathrm{f}}/X$, therefore,

$$\begin{aligned}\frac{dZ_{\mathrm{d}}}{Z_{\mathrm{d}}} &= \mu_{Z_{\mathrm{d}}}dt + \vec{\sigma}_{Z_{\mathrm{d}}} \bullet d\vec{W}_t^P, \\ &= r_{\mathrm{d}}dt + \vec{\sigma}_{Z_{\mathrm{d}}} \bullet d\vec{W}_t^{Q_{\mathrm{d}}}.\end{aligned}$$

We also have (by the lemma of Ito):

$$\begin{aligned}\frac{dZ_{\mathrm{d}}}{Z_{\mathrm{d}}} &= \frac{dX_{\mathrm{f}}}{X_{\mathrm{f}}} - \frac{dX}{X} - \left\langle \frac{dX_{\mathrm{f}}}{X_{\mathrm{f}}} - \frac{dX}{X}, \frac{dX}{X} \right\rangle \\ &= [\mu_{X_{\mathrm{f}}} - \mu_X - (\vec{\sigma}_{X_{\mathrm{f}}} - \vec{\sigma}_X) \bullet \vec{\sigma}_X]dt + (\vec{\sigma}_{X_{\mathrm{f}}} - \vec{\sigma}_X) \bullet d\vec{W}_t^P.\end{aligned}$$

Under some technical hypotheses — that we can assume in Finance — there is only a possible decomposition of an Ito process so, remembering that $\mu_{X_{\mathrm{f}}} = r_{\mathrm{f}} + \vec{\lambda}_{\mathrm{f}} \bullet \vec{\sigma}_{X_{\mathrm{f}}}$:

$$\begin{aligned}\mu_{X_{\mathrm{f}}} - \mu_X - (\vec{\sigma}_{X_{\mathrm{f}}} - \vec{\sigma}_X) \bullet \vec{\sigma}_X &= \mu_{Z_{\mathrm{d}}}, \\ \vec{\sigma}_{X_{\mathrm{f}}} - \vec{\sigma}_X &= \vec{\sigma}_{Z_{\mathrm{d}}},\end{aligned}$$

so

$$\begin{aligned}r_{\mathrm{f}} + \vec{\lambda}_{\mathrm{f}} \bullet \vec{\sigma}_{X_{\mathrm{f}}} - \mu_X - (\vec{\sigma}_{X_{\mathrm{f}}} - \vec{\sigma}_X) \bullet \vec{\sigma}_X &= r_{\mathrm{d}} + \vec{\lambda}_{\mathrm{d}} \bullet \vec{\sigma}_{Z_{\mathrm{d}}}, \\ \vec{\sigma}_{X_{\mathrm{f}}} - \vec{\sigma}_X &= \vec{\sigma}_{Z_{\mathrm{d}}},\end{aligned}$$

hence

$$r_{\mathrm{f}} + \vec{\lambda}_{\mathrm{f}} \bullet \vec{\sigma}_{X_{\mathrm{f}}} - \mu_X - (\vec{\sigma}_{X_{\mathrm{f}}} - \vec{\sigma}_X) \bullet \vec{\sigma}_X = r_{\mathrm{d}} + \vec{\lambda}_{\mathrm{d}} \bullet (\vec{\sigma}_{X_{\mathrm{f}}} - \vec{\sigma}_X), \quad (3.1)$$
$$\vec{\sigma}_{X_{\mathrm{f}}} - \vec{\sigma}_X = \vec{\sigma}_{Z_{\mathrm{d}}}.$$

In the same way, we have a foreign quantity defined by $Z_{\mathrm{f}} = X.X_{\mathrm{d}}$ therefore (by the lemma of Ito)

$$\begin{aligned} \frac{dZ_{\mathrm{f}}}{Z_{\mathrm{f}}} &= \frac{dX_{\mathrm{d}}}{X_{\mathrm{d}}} + \frac{dX}{X} + \left\langle \frac{dX_{\mathrm{d}}}{X_{\mathrm{d}}}, \frac{dX}{X} \right\rangle \\ &= (\mu_{X_{\mathrm{d}}} + \mu_X + \vec{\sigma}_{X_{\mathrm{d}}} \bullet \vec{\sigma}_X)dt + (\vec{\sigma}_{X_{\mathrm{d}}} + \vec{\sigma}_X) \bullet d\vec{W}_t^P. \end{aligned}$$

Then

$$\mu_{X_{\mathrm{d}}} + \mu_X + \vec{\sigma}_{X_{\mathrm{d}}} \bullet \vec{\sigma}_X = \mu_{Z_{\mathrm{f}}}, \quad \vec{\sigma}_{X_{\mathrm{d}}} + \vec{\sigma}_X = \vec{\sigma}_{Z_{\mathrm{f}}}$$

so

$$r_{\mathrm{d}} + \vec{\lambda}_{\mathrm{d}} \bullet \vec{\sigma}_{X_{\mathrm{d}}} + \mu_X + \vec{\sigma}_{X_{\mathrm{d}}} \bullet \vec{\sigma}_X = r_{\mathrm{f}} + \vec{\lambda}_{\mathrm{f}} \bullet \vec{\sigma}_{Z_{\mathrm{f}}}$$
$$\vec{\sigma}_{X_{\mathrm{d}}} + \vec{\sigma}_X = \vec{\sigma}_{Z_{\mathrm{f}}}$$

hence

$$r_{\mathrm{d}} + \vec{\lambda}_{\mathrm{d}} \bullet \vec{\sigma}_{X_{\mathrm{d}}} + \mu_X + \vec{\sigma}_{X_{\mathrm{d}}} \bullet \vec{\sigma}_X = r_{\mathrm{f}} + \vec{\lambda}_{\mathrm{f}} \bullet (\vec{\sigma}_{X_{\mathrm{d}}} + \vec{\sigma}_X) \quad (3.2)$$
$$\vec{\sigma}_{X_{\mathrm{d}}} + \vec{\sigma}_X = \vec{\sigma}_{Z_{\mathrm{f}}}$$

Let us look at Eqs. (3.1) and (3.2) together:

$$r_{\mathrm{f}} + \vec{\lambda}_{\mathrm{f}} \bullet \vec{\sigma}_{X_{\mathrm{f}}} - \mu_X - (\vec{\sigma}_{X_f} - \vec{\sigma}_X) \bullet \vec{\sigma}_X = r_{\mathrm{d}} + \vec{\lambda}_{\mathrm{d}} \bullet (\vec{\sigma}_{X_{\mathrm{f}}} - \vec{\sigma}_X) \quad (3.1)$$
$$r_{\mathrm{d}} + \vec{\lambda}_{\mathrm{d}} \bullet \vec{\sigma}_{X_{\mathrm{d}}} + \mu_X + \vec{\sigma}_{X_{\mathrm{d}}} \bullet \vec{\sigma}_X = r_{\mathrm{f}} + \vec{\lambda}_{\mathrm{f}} \bullet (\vec{\sigma}_{X_{\mathrm{d}}} + \vec{\sigma}_X) \quad (3.2)$$

If we add them together we deduce:

$$\begin{aligned} &\vec{\lambda}_{\mathrm{f}} \bullet \vec{\sigma}_{X_{\mathrm{f}}} + \vec{\lambda}_{\mathrm{d}} \bullet \vec{\sigma}_{X_{\mathrm{d}}} - (\vec{\sigma}_{X_{\mathrm{f}}} - \vec{\sigma}_X) \bullet \vec{\sigma}_X + \vec{\sigma}_{X_{\mathrm{d}}} \bullet \vec{\sigma}_X \\ &\quad = \vec{\lambda}_{\mathrm{d}} \bullet (\vec{\sigma}_{X_{\mathrm{f}}} - \vec{\sigma}_X) + \vec{\lambda}_{\mathrm{f}} \bullet (\vec{\sigma}_{X_{\mathrm{d}}} + \vec{\sigma}_X), \end{aligned}$$

that is,

$$\begin{aligned} &\vec{\lambda}_{\mathrm{f}} \bullet (\vec{\sigma}_{X_{\mathrm{f}}} - \vec{\sigma}_{X_{\mathrm{d}}} - \vec{\sigma}_X) + \vec{\lambda}_{\mathrm{d}} \bullet (\vec{\sigma}_{X_{\mathrm{d}}} - \vec{\sigma}_{X_{\mathrm{f}}} + \vec{\sigma}_X) \\ &\quad + \vec{\sigma}_X \bullet (\vec{\sigma}_X - \vec{\sigma}_{X_{\mathrm{f}}} \bullet + \vec{\sigma}_{X_{\mathrm{d}}}) = 0, \end{aligned}$$

that is

$$(\vec{\sigma}_{X_{\mathrm{d}}} - \vec{\sigma}_{X_{\mathrm{f}}} + \vec{\sigma}_X) \bullet (\vec{\lambda}_{\mathrm{d}} + \vec{\sigma}_X - \vec{\lambda}_{\mathrm{f}}) = 0.$$

We assume that whatever the vectors $\vec{\sigma}_{X_{\mathrm{d}}}$ or $\vec{\sigma}_{X_{\mathrm{f}}}$ there exist some assets that have them as "volatility" vectors, so $\vec{\lambda}_{\mathrm{d}} + \vec{\sigma}_X - \vec{\lambda}_{\mathrm{f}}$ is a vector orthogonal to the whole space. Hence, it is the null vector and we found: $\vec{\sigma}_X = \vec{\lambda}_{\mathrm{f}} - \vec{\lambda}_{\mathrm{d}}$. If we use that $\vec{\lambda}_{\mathrm{d}} + \vec{\sigma}_X = \vec{\lambda}_{\mathrm{f}}$ in Eq. (3.2) we find:

$$\begin{aligned}\mu_X &= r_{\mathrm{f}} + \vec{\lambda}_{\mathrm{f}} \bullet (\vec{\sigma}_{X_{\mathrm{d}}} + \vec{\sigma}_X) - r_{\mathrm{d}} - \vec{\lambda}_{\mathrm{d}} \bullet \vec{\sigma}_{X_{\mathrm{d}}} - \vec{\sigma}_{X_{\mathrm{d}}} \bullet \vec{\sigma}_X \\ &= r_{\mathrm{f}} - r_{\mathrm{d}} + \vec{\lambda}_{\mathrm{f}} \bullet (\vec{\sigma}_{X_{\mathrm{d}}} + \vec{\sigma}_X) - \vec{\sigma}_{X_{\mathrm{d}}} \bullet (\vec{\lambda}_{\mathrm{d}} + \vec{\sigma}_X) \\ &= r_{\mathrm{f}} - r_{\mathrm{d}} + \vec{\lambda}_{\mathrm{f}} \bullet \vec{\sigma}_X.\end{aligned}$$

As we know the drift and the volatility vector of the exchange rate we can now write:

$$\begin{aligned}\frac{dX}{X} &= (r_{\mathrm{f}} - r_{\mathrm{d}} + \vec{\lambda}_{\mathrm{f}} \bullet \vec{\sigma}_X)dt + \vec{\sigma}_X \bullet d\vec{W}_t^P \\ &= (r_{\mathrm{f}} - r_{\mathrm{d}})dt + \vec{\sigma}_X \bullet d\vec{W}_t^{Q_{\mathrm{f}}}.\end{aligned}$$

Looking at the Brownian motions:

$$\begin{aligned}d\vec{W}_t^{Q_{\mathrm{f}}} - d\vec{W}_t^{Q_{\mathrm{d}}} &= (d\vec{W}_t^P + \vec{\lambda}_{\mathrm{f}}dt) - (d\vec{W}_t^P + \vec{\lambda}_{\mathrm{d}}dt) \\ &= \vec{\sigma}_X dt.\end{aligned}$$

To sum up:

$$\frac{dX}{X} = (r_{\mathrm{f}} - r_{\mathrm{d}})dt + \vec{\sigma}_X \bullet d\vec{W}_t^{Q_{\mathrm{f}}} \quad \text{and} \quad d\vec{W}_t^{Q_{\mathrm{f}}} = d\vec{W}_t^{Q_{\mathrm{d}}} + \vec{\sigma}_X dt.$$

Final remark on some technical conditions: we mentioned several times these kinds of conditions, stating that we could reasonably assume them.

(1) *Condition on smoothness of functions*: Actually we can only program functions that are pieces of polynomials. Even functions such as the exponential function are coded through a polynomial approximation.
Therefore, in practice, the functions that we can use are functions that we can differentiate as often as we want, except on maybe a finite number of exceptional points.

(2) *Condition on the volatility function*: Apart the condition of smoothness, we need to have martingales processes.

In all practical cases the norm $\|\vec{\sigma}_s\|$ is a function:

(a) That is never negative.
(b) With an upper bound: in a trading room, everybody will probably agree that a volatility will not move above 100 billion %. Anyway, if the volatility ever moved beyond this level, we clearly would have other concerns that the validity of some mathematical assumptions and the accuracy of our pricing.
(c) That we can differentiate as often as we want, as we already saw, except maybe on few exceptional points.
(d) That is zero after some time (whatever the firm, we can boldly assume that it will have disappeared in let us say 10 billion years).

A mathematical consequence of these constraints is that the mathematical expectation of $\int_0^{+\infty} \sigma^2 dx$ is finite. This will also allow us to apply Girsanov's theorem.

3.4. Summary

What is interesting in finance over the last 30 years is the spectacular development of mathematical tools applied to the derivatives industry. We have first reviewed the evolution of the modeling approach with two different methods to solve the "hedging portfolio" problem: the PDE and the probability approach.

We have discussed how we can analyze the quality of a model in terms of pricing but also hedging. An appropriate model should not only fit the market data in terms of price, but also lead to realistic value for its Greeks. In addition, it is important to look at the estimation of the model parameters referred to as the calibration problem. The calibration may result in unstable parameters and indicate poor modeling assumptions or the non-existence of a robust solution.

We should also keep in mind that the development of modeling tools contains some operational risk and contributes to operational risk with more and more sophisticated models that are harder to validate.

We conclude on the fact that the absence of arbitrage implies the existence of a risk neutral probability and the corresponding risk free rate.

References

[1] Davis EL (2005). *Operational Risk.* London, UK: Risk Books.
[2] Harrisson JM and Kreps D (1979). Martingales and arbitrage in multi-period securities market. *Journal of Economy Theory,* **20**, 381–408.
[3] Karatzas I and Shreve SE (1991). *Brownian Motion and Stochastic Calculus.* New York: Springer.
[4] Musiela M and Rutkowski M (2005). *Martingale Methods in Financial Modelling.* Berlin: Springer-Verlag.
[5] Revuz D and Yor M (1994). *Continuous Martingales and Brownian Motion.* Berlin: Springer-Verlag.

Chapter 4

THE BLACK–SCHOLES MODEL

In this chapter we tackle the famous Black–Scholes (BS) model that has had an important influence in the world of finance. As a first step, we introduce the model with some assumptions. We explain how to use it to price European call and put options on a non-dividend paying option. We analyze the different hedging strategies and then introduce the price derivatives which measure different dimensions of the risk in an option position. Finally, we discuss about the extensions of the BS model.

4.1. Model History

In 1973 Fischer Black, a 31-year-old independent finance contractor, and Myron Scholes, a 28-year-old assistant professor of finance, at MIT had written a first draft of a paper that presented an analytical model to determine the fair value for European call option. For his PhD degree in applied mathematics, Scholes worked on a model to valuate securities and his colleagues worked on options, not very liquid at this moment. They submitted their article, The pricing of options and corporate liabilities, to the *Journal of Political Economy* and the *Review of Economics and Statistics* for publication, which was rejected by both journals. After some revision advised by Merton Miller (Nobel Laureate from the University of Chicago) and Eugene Fama, of the University of Chicago, it was finally accepted by the

Journal of Political Economy. At the same time options on stocks started to be traded on organized exchanges and in 1975 the BS formulae was adopted.

In fact, the mathematical finance was born 73 years earlier when Louis Bachelier successfully defended his thesis, *Theory de la Speculation, supported* by Henri Poincaré, the "Pope" of mathematics at that time. For this purpose he also introduced the Brownian motion 5 years before Einstein.

4.2. Initial Problems

The main issue in the financial risk management is the pricing of a contract. In an option contract the two parties carry different risks:

(1) the buyer limits his risk to the premium;
(2) the seller of a call option fears a sharp increase of the stock price before the maturity of the contract.

However, the risk of the underlying price is the result of the everyday price change (either increase or decrease) that can be easily monitored. So the trader of the option can take advantage of this information to build a strategy minimizing the risk, using the premium collected from the buyer of the option.

For instance, we can notice that the price of a call (respectively put) is positively (respectively negatively) correlated to the underlying price. It implies that a trivial (but costly) hedge for selling a call option can be to buy the underlying stock and for put to sell it. But as we will see, this strategy is not the best and buying the full stock to hedge the call option is for sure a hedging strategy but this strategy hedges too much. Instead, one can enter into a dynamic hedging strategy much cheaper and still offsetting the full risk of the option. We will see this in greater details later in this chapter. And this was the mastermind idea of Black and Scholes.

Black and Scholes suggested "creating a self-financing portfolio" which replicates at the expiry the payoff of the option. A portfolio is called self-financing if at any time the change in its value is due to change in market prices and not to any redistribution:

$$\begin{aligned} V(t) &= \sum_{i=1}^{N} q_i(t)S_i(t), \\ dV(t) &= \sum_{i=1}^{N} q_i(t)dS_i(t). \end{aligned} \tag{4.1}$$

It also means that no new assets will be added in the portfolio without selling some others.

$$0 = \sum_{i=1}^{N} dq_i(t)S_i(t). \tag{4.2}$$

Let us use this proposal to calculate options. For more generality the payoff of the European option will be note $\Phi(S(T))$. It will expire at a future time T. The replication portfolio is composed with the underlying stocks and money markets. When an investor buys a money market, he invests 1 euro and receives every day the risk free rate.

$$V(t) = \delta(t)S(t) + \delta_0(t)S_0(t). \tag{4.3}$$

The money market dynamic is:

$$dS_0(t) = S_0(t)r_t dt.$$

The money market value is:

$$S_0(t) = S_0(0)\mathrm{e}^{-\int_0^t r(s)ds}, \tag{4.4}$$

where $S(t)$ is the price of the underlying stock and $S_0(t)$ is the money market price. The present value $V(0)$ of the portfolio is exactly the price of the option and $\delta(t)$ defines the hedging strategy at the time t to apply until the maturity of the option. It is the number of underlying stocks in our hedging portfolio. At expiry T, the value of the portfolio is exactly equal to the option payoff $V(0) = \Phi(S(T))$. The two expressions $(V(t), \delta(t))$ are the solutions to our problem.

4.3. Model Hypothesis

4.3.1. *Lognormal distribution and constant volatility*

The main assumption of the BS model is the lognormal distribution of the underlying stock price. It means the logarithm of the stock price is normally distributed.

$$S(t) = S(0)e^{\left(\mu - \frac{\sigma^2}{2}\right)t + \sigma t z}, \quad z \sim N(0,1). \tag{4.5}$$

This property can be rewritten with a diffusion equation:

$$dS(t) = \mu S(t)dt + \sigma S(t)dW_t. \tag{4.6}$$

Then, if you apply the Itô lemma, we get:

$$d\ln(S(t)) = \left(\mu - \frac{\sigma^2}{2}\right)dt + \sigma dW_t, \tag{4.7}$$

where σ, the volatility term, is constant. It can also be a known function of time (deterministic function), in this case:

$$S(t) = S(0)\exp\left\{\int_0^t \left[\mu - \frac{1}{2}\sigma^2(t)\right]dt + \sqrt{\int_0^t \sigma^2(s)ds}\right. \tag{4.8}$$

An important probabilistic property of the stock price is:

$$E(S(t)) = S(0)e^{\mu t}. \tag{4.9}$$

At this moment, we suppose that the drift term μ is constant, but later in this chapter, we will see that we can easily relax this hypothesis.

In the previous example (Fig. 4.1), we can observe the lognormal distribution for a mean of 1 and standard deviation of 20%.

4.3.2. *Assumptions*

To simplify the modeling, we made some assumptions, which are not very realistic:

(1) The underlying stock price follows a lognormal distribution driven by a single Brownian motion with a constant σ and μ.

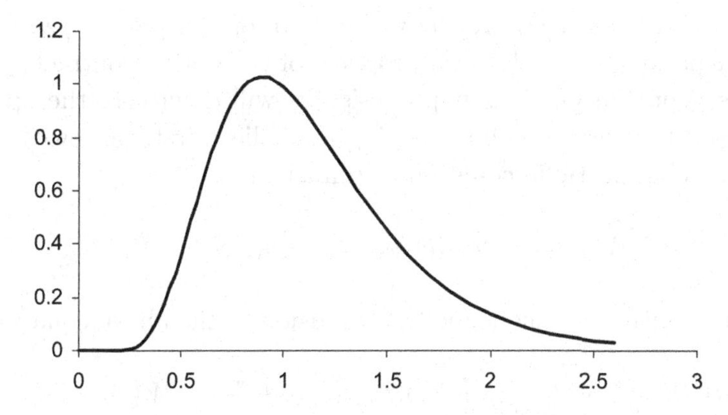

Figure 4.1: Lognormal distribution.

(2) Short selling is allowed without any limits. It means that we can borrow money to buy a stock without any limits.
(3) There are no transaction costs. In reality, one should account for the spread between bid and ask price, making re-hedging expensive.
(4) The underlying stock does not pay any dividends until the maturity of the option. Later, we will see that we can easily relax this assumption.
(5) The stock is continuously traded so the hedging portfolio can be rebalanced at any time. This is wrong, because exchanges are not usually opened 24 hours a day. Moreover, transaction costs force traders to do hedging at discrete time.
(6) The risk free rate is assumed to be constant. We can get equivalent result if it is a known function time. In reality, interest rates are stochastic.
(7) There is no arbitrage opportunity on the market. This hypothesis is the foundation of the pricing theory. It allows us to calculate a single price for a derivative that does not depend on market expectations.

4.4. Black–Scholes Partial Differential Equation

$V(t)$ is a function of different parameters:

$$V(t, S(t); \sigma, \mu; T; r). \tag{4.10}$$

In fact, just t and $S(t)$ are the variables of option price; σ, μ, r are market parameters and T is a property of the option contract.

Based on the previous hypothesis, we will determine the option price and the corresponding hedging portfolio. First, we rewrite the self-financing portfolio equation as diffusion:

$$dV(t) = \delta(t)dS(t) + (V(t) - \delta(t)S(t))rdt. \tag{4.11}$$

Let us introduce the lognormal diffusion in the above equation:

$$dV(t) = (rV(t) + (\mu - r)\delta(t)S(t))dt + \delta(t)\sigma dW(t). \tag{4.12}$$

Then, we apply the Itô Lemma to the price option $V(t, S(t))$:

$$\begin{aligned} dV(t) = {} & \left(\frac{dV(t)}{dt} + \mu S(t)\frac{dV(t)}{dS} + \frac{1}{2}S(t)^2\sigma^2\frac{d^2V(t)}{dS^2}\right) dt \\ & + \sigma S(t)\frac{dV(t)}{dS}dW(t). \end{aligned} \tag{4.13}$$

If we want the portfolio to replicate perfectly the part of the option linked to the underlying stock, we need to have:

$$\delta(t) = \frac{dV(t)}{dS} \tag{4.14}$$

and the price of the option is the solution of the BS partial differential equation (BS PDE):

$$\frac{dV(t)}{dt} + rS(t)\frac{dV(t)}{dS} + \frac{1}{2}S(t)^2\sigma^2\frac{d^2V(t)}{dS^2} = rV(t) \tag{4.15}$$

with the boundary conditions:

$$V(T, S(T)) = \Phi(S(T)). \tag{4.16}$$

In the previous partial differential equation we can notice the drift μ has been eliminated. It means the expectation of the evolution underlying stock price does not affect the option price.

The Feynman–Kac formula shows that the solution BS PDE is

$$V(t, S(t)) = E[e^{-rT}\Phi(S_T)], \tag{4.17}$$

where

$$dS(t) = rS(t)dt + \sigma S(t)d\tilde{W}(t) \tag{4.18}$$

(see also Appendix A.3). We say that we price the option under the risk neutral probability. $\tilde{W}$ is a Brownian movement under this probability (see also Appendix A.6).

4.5. Black–Scholes Formulae for Call and Put Options

Now let us apply the general result developed in the previous section to the particular payoff:

(1) the call: $\Phi(S(T)) = (S(T) - K)_+$
(2) the put: $\Phi(S(T)) = (K - S(T))_+$

Under the risk neutral probability:

$$S(T) = S(0)e^{rT - \frac{1}{2}\sigma^2 T + \sigma W(T)}. \tag{4.19}$$

For the call option we can divide the price equation into two parts:

$$C(t, S(t)) = e^{-rT}E\big(S(T)1_{\{S(T)\geq K\}}\big) - e^{-rT}KE\big(1_{\{S(T)\geq K\}}\big). \tag{4.20}$$

The second part is easy to calculate with using the normal cumulative:

$$E(1_{\{S(T)\geq K\}}) = N(d_2), \tag{4.21}$$

where

$$d_2 = \frac{\ln(S(0)/K) + rT}{\sigma\sqrt{T}} - \frac{1}{2}\sigma\sqrt{T}. \tag{4.22}$$

The second part can be calculated by integration with variable change or by applying the Girsanov theorem with the following exponential martingale:

$$\frac{dQ^S}{dQ} = \frac{e^{-rT}S(T)}{S(0)} = e^{-\frac{1}{2}\sigma^2 T + \sigma W(T)} \tag{4.23}$$

(see also Appendix A.2). We get the final formulae:

$$\mathrm{Call}(0,T,K) = S(0)N(d_1) - K\mathrm{e}^{-rT}N(d_2), \tag{4.24}$$

where

$$d_1 = d_2 + \sigma\sqrt{T} = \frac{\ln(S(0)/K) + rT}{\sigma\sqrt{T}} + \frac{1}{2}\sigma\sqrt{T}. \tag{4.25}$$

We get a similar formula for the put option with the same arguments:

$$\mathrm{Put}(0,T,K) = K\mathrm{e}^{-rT}N(-d_2) - S(0)N(-d_1). \tag{4.26}$$

For at the money option, when $S(0)$ is equal to K, we can assume that $\sigma\sqrt{T}$ is really small and approximate $N(.)$ by its limited expansion then we get:

$$\mathrm{Call}(0,T,S) = 0.4S\sigma\sqrt{T}. \tag{4.27}$$

4.6. Implied Volatility and Smile

The BS model assumes that the volatility parameters are constant and deterministic. But if we observe option price in the market we will get different volatility value for different options on the same underlying assets with different time of expiry or strike value. Its quite convenient to extract the "implied volatility" from an option price. The call or put prices are strictly increasing functions of the volatility so that they can be easily reversed.

The BS model can be extended with known function of time as volatility. We can then fit the price of an option for all its expiry dates. The volatility smile is a plot of the implied volatility of an option as a function of its strike price. The existence of a non-flat smile curve proves that the future distribution of the underlying asset is not lognormal.

4.7. Hedging Strategies

Up to now, we use the same example: a 5-year European call with a 110-euros strike. The present value of the stock is 100 euros and we assume the interest remains constant equal to 5% until the end of maturity of the option and the market volatility is 20%. The present value of the option is 24.54 euros.

4.7.1. *Naked and cover positions*

A simple way to cover the risk associated to an option is doing nothing. This strategy is also called "naked position". In the example of a European call, this approach is perfect when the option ends in the money. Then the trader buys the stock at the strike price and sells it at higher price in the market. But if the stock price is out of the money at the maturity, the trader looses the difference between the market price and the strike. In an opposite point view, the trader can buy the stock at the issue of the option and deliver it to the client at the maturity of the option. This strategy is also called a "covered position" and is the more expensive one. Those two strategies give a good range of a call option price:

$$(S-K)_+ \leq \text{Call}(S,T,K) \leq S. \tag{4.28}$$

4.7.2. *Delta hedging*

The BS model advises the trader to buy $\Delta_{\text{BS}} = N(d_1)$ unit of the underlying stock and to borrow the money to realize this transaction. At the maturity of the option stock is delivered to the client if the price is out of the money and nothing if the price is in the money (Table 4.1). The fact that this strategy costs some money is due to two reasons. On the one hand the trader buys the stock at expensive price and sells it at cheap price. On the other hand he has to pay interest for his loan. With a Monte Carlo, we calculate the average cost of this strategy and naturally the option price is inside the confident interval. We also graph the distribution of cost in Fig. 4.2.

Price	24.67
StdDev	0.09

4.7.3. *Stop and loss strategy*

Another well-known idea of hedging is the "stop and loss" strategy. For instance, the seller of a call option buys the stock, when

Table 4.1: Delta hedging scenario.

	OTM scenario				ITM scenario			
T	Stock	Delta	Purchase	Interests	Stock	Delta	Purchase	Interests
0	100	0.72	71.55	72	100	0.72	71.55	72
0.25	119	0.83	0.11	86	87	0.59	−0.13	61
0.5	112	0.78	−0.04	82	93	0.64	0.05	67
0.75	116	0.80	0.02	85	83	0.52	−0.12	58
1	127	0.86	0.05	93	83	0.50	−0.02	57
1.25	137	0.89	0.04	99	84	0.49	−0.01	57
1.5	124	0.83	−0.06	93	105	0.70	0.21	79
1.75	135	0.88	0.05	101	135	0.89	0.19	105
2	151	0.94	0.05	110	129	0.85	−0.03	103
2.25	136	0.89	−0.05	105	126	0.84	−0.02	102
2.5	151	0.94	0.05	114	136	0.89	0.05	110
2.75	150	0.94	0.00	116	123	0.82	−0.07	103
3	197	0.99	0.05	128	126	0.83	0.02	106
3.25	222	1.00	0.00	130	116	0.74	−0.09	97
3.5	195	1.00	0.00	132	103	0.57	−0.17	80
3.75	190	1.00	0.00	133	98	0.45	−0.12	70
4	217	1.00	0.00	136	93	0.32	−0.14	58
4.25	216	1.00	0.00	137	96	0.32	0.00	59
4.5	219	1.00	0.00	139	91	0.14	−0.18	43
4.75	225	1.00	0.00	141	81	0.00	−0.14	33
5	229	1.00	0.00	143	74	0.00	0.00	33
	Final cost			33				33
	Discounted final cost			25				26

its price is upper than the strike level and sells it in the opposite case. A naïve analysis will tell us that this strategy does not cost anything. However, this result is inconsistent with the previous equation, the price of an option should be greater than its intrinsic value.

There are two reasons to explain this paradox: First, in this strategy the stock is not bought or sold at the same time so each flows has to be discounted. Second, this strategy is done at discrete time so the purchase price is smaller than the sales price. Consequently, the trader looses some money for each transaction. This second issue is the most important.

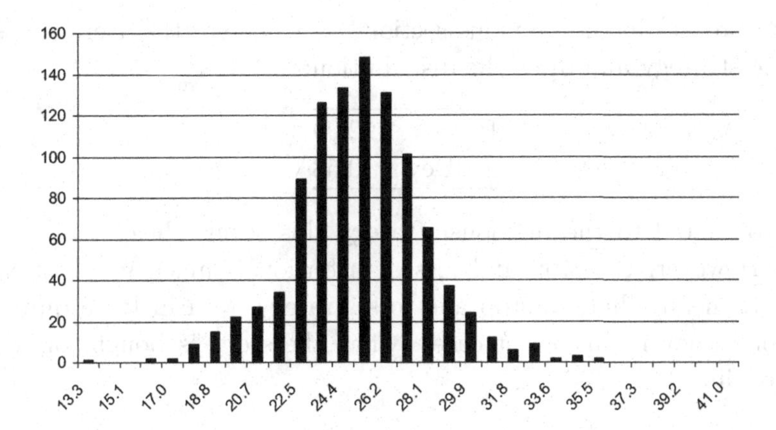

Figure 4.2: Delta hedging cost distribution.

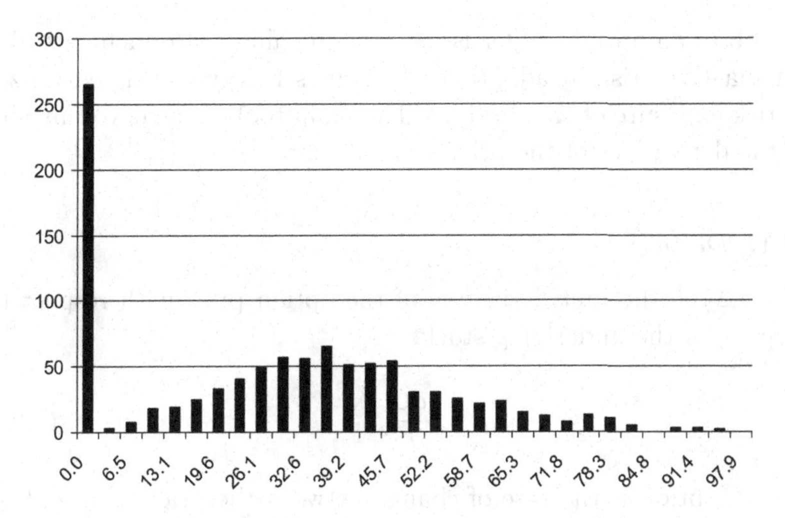

Figure 4.3: Stop and loss cost distribution.

Practically, the trader defines threshold ε around the strike value. He buys the stock when the stock price is greater than $X+\varepsilon$ and sells it when it is lower than $X-\varepsilon$. If the trader checks the price more frequently, he can reduce the threshold. But in this case, the number of transactions increases. When $\varepsilon \to 0$, the number of transaction tends to infinity (Figure 4.3).

As we did in the previous section, we calculate the average cost of the strategy and draw the distribution:

Price	22.99
StdDev	0.58

Compared to the previous strategy this seems cheaper in average. However, the stop and loss distribution is much more spread out. It means that, in stop and loss strategy, we can loose quite a lot of money in the worst cases (when the stock is bought or sells many times).

4.8. The Derivative Function of the Price or Greeks

The job of an option trader is not just to valuate products but also to manage the risk by adopting a hedging strategy which neutralizes the risk exposure of the product. The main tools to achieve this aim are the derivatives of the price.

4.8.1. *Delta*

Delta (Δ) is the first derivative of the option price with respect to the price of the underlying stock.

$$\frac{\partial C}{\partial S}. \tag{4.29}$$

Δ of an option is the rate of change between the price of an option and its underlying price. It shows the investor, who sold an option, to buy or sell what quantity of the underlying stocks. This strategy succeeds for a small change in the price of the underlying stocks.

In BS model, we get this formula for an European call:

$$\Delta_{\mathrm{BS}} = N(d_1). \tag{4.30}$$

In the example, we need to buy 0.72 unit of underlying stock to be hedged at $t = 0$.

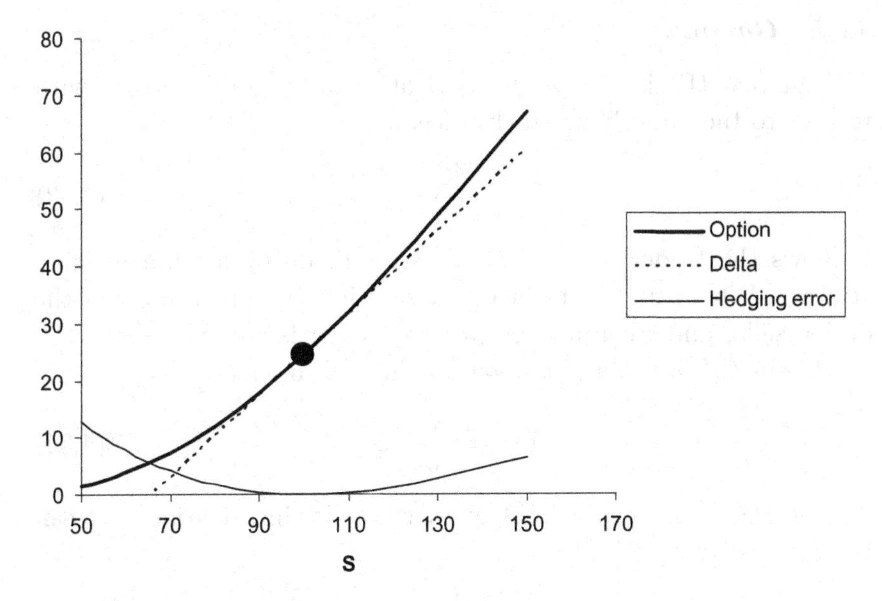

Figure 4.4: Delta heding.

For an European put the formula is:

$$\Delta_{\mathrm{BS}} = N(d_1) - 1 \leq 0. \tag{4.31}$$

So for put options, we need to sell the underlying stock. This is quite natural for an option which guaranties a selling price of the underlying stock (Fig. 4.4).

4.8.2. *Theta*

The theta (Θ) is the first partial derivative of the option price with respect to the maturity time. The theta is usually negative because the price of an option is less valuable when the time is passing. It does not make any sense to be hedged against the passage of time because there is no uncertainty in this phenomenon.

Here is the formula of the theta for an European call in the BS model:

$$\Theta_{\mathrm{BS}} = -\frac{SN'(d_1)}{2\sqrt{T}} - rK\mathrm{e}^{-rT}N(d_2). \tag{4.32}$$

4.8.3. *Gamma*

The gamma (Γ) is the second derivative of the option price with respect to the underlying stock price.

$$\frac{\partial^2 C}{\partial S^2}. \tag{4.33}$$

Γ shows the trader an idea of the error made by a delta hedging strategy. When the Γ is positive the hedging error is in favor of the option seller and when it is negative the error is adverse.

Here is the formula of the gamma in a BS model:

$$\Gamma_{\mathrm{BS}} = \frac{N'(d_1)}{S\sigma\sqrt{T}} \tag{4.34}$$

The BS PDE (Eq. (4.15)) gives the following relation between price derivatives:

$$\Theta + rS\Delta + \frac{1}{2}\sigma^2 S^2 \Gamma = rV, \tag{4.35}$$

when the option is delta hedged the equation become:

$$\Theta + \frac{1}{2}\sigma^2 S^2 \Gamma = rV. \tag{4.36}$$

This equation shows us that when the theta is negative the gamma should be positive.

4.8.4. *Vega*

From now, we will assume that the volatility is constant, but we can also calculate the sensitivity of the product with the volatility. This greek is called Vega (V):

$$\frac{\partial C}{\partial \sigma}. \tag{4.37}$$

We can estimate the delta hedge error (using this value) when we misevaluate the volatility parameter.

In the BS model we get this formula for an European call:

$$V = S\sqrt{T}N(d_1) \tag{4.38}$$

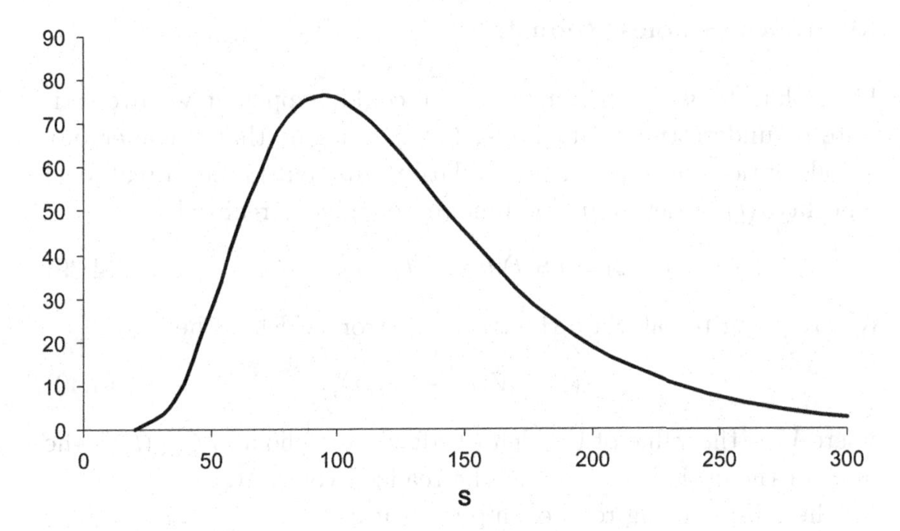

Figure 4.5: Vega function of the underlying price.

If we look at the Vega function of the underlying price (Fig. 4.5), we can notice that the option price is affected by the volatility parameter when the option is closed to the money. Deep out or in the money options are quite independent of the volatility level.

In the BS formula, we can notice that the Vega increases with the maturity of the deal, which is consistent with the intuition that the uncertainty increase with the time. Table 4.2 summarizes the Greeks or price derivatives for an European call and put.

Table 4.2: BS Greeks.

Greeks	European call	European put
Delta	$N(d_1)$	$N(d_1) - 1$
Gamma	$\frac{N'(d_1)}{S\sigma\sqrt{T}}$	$\frac{N'(d_1)}{S\sigma\sqrt{T}}$
Theta	$-\frac{SN'(d_1)}{2\sqrt{T}} - rKe^{-rT}N(d_2)$	$-\frac{SN'(d_1)}{2\sqrt{T}} + rKe^{-rT}N(-d_2)$
Vega	$S\sqrt{T}N(d_1)$	$S\sqrt{T}N(d_1)$

4.9. Black–Scholes Robustness

The volatility is uncertain. So what could happen if we overestimate or underestimate its value. Let us imagine that a trader buy or sells options with a constant volatility σ_{BS} but in fact a real local volatility$\sigma(t)$, a deterministic function of time, is realized.

$$dS(t) = rS(t)dt + \sigma(t)S(t)dW(t). \tag{4.39}$$

We are going to calculate the tracking error, which is the

$$e(t) = V(t) - C_{\mathrm{BS}}(t), \tag{4.40}$$

where V_t is the value of the delta hedged portfolio and $C_{\mathrm{BS}}(t)$ is the value of the option priced with the realized volatility.

The self-financing relationship gives us:

$$dV(t) = rV(t)dt + \frac{dC_{\mathrm{BS}}(t)}{dS}(dSt - rStdt). \tag{4.41}$$

We apply the Itô lemma on $C_{\mathrm{BS}}(t)$:

$$dC_{\mathrm{BS}}(t) = \left(\frac{dC_{\mathrm{BS}}(t)}{dt} + rS(t)\frac{dC_{\mathrm{BS}}(t)}{dS} + \frac{1}{2}S(t)^2\sigma^2(t)\frac{d^2C_{\mathrm{BS}}(t)}{dS^2}\right)dt + \sigma(t)S(t)\frac{dC_{\mathrm{BS}}(t)}{dS}dW(t). \tag{4.42}$$

If we introduce the BS PDE in the above equation:

$$\frac{dC_{\mathrm{BS}}(t)}{dt} + rS(t)\frac{dC_{\mathrm{BS}}(t)}{dS} + \frac{1}{2}S(t)^2\sigma_{\mathrm{BS}}^2\frac{d^2C_{\mathrm{BS}}(t)}{dS^2} = rC_{\mathrm{BS}}(t), \tag{4.43}$$

then we get:

$$dC_{\mathrm{BS}}(t) = \left(rC_{\mathrm{BS}}(t) + \frac{1}{2}S(t)^2(\sigma^2(t) - \sigma_{\mathrm{BS}}^2)\frac{d^2C_{\mathrm{BS}}(t)}{dS^2}\right)dt + \sigma(t)S(t)\frac{dC_{\mathrm{BS}}(t)}{dS}dW(t). \tag{4.44}$$

Then we obtain the ordinary differential equation for the tracking error:

$$de(t) = \left(re(t) + \frac{1}{2}S(t)^2(\sigma_{\mathrm{BS}}^2 - \sigma^2(t))\frac{d^2C_{\mathrm{BS}}(t)}{dS^2}\right)dt \tag{4.45}$$

By integration, we get:

$$e(T) = \int_0^T \mathrm{e}^{-r(T-t)} \frac{1}{2} S(t)^2 (\sigma_{\mathrm{BS}}^2 - \sigma^2(t)) \frac{d^2 C_{\mathrm{BS}}(t)}{dS^2} dt. \tag{4.46}$$

It means that when a trader does not know the value of the volatility parameter and want to use the BS model. He has to overestimate its value for gamma positive option and to underestimate its value for gamma negative options in order to obtain a positive tracking error.

4.10. Black–Scholes Extensions

4.10.1. *Volatility: deterministic function of time*

In the "classical" BS model we assume that the volatility parameter is a constant parameter. We can extend the model with volatility, which is a deterministic function of time.

$$dS(t) = \mu S(t) dt + \sigma(t) S(t) dW_t. \tag{4.47}$$

With the same argument we get the following formulae for option:

$$\mathrm{Call}(0, T, K) = S(0) N(d_1) - K\mathrm{e}^{-rT} N(d_2), \tag{4.48}$$

$$d_1 = d_2 + \Sigma\sqrt{T} = \frac{\ln(S(0)/K) + rT}{\Sigma\sqrt{T}} + \frac{1}{2}\Sigma\sqrt{T}, \tag{4.49}$$

$$\Sigma = \int_0^T \sigma(t) dt. \tag{4.50}$$

This model is really convenient when we want to valuate several options on the same underlying with the same maturity.

4.10.2. *Dividend*

In the classical BS model, we assume that the stock does not pay any dividend. In practice the dividend are paid at precise time. When the payment occurs, the price drops down to the amount of the dividend. Here we assume that the dividend is paid continuously with a rate d.

Under the risk neutral probability the diffusion of the underlying becomes:

$$dS(t) = (r - d)S(t)dt + \sigma(t)S(t)dW_t. \tag{4.51}$$

Then we get the following formulae for the European call by replacing $S(0)$ by $S(0)e^{-dT}$ in BS formula:

$$\text{Call}(0, T, K) = S(0)e^{-dT}N(d_1) - Ke^{-rT}N(d_2) \tag{4.52}$$

$$d_1 = d_2 + \sigma\sqrt{T} = \frac{\ln(S(0)/K) + (r - d)T}{\sigma\sqrt{T}} + \frac{1}{2}\sigma\sqrt{T} \tag{4.53}$$

These results were first obtained by Merton in 1973.

4.10.3. *Normal and shifted lognormal models*

There exists a wide variety of model to handle non-lognormal distribution. For instance, the shifted lognormal and the normal model are easy to use and are some alternatives. A normal model is given by for the forward asset by:

$$dS(t) = \sigma(t)dW_t. \tag{4.54}$$

The pricing formula is easy to derive and is given for call and put options respectively by:

$$C_t = (S_0 - K)N(d_1) + \text{Vol}.n(d_1), \tag{4.55}$$

$$P_t = (K - S_0)N(-d_1) + \text{Vol}.N(-d_1), \tag{4.56}$$

where

$$d_1 = \frac{S_0 - K}{\text{Vol}}, \tag{4.57}$$

$$\text{Vol} = \sqrt{\int_0^T \sigma^2(s)ds} \tag{4.58}$$

and $n(x) = \frac{1}{\sqrt{2\pi}}e^{-x^2/2} = N'(x)$ is the derivative function of the cumulative function.

A shifted lognormal model is given by:

$$dS(t) = \sigma(t)(S(t) + \mu)dW_t. \tag{4.59}$$

The pricing formula for call and put options are given respectively by

$$C_t = (S_0 + \mu)N(d_1) - (K + \mu)N(d_2), \tag{4.60}$$
$$P_t = (K + \mu)N(-d_2) - (S_0 + \mu)N(-d_1), \tag{4.61}$$

where

$$d_1 = \frac{\ln (S_0 + \mu)/(\mathrm{K} + \mu)}{\sigma\sqrt{T}} + \frac{1}{2}\sigma\sqrt{T}, \tag{4.62}$$

$$d_2 = \frac{\ln (S_0 + \mu)/(\mathrm{K} + \mu)}{\sigma\sqrt{T}} - \frac{1}{2}\sigma\sqrt{T}. \tag{4.63}$$

4.11. Summary

We started this chapter by examining the approach of Black and Scholes. A self-financing portfolio is composed with the underlying stocks and money market which replicates at the expiry the payoff of the option. The BS results are based on several hypotheses, the main one being the lognormal distribution with constant volatility. Thanks to these assumptions, we obtained the differential partial equation solved by the option price. Finally, the Feyman–Kac formula gives us the price of the option at any time. This approach can always be used because the derivative and the stock price are both dependant of the same uncertainty source. We applied these results to price European call and put options

The implied volatility is the volatility that is used in BS pricing formula in order to fit the market price. We observe a smile of volatility that proves that it is a function of expiry and strike value.

The issue at stake for a trader is not only to buy or sell instruments but to hedge his position. We have taken at first a look at some hedging strategies. They are not perfect because they required to continually make transactions. Greeks gives an idea about how to neutralize the risk exposure, they give a lot of information about how the price of an option will change as market conditions change: they are tools for the option sensitivity measure.

The BS results can be extended to cover options with a deterministic function of time volatility, to cover options with a dividend paying stock and to non-lognormal distributions.

References

[1] Black F and Scholes M (1973). The pricing of options and corporate liabilities. *Journal of Political Economy*, **81**(3), 637–654.

[2] Engle RF and Mezrich J (1995). *Grappling with GARCH.* London, UK: Risk Books.

[3] Engle RF and Mezrich J (1996). *GARCH for Groups.* London, UK: Risk Books.

[4] Fama EF (1965). The behavior of stock market prices. *Journal of Business*, **38**, 34–105.

[5] Hull J (2002). *Options, Futures, and Other Derivative Securities*, 5th edn. Englewood Cliffs, NJ: Prentice Hall.

[6] Merton RC (1974). Theory of rational option pricing. *Bell Journal of Economics and Management Science*, **4**, 141–183.

[7] Neftci S (1996). *Introduction to Mathematics of Financial Derivatives.* New York: Academic Press.

[8] Smith CW (1976). Option pricing: a review. *Journal of Financial Economics*, **3**, 3–51.

[9] Taleb NM (1996). *Dynamic Hedging: Managing Vanilla and Exotic Options.* New York: Wiley Finance.

Chapter 5

FIXED INCOME BASIS

The fixed income markets are among the world's most complex markets. In this introductory chapter, we will provide a brief summary of fixed income basis interest rate products. It focuses on fundamentals of derivative products to discover characteristics, particularities and valuations of a basis instruments and useful tools to build interest rate curve (i.e. forward rate, Euro future, swap rate, duration, ...), interest rate swap and vanilla options (cap, floor, swaption, option on bond, spread option, ...).

5.1. Simple Instruments

5.1.1. *Markets conventions*

Each financial market specifies its own conventions for day count, date roll, amortization structures, etc.

Day-count convention: A day-count convention is a method used to calculate the fraction of a year between two dates. It is mainly used to determine the accrued interests. The notation used is d/y where d is the assumed number of days in a month and y is the assumed number of days in a year. There exist three main day-count conventions:

- actual/actual
- 30/360
- actual/360

For example, an actual/360 day count basis converts the number of days between the start day and the end day to a time value δ = days/360. The interests earned during the interest period are calculated as follows:

$$\delta \times \text{annual interest.}$$

For sterling bonds these are assumed to be the actual number of days in each month and 365 days in each year, known as the "actual/actual" convention. For US$ bonds, the convention is "30/360"; each month is assumed to have 30 days and each year 360 days. Actual/360 is commonly used for money market rates.

Date-roll convention: This convention determines the rescheduling of dates when they fall on non-business days. Examples of date-roll conventions:

(1) *Following*: If the date falls on a non-business day, the date will roll forward to the next business day.
(2) *Modified following*: If the date falls on a non-business day, the date will roll to the first following business day. However, if the first following business day falls in the next calendar month, the date will roll back to the first.

Table 5.1 summarizes the Euro market conventions (1998).[1]

5.1.2. *LIBOR rates*

LIBOR stands for London interbank offered rate. It is a short-term interest rate that represents the simple interest rate at which banks are willing to lend to each other. As a practical matter, it is the rate other banks must pay to borrow. LIBOR rate exist for 1-, 3-, 6- and 12-month maturities. It is calculated on an actual over 360 days basis, except for GBP, which uses an actual over 365 day basis. The LIBOR rate is usually used as the "risk-free rate" rather than the treasury

[1]Table 5.1 is retrieved from the ISDA website www.isda.org.

Table 5.1: Euro market conventions.

Euro money markets	Day count basis: actual/360 Settlement basis: spot (two day) standard Business days: TARGET operating days should form the basis for euro business days
Euro swap markets	Floating day count basis: actual/360 Fixed rate day count basis: 30/360 Business days: TARGET operating days should form the basis for euro business days Fixing period: two day rate fixing convention Coupon frequency: annual
Euro bond markets	Day count basis: actual/actual Quotation basis: decimals rather than fractions Business days: TARGET operating days should form the basis for euro business days Coupon frequency: annual Settlement dates: the standard for internationally traded cross-border transactions for the euro should remain on a $T+3$ business day cycle
Euro foreign exchange markets	Settlement timing: spot convention, with interest accrual beginning on the second day after the deal has been struck Quotation: "certain for uncertain" (i.e. 1 euro = x foreign currency units) Reference rate: the ECB (or NCBs) should be responsible for the publication of daily closing reference rates

rate to evaluate derivatives. LIBOR is, by convention, the liquidity and credit risk base point for all swap-market securities: swaps, swaptions, Bermuda swaptions, exotics, etc.

5.1.3. *Repo rate*

A repurchase agreement (or repo) is an agreement between two parties whereby one party sells the other a security at a specified price with a commitment to buy the same security or equivalent back at a later date for another specified price. Most repos are overnight transactions. Long-term repos — called term repos — can extend for a month or more.

5.1.4. *Duration*

Duration is a measure of the average (cash-weighted) term-to-maturity value of a bond. It measures bond's sensitivity towards exposure to parallel shifts in the spot interest rate curve. The weights are the present values of payments, using bond's yield-to-maturity value as discount rate.

A bond with zero coupon will have duration exactly equal to its maturity because it pays no coupons. However, a non-zero coupon bond will have duration shorter than its maturity.

Definition[2]: Suppose that a portfolio whose cash flows are all fixed c_i at time $t_i (1 \leq i \leq n)$. The price B and the present value $pv(c_i)$ are related by

$$B = \sum_{1}^{n} pv(c_i). \tag{5.1}$$

The duration D of the bond[3] is defined as

$$D = \frac{\sum_{1}^{n} t_i pv(ci)}{B},$$

which can be written as:

$$D = \sum_{1}^{n} t_i \left[\frac{pv(c_i)}{B} \right], \tag{5.2}$$

where the present value $pv(c_i) = c_i e^{-yt_i}$ and y is the yield-to-maturity value.

Duration D appears then as a weighted average of coupon payment dates. The weight applied to each payment date is the ratio between the present value of the coupon and the bond current price.

---

[2]Same definition as that is given in Hull (2002).

[3]The name duration should make more sense, as should the fact that duration is measured in years. When duration is calculated in this way, it is referred as the bond's *Macaulay duration.*

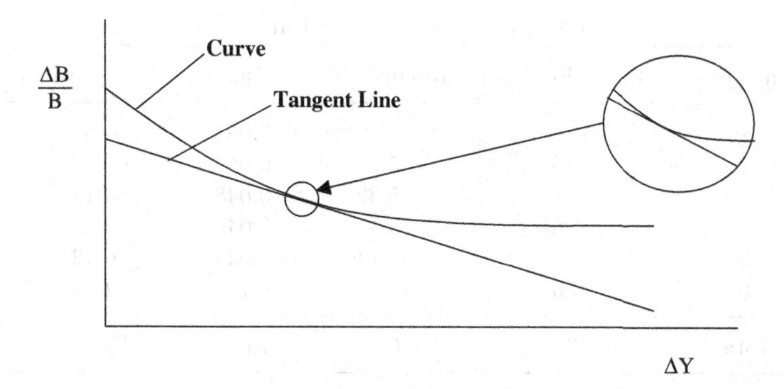

Figure 5.1: Duration is the opposite of slope of a line drawn tangent to the price yield curve and shows the % change in bond price for a given change in yield.

From Eq. (5.7)

$$\frac{\partial B}{\partial y} = -\sum_{1}^{n} t_i pv(c_i) = -DB. \tag{5.3}$$

We have then

$$D \approx -\frac{1}{B}\frac{\Delta B}{\Delta y}. \tag{5.4}$$

This equation shows the percentage change in the price of bond for a given change in yield-to-maturity value. This change can be illustrate as shown in Fig. 5.1.

Example: Consider a bond with €100 face value and 6% coupon. The yield-to-maturity value is 4% per year with continuous compounding and it matures in 6 years. We suppose that coupon payments are made every year. The bond thus pays €6 a year from now per year and €106 in 6th year. In order to compute duration, we calculate the present values and weights at each date. As Table 5.2 shows, the present value at time $t_i (1 \le i < 6)$ is €6*$e^{-0.04^* t_i}$ and at time $t_{i=6}$ is €106*$e^{-0.04^* t_i}$, and the weights are calculated by dividing present values by their sum. So the duration is:

$$D = \frac{5.76^*1 + 5.53^*2 + 5.32^*3 + 5.11^*4 + 4.91^*5 + 83.38^*6}{5.76 + 5.53 + 5.32 + 5.11 + 4.91 + 83.38} = 5.25.$$

Table 5.2: Calculation of duration.

Time (year)	Cash flow (€)	Present value	Weight	Time × weight
1	6	5.76	0.052	0.052
2	6	5.53	0.050	0.10
3	6	5.32	0.048	0.14
4	6	5.11	0.046	0.18
5	6	4.91	0.044	0.22
6	106	83.38	0.75	4.54
Total	136	110.01	1.00	5.25

The equality (5.4) is correct only if interest rates are continuously compounded. In practice, we often calculate all present values with a discontinuously compounded yield y. If this is done, Eq. (5.3) should be slightly modified and gives *modified duration.*

Modified duration: This, Modified duration is a measure of the bond's price sensitivity to interest rate movements. It is defined as follows:

$$D^* = \frac{D}{1 + y/n},$$

where D and y are already defined and y is expressed with a compounding frequency n times per year. Then,

$$\frac{\Delta B}{B} \approx -D^* \Delta y.$$

5.2. Bootstrapping and Curve Fitting

Zero curves (spot and forward rate, discount factor, and discount rates curves) are the corner stone of securities valuation. They are built from current cash deposit rates, Euro futures prices (actually FRA's), treasury yields, and interest rate swap spreads. The method used is called bootstrapping". This involves deriving each new point on the curve from previously determined points (hence the definition, "bootstrapping").

First, the zero-coupon yield curve that is to say $T \to R(0,T)$ is constructed by fitting to a given portfolio of market instruments.

Basically, we start by fitting to the deposit quotes, then the forwards and then move on to the swap rates. We here need to use the following pricing formulae:

$$\text{Deposit rate} \quad R(0,T) = \frac{1}{T}\left(\frac{1}{B(0,T)} - 1\right),$$

$$\text{Forward rate} \quad F_k(0) = \frac{1}{T_{k+1} - T_k}\left(\frac{B(0,T_k)}{B(0,T_{k+1})} - 1\right),$$

$$\text{Par swap rate} \quad S(0) = \frac{1 - B(0,Tn)}{\sum_i \delta_i B(0,T_i)}.$$

In this bootstrapping exercise, we write $B(0,T) = \mathrm{e}^{-R(0,T)T}$ where R is assumed piecewise linear. Gaps are filled by the assumption of piecewise linearity. Curve fitting is the way we bridge the gaps for maturities for which we do not have market data. There exist many approaches to curve fitting: bootstrapping piecewise flat in forward space, smoothing splines, local quadratic splines, tension splines, etc. These methods generally involve global optimization. A smooth curve is generally preferable. However, one problem is that smooth curves typically create spill over effects in risk calculations, so that a 5-year swap will have sensitivities to 6-, 7-year, etc. So they are not always popular.

5.3. Introduction to Swaps

5.3.1. *Vanilla swaps*

A swap[4] is an exchange of cash flows between two companies in the future for a predetermined period on prescribed dates. A vanilla interest rate swap (IRS) stays the most popular kind of swap. In this, the parties must exchange a floating interest rate for specified terms. Generally the LIBOR is the floating rate index. The interest rate swap is equivalent to an exchange of cash flows from two bonds. In fact, receiving a fixed rate is akin to holding a bond, while payment of a floating rate resembles the funding costs of this long bond position. A *payer* IRS is a swap where the customer pays the fixed interest

[4]For more documentation on swaps see Selic.

rate and when the customer receives the fixed interest rate is called *receiver* swap.

Purpose of the product: The IRS enables a company to change its floating rate debt to fixed rate, or vice versa. It can be used as a hedge to lock in the future cost of borrowing or to lock in the future rate of return on an investment. With IRS, there is no principal exchanged, there can be no foreign exchange on that principal, and as it can be reversed at any time, this last target market is a major advantage when comparing to alternative strategies. Swaps are the most common form of interest rate derivative.

Pricing: Since the interest swap is equivalent to an exchange of cash flows from two bonds, its value is the difference between the present value of the two bonds. If an investor receives fixed interest rate and pays floating, then the value of the swap contract is:

$$V_{\mathrm{S}} = V_{\mathrm{fixed}} - V_{\mathrm{float}},$$

where V_{fixed} is the present value of the fixed-coupon bond (or fixed interest leg of the swap), and V_{float} is the present value of the floating-coupon bond (or floating leg). When issued, the swap value is equal to zero.

We can then define the *swap rate* as the fixed payment that makes $V_{\mathrm{fixed}} = V_{\mathrm{float}}$.

Let us consider a receiver swap where the floating leg is indexed on a LIBOR with payment dates $\{T_1, \ldots, T_n\}$. Then the swap rate is given by the formula $S(0) = (1 - B(0, Tn))/\sum_i \delta_i B(0, T_i)$.

Example: In Fig. 5.2, the measuring of value of a swap is illustrated.[5]

5.3.2. *Cross-currency swaps*

Description: A cross-currency swap is an IRS where each leg of the swap is described in a different currency. The interest rates of the two legs can be both fixed rate, both floating rate or one fixed and

[5]The example is retrieved from an article of the Bank of Montreal (1998) "what are your swaps worth?".

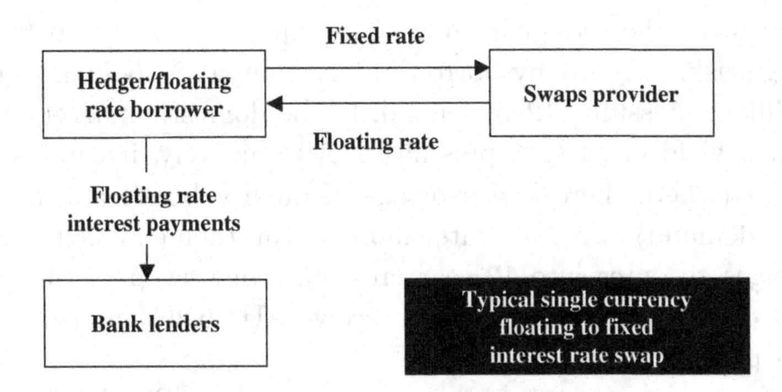

Figure 5.2: Swap mechanism.

one floating. It has two principal amounts one for each currency that are usually exchanged both at the beginning and the maturity of the swap. Normally, the exchange rate used to predetermine the two principals is then prevailing spot rate even for delayed start transaction. The investors can either agree to use the forward exchange rate or agree to set the rate two business days prior to the start of the deal.

A cross-currency swap where both legs are relying on floating rates is termed *floating-for-floating* currency swap and it is part of the basis swap product family (Fig. 5.3).

Purpose of the product: A cross-currency swap enables us to switch loans, deposits, assets or liabilities to a required currency. Like all

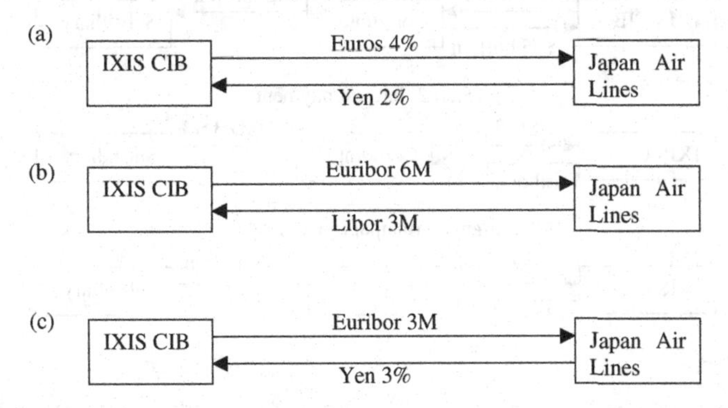

Figure 5.3: The currency swap options for (a) fixed-for-fixed; (b) floating-for-floating; and (c) fixed-for-floating swaps.

FX Forwards, the cross-currency swap exposes the user to foreign exchange risk. A company borrower, for example, is looking to raise €35 million by issuing 12-year bond. In the domestic market, it can issue at a yield of LIBOR plus 35 bp. Alternatively, it can issue in Australia (where there is a shortage of quality bonds and a high-investor demand) at a fixed rate 6.80%. It can then exploit potential advantages to enter into 12-year cross-currency swap for principal amount of €35 million agreeing to receive AU$ 6.80% and pay US$ LIBOR plus 25pb.

Cross-currency swap has most advantage. The following example illustrates the above swap options.

Example: A subsidiary of French corporate in the USA would like to finance its local project. It asks its parent company to grant it an internal loan in US$ for 5 years at a fixed rate of 3.6% for a total principal $45 million. The corporate receives interest payments in US$ from its subsidiary and pays interests in Euro to the bank. It decides to enter into a cross-currency swap with a financial institution in order to eliminate both interest and currency risks. We suppose that the parent corporate pays 3.6% in US$ and receives Euribor + 20 bp for a principal amount of €35 millions. The parties agree to exchange principal amount at the outset of the swap.

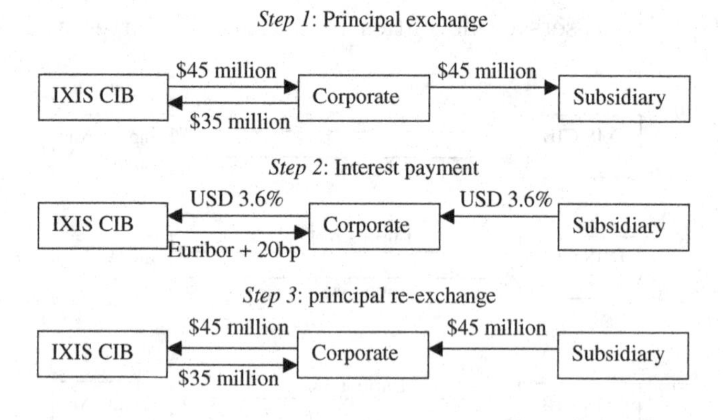

Pricing: A cross-currency swap can be viewed as a basket of two interest rate mono-leg swaps. Each swap is evaluated in its proper currency that is to say each leg is calculated using the prevailing zero-coupon yield curve of its currency. At the outset of the swap, the present value of one leg must be equal to the present value of the other leg both expressed in domestic currency.

In general, the value of swap V_{ccs} where we receive domestic currency and pay foreign currency is as follows:

$$V_{\text{ccs}} = B_{\text{Domestic}} - FX_{\text{spot}} B_{\text{Foreign}},$$

where B_{Domestic} is the domestic leg value, calculated in the domestic currency, B_{Foreign} is the foreign leg value, calculated in the foreign currency, and FX_{spot} is the spot exchange rate. We notice that the price and value of a cross-currency swap is influenced by the following:

the zero-coupon yield curve on domestic currency,
the zero-coupon yield curve on foreign currency,
the spot exchange rate.

By a similar reasoning, the value of a swap V^*_{ccs} where we receive foreign currency and pay domestic one is:

$$V_{\text{ccs}} = FX_{\text{spot}} B_{\text{Foreign}} - B_{\text{Domestic}}$$

Summary: A cross-currency swap can be cheaper than the cash markets (simple documentation, issuing currency bonds directly, ...). It can be replicated on-balance-sheet instruments, loans and deposits in different currencies. This explains the necessity of exchanging principal at expiry date.

5.3.3. *Forward starting swaps*

Description: A forward starting swap is an IRS that does not start immediately but on a future date.

Purpose of the product: A forward starting swap allows a company to profit from favorable short-term rates movements while locking in the forward financing costs or investment yields. It can be of a great benefit if the market anticipates a steepening yield curve. In this case, hedgers can make profit by swapping their fixed debt to a lower floating rate spread if they purchase a forward starting swap instead of entering immediately into a swap.

5.3.4. *LIBOR-in-arrears swaps*

In a plain vanilla IRS, the floating interest rate is observed at the start of a period, and paid at the end of that period. In a LIBOR-in-arrears swap/note, the floating rate is observed and paid at the end of the period. For example, in a LIBOR-in-arrears swap with semi-annual resets, the 6-month LIBOR rate from time t_i to t_{i+1} is used to calculate the coupon payment at time t_i.

This mismatch in payment timing makes a difference between the present value of a floating cash flow in a LIBOR-in-arrears swap and the present value of the forward rate for this date. This phenomenon is called *convexity adjustment.*

Pricing: On account of convexity adjustment the valuation of LIBOR-in-arrears swaps should take into account forward rate correction. This correction depends on the volatilities of interest rates. In other words, it is based on the following mathematics' assumption for expected forward rate at time T:

$$F_i + \frac{F_i^2 \sigma_i^2 \tau_i t_i}{1 + \tau_i F_i},$$

where, F_i is the forward LIBOR rate value resetting at t_i and ending at t_{i+1}, σ_i is the corresponding implied volatility and τ_i is the interest rate period. For more details on pricing techniques see Section 5.4.

5.3.5. *CMT and CMS swaps*

A constant maturity swap (CMS)[6] is an agreement in which one leg is indexed to a market swap rate of fixed maturity. The other leg can be fixed, as is most common, or floating (LIBOR-based).

[6]Definition and example retrieved from Carmon and Durrleman (2003).

A simple example is a 2-year quarterly reset swap in which one leg pays LIBOR and pays receives the 5-year CMS rate. It is a constant maturity swap because the 5-year term is from each reset date, rather from the swap's inception.

Constant maturity swaps are part of a larger class of swaps known as yield curve swaps. For details on constant maturity treasury (CMT) and CMS swap valuation, see Section 5.4. Further details can be found in Benhamou (2000), Amblard and Lebuchoux (2000), Selic and Hagan.

5.4. Convexity Adjustment

5.4.1. *Convexity correction*

Some derivatives payoffs, as bond or IRS, have their prices which are convex functions with respect to the value of their underlying. Mathematically, the convexity property of a decreasing function in a finite interval, means that this function is "lower" that the straight line joining the bounds of this interval (Fig. 5.4). This is written mathematically as:

$$\forall 0 \leq \lambda \leq 1, \quad \forall x, \quad y : f(\lambda x + (1-\lambda)y) \leq \lambda f(x) + (1-\lambda) f(y). \tag{5.5}$$

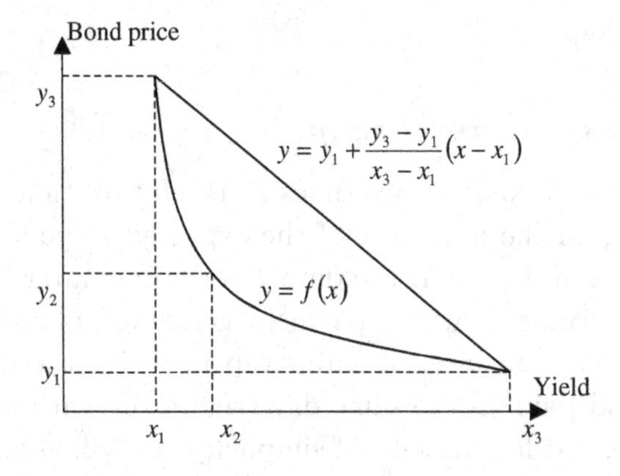

Figure 5.4: The convex bond price-yield function.

The Black–Scholes pricing of a European option whose payoffs underlying is a linear function of the bond or the swap yield requires an adjustment from a lack of convexity value. Examples of some derivatives whose payoffs are linear of the swap or the bond yield are the CMS and the CMT swaps.

The convexity adjustment concerns also the values of some derivative payoffs which are not paid at their underlying fixing time. This is the case, for instance, of the LIBOR-in-arrears swap. The expected value of a future floating rate is the forward rate only under its "natural probability" that is, the one corresponding to its natural payoff. As soon as the forward is not paid in its natural payoff, the expected value of the future floating rate is equal to the rate forward with an adjustment (adjustment corresponding to the change of probability).

More rigorously, the convexity adjustment comes from the Jensen inequality in probability, implying that the expectation of a convex function f of a random variable X is at least greater that the function taking the expectation of this variable. Explicitly we have:

$$f(E[X]) \leq E[f(X)], \tag{5.6}$$

under the same probability measure. This inequality holds also for conditional expectations.

The detailed principle of the calculation of the different corrections for each of the products mentioned above are discussed in the following section.

5.4.2. *LIBOR-in-arrears swap*

Let X be a random asset which fixes at the future time $T_1 > 0$. We are interesting on the difference of the expected value at time $t = 0$ of a payoff of X if it paid at the date T_1 or a date after $T_2 > T_1$. An example of instruments whose payment is delayed is the LIBOR-in-arrears swap. It is an interest rate swap in which the floating rate is observed and paid on the same day (rigorously with two business day difference, but for the sake of simplicity, we will assume it is the same day).

Let us consider a LIBOR forward rate (the zero-coupon forward bond respect), seen at time t between the time T_1 and T_2, denoted by $L(t, T_1, T_2)$ (by $B(t, T_1, T_2)$ respect). We know that the relation between the zero-coupon bond forward and the LIBOR forward bond respectively is model free and is written as:

$$B(t, T_1, T_2) = \frac{1}{1 + (T_2 - T_1)L(t, T_1, T_2)}. \tag{5.7}$$

Let us denote also by:

$\Delta_{1,2} = T_2 - T_1$: the forward tenor of $L(t, T_1, T_2)$,
$B(t, T)$: the zero-coupon bond seen at time t maturing at time T,
$\boldsymbol{E}_i$: the forward expected value in the $B(0, T_i)$ risk neutral world (respecting to the $B(0, T_i)$ numeraire).

Using standard change of measure between the T_1- and the T_2-forward neutral measure, we have (see Chapter 7f):

$$\boldsymbol{E}_1[L(T_1, T_1, T_2)] = \boldsymbol{E}_2\left[L(T_1, T_1, T_2)\frac{B(T_1, T_1)/B(0, T_1)}{B(T_1, T_2)/B(0, T_2)}\right], \tag{5.8}$$

which can be written as:

$$\boldsymbol{E}_1[L(T_1, T_1, T_2)] = \boldsymbol{E}_2\left[L(T_1, T_1, T_2)\frac{B(T_1, T_1)/B(T_1, T_2)}{B(0, T_1)/B(0, T_2)}\right]. \tag{5.9}$$

Relation (5.7) will lead to:

$$\boldsymbol{E}_1[L(T_1, T_1, T_2)] = \boldsymbol{E}_2\left[L(T_1, T_1, T_2)\frac{1 + \Delta_{1,2}L(T_1, T_1, T_2)}{1 + \Delta_{1,2}L(0, T_1, T_2)}\right], \tag{5.10}$$

from where, we have immediately:

$$\boldsymbol{E}_1[L(T_1, T_1, T_2)] = \frac{L(0, T_1, T_2) + \Delta_{1,2}\boldsymbol{E}_2[L^2(T_1, T_1, T_2)]}{1 + \Delta_{1,2}L(0, T_1, T_2)}. \tag{5.11}$$

Supposing that at each date $t > 0$, the LIBOR ratio $L(t, T_1, T_2)/L(0, T_1, T_2)$ is a lognormal of parameters $\mathrm{LN}(0, t\sigma_t^2)$ under the

$B(0, T_2)$ risk neutral world (geometric Brownian motion, see Chapter 7a), we can calculate its second moment using its variance.

We find finally its T_1-forward value:

$$\boldsymbol{E}_1[L(T_1, T_1, T_2)] = L(0, T_1, T_2) + \frac{\Delta_{1,2} L^2(0, T_1, T_2)(e^{T_1 \sigma^2 T_1} - 1)}{1 + \Delta_{1,2} L(0, T_1, T_2)}. \tag{5.12}$$

where σ_{T_1} represents the volatility of the interest rate at T_1, implied from the caplet prices. By considering a first-order Taylor[7] development for $\sigma_{T_1}^2 T_1$ around 0:

$$\boldsymbol{E}_1[L(T_1, T_1, T_2)] = L(0, T_1, T_2) + \frac{\Delta_{1,2} L^2(0, T_1, T_2) \sigma_{T_1}^2 T_1}{1 + \Delta_{1,2} L(0, T_1, T_2)}. \tag{5.13}$$

This is the simplest convexity adjustment case, that is of LIBOR-in-arrears due to the probability measure changing. We give in the following sections others examples of convexity adjustment on the CMS and CMT swap which is due to the convexity of the payoff function of the underlying.

5.4.3. *CMS/CMT/spread lock*

We are interested in correcting the CMS rate and the CMT yield from the convexity lag (Benhamou 2000(b)). And we want to do this using the forward-swap value. The price of a swap P maturing at time T_n with periodicity δ is a non-linear function of its rate S, given by:

$$P = f(S) = \sum_{i=1}^{n} \frac{\delta S_0}{(1 + \delta S)^i} + \frac{1}{(1 + \delta S)^n}. \tag{5.14}$$

We can explain this formula by considering δS as the yield of a coupon bond with coupon dates $T_1, T_2, \ldots, T_n$ and coupon δS_0. And here we have a bond with periodic coupons, $T_{i+1} = T_i + \delta$, and a face value of 1.

[7] $e^x = 1 + x \approx 1$ for x in the vicinity of 0 (for small x).

Let us denote by S_t the value of the swap S at time t, f the payoff convex function of the swap rate: $P = f(S)$ defined in the above equation, and $\sigma^S t^2$ the variance of the process S_t.

We would work under the P-risk-forward neutral probability, that is, under which its expected value is equal to its initial condition P_0:

$$\boldsymbol{E}^*[P_T] = \boldsymbol{E}^*[f(S_T)] = P_0 = f(S_0). \tag{5.15}$$

The inequality above comes from the Jensen result (6.i.1). We search then to obtain the "real" expected swap rate under the risk-forward neutral probability of the payoff. We need to determinate or approximate as better as possible the expected value of S_T, from S-risk-forward neutral value S_0:

$$\boldsymbol{E}^*[S_T] = S_0 + (\boldsymbol{E}^*[S_T] - S_0). \tag{5.16}$$

The best linear approximation of $f(S)$ is its Taylor development around S_0:

$$f(S_T) \approx f(S_0) + f'(S_0)(S_T - S_0) + \frac{1}{2} f''(S_0)(S_T - S_0)^2. \tag{5.17}$$

The expected correction of S is equal to:

$$\begin{aligned}\boldsymbol{E}^*[S_T - S_0] = \frac{1}{f'(S_0)} \Big\{ & \boldsymbol{E}^*[f(S_T) - f(S_0)] - \frac{1}{2} f''(S_0) \\ & \times \boldsymbol{E}^*[(S_T - S_0)^2] \Big\}. \end{aligned} \tag{5.18}$$

So taking in account the relation (1.3.2) and the value of the variance of S:

$$\boldsymbol{E}^*[S_T - S_0] = -\frac{f''(S_0)}{2f'(S_0)} (\sigma^S)^2 T = -\frac{f''(S_0)}{2f'(S_0)} S_0^2 \cdot \sigma^2 T \tag{5.19}$$

This correction is called *convexity adjustment* and it is always positive.[8]

---

[8] f convex $\Rightarrow f''(x) > 0, f$ is decreasing $\Rightarrow f'(x) < 0$.

For example, applying this on the CMS swap f function defined in Eq. (1.3.1) gives:

$$\boldsymbol{E}^*[S_T - S_0] = \left(1 - \frac{n\delta S_0}{(1+\delta S_0)((1+\delta S_0)^n - 1)}\right)\sigma^2 T. \tag{5.20}$$

The CMT yield C being a linear function if the CMS rate S:

$$C = S - \text{SpreadLock}, \tag{5.21}$$

it can be corrected from the convexity adjustment:

$$\boldsymbol{E}^*[C_T - C_0] = -\frac{g''(C_0)}{2g'(C_0)}(\sigma^g)^2 T = -\frac{g''(S_0)}{2g'(S_0)}(\sigma^g)^2 T, \tag{5.22}$$

where $g(C)$ is the price of the CMT yield C. Generally, the adjustment convexity comes often from the fact that a payoff and its underlying are not martingale under the same probability measure; in more usual words, the two do not have the same P-risk-forward neutral probability.

5.4.4. *Replication*

The replication constitutes an alternative to the closed forms for convexity correction to compute forward payoff value. The replication method has the advantage to be approximation free, in the sense that it is not based on Taylor expansion and similar techniques. It is also model free in the sense that it does not require some modeling assumption about the distribution to compute the convexity correction.

The replication method consists in "replicating" the payoff value of an underlying with some other assets. The first example of the replication method shows how to replicate a CMS statically with swaptions. More generally, let us consider a derivative whose payoff is a function f of the underlying S. We want to write the payoff $f(S)$ by means of European options.

We can first remark that we can write any twice derivable function f (supposed here positive) as a continuous sum using Dirac delta function as:

$$f(S) = \int_0^\infty f(k)\delta(S-k)dk$$
$$= \int_0^K f(k)\delta(S-k)dk + \int_K^\infty f(k)\delta(S-k)dk. \quad (5.23)$$

The decomposition of the second equality holds for any positive "transition" value K. Integrating twice by parts the two terms in Eq. (5.23) leads to:

$$f(S) = f(K)1_{\{S<K\}} - f'(k)(k-S)^+|_0^K$$
$$+ \int_0^K f''(k)(k-S)^+dk$$
$$+ f(K)1_{\{S\leq K\}} - f'(k)(S-k)^+|_K^\infty$$
$$+ \int_K^\infty f''(k)(S-k)^+dk. \quad (5.24)$$

Payoff function $(S-k)^+$ and $(k-S)^+$ are precisely the ones of European calls or puts option with underlying S and strike k.

By considering $(k-S)^+ - (S-k)^+ = k - S$ we have the following interesting result:

$$f(S) = f(K) + f'(K)(S-K) + \int_0^K f''(k)(k-S)^+dk$$
$$+ \int_K^\infty f''(k)(S-k)^+dk. \quad (5.25)$$

The expansion in Eq. (5.25) means that we can replicate or reconstitute any "smooth" payoff of an underlying by a infinity sum of call and put options with successive strikes. The most important remark is that there is no restriction on the underlying price process (Fig. 5.5).

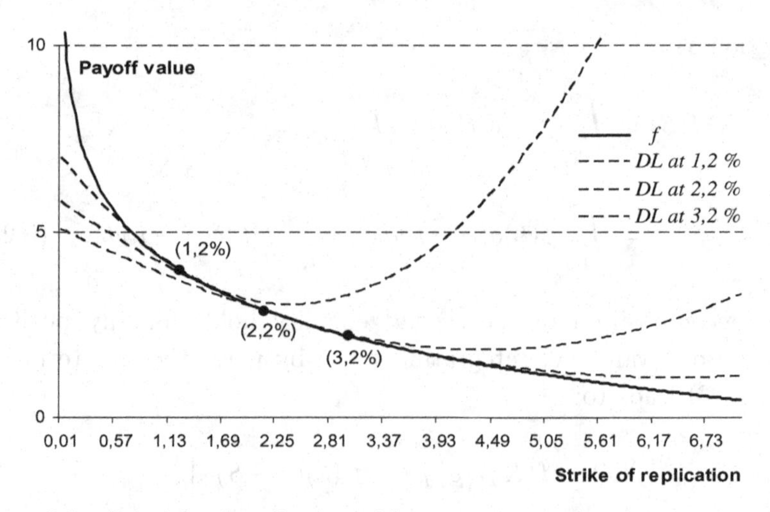

Figure 5.5: Principle of replication with tangent approximation at successive replication strike options.

In order to implement this, we need to discretize Eq. (5.25). We use a polynomial approximation of the two integrals:

$$f(S) \approx f(K) + f'(K)(S-K) + \sum_{i=0}^{N} f''(k_i)(k_i - S)^+ \delta_i + \sum_{i=l}^{N} f''(k_i)(S-k_i)^+ \delta_i, \quad (5.26)$$

The points $\{k_i, i = 0, N+1\}$ chosen such that $k_0 = 0$ and $k_l = K$ and the weights are defined by $\delta_i = k_{i+1} - k_i, i = 1, N$.

For example, we will consider an application of the technique on option payoff $f(x) = (x-K)^+$ on a CMS underlying S_T. We will calculate the forward value of a caplet CMS by a finite sum of payer and receiver swaptions:

$$\boldsymbol{E}^T[f(S_T)] = f(S_0) + \sum_{i=1}^{N_1} \alpha_i^{(1)} \boldsymbol{E}^T[A(S_T)(k_i^{(1)} - S_T)^+] + \sum_{i=1}^{N_2} \alpha_i^{(2)} \boldsymbol{E}^T[A(S_T)(S_T - k_i^{(2)})^+], \quad (5.27)$$

where A is the annuity of yield, and S_0 is the forward value of the CMS S_T: $\boldsymbol{E}^T[S_T] = S_0$.

The problem is how to determine the different weights $\{\alpha_i^{(1)}, i = 1, \ldots, N_1\}$, $\{\alpha_i^{(2)}, i = 1, \ldots, N_2\}$ and the strikes $\{k_i^{(1)}, i = 1, \ldots, N_1\}$, $\{k_i^{(2)}, i = 1, \ldots, N_2\}$. Let us consider the function:

$$\begin{aligned}\phi(x) &= \frac{f(x) - f(S_0)}{A(x)} \\ &= \sum_{i=1}^{N_1} \alpha_i^{(1)} (k_i^{(1)} - S_T)^+ + \sum_{i=1}^{N_2} \alpha_i^{(2)} (S_T - k_i^{(2)})^+, \qquad (5.28)\end{aligned}$$

and by supposing that it is derivable anywhere its second-order Taylor development order gives:

$$\begin{aligned}\phi(x) = \phi'(S_0)(x - S_0) + \frac{1}{2}\phi''(S_0)(x - S_0)^2 \\ + o(x - S_0)^2. \qquad (5.29)\end{aligned}$$

We take the firsts strikes as the forward CMS value $k_1^{(1)} = k_2^{(1)} = S_0$. We distinguish two cases:

$$x > S_0 \Rightarrow \phi(x) = \sum_{i=1}^{N_2} \alpha_i^{(2)} (S_T - k_i^{(2)})^+,$$

then necessary from (5.27) $\alpha_1^{(2)} = \phi'(S_0)$,

$$x < S_0 \Rightarrow \phi(x) = \sum_{i=1}^{N_1} \alpha_i^{(1)} (k_i^{(1)} - S_T)^+,$$

then necessary from (5.27)$\alpha_1^{(1)} = -\phi'(S_0)$.

In order to determine the others parameters, we use from Eq. (5.27) that the function ϕ can be approximated by the tangent parabola $x \mapsto (x - S_0)^2/2$ in S_0. We reconstitute by a successive algorithm this function for the equal-spaced values of strikes $\{k_i^{(j)}, i = 1, \ldots, N_j, j = 1, 2\}$:

$$x > S_0 \colon Z_1 = k_1^{(1)} = S_0,\ \alpha_1^{(1)} = \phi'(S_0),$$

$$\begin{aligned} i = 1\colon \quad & Z_{i+1} = Z_i + \Delta/2 \\ i \geq 2\colon \quad & k_{i+1}^{(1)} = k_i^{(1)} + \Delta,\ \ Z_{i+1} = Z_i + \Delta, \\ & \alpha_{j+1}^{(1)} = \phi'(Z_{i+1}) - \sum_{j=1}^{i} \alpha^{(1)}{}_j, \end{aligned}$$

$$x < S_0 \colon Z_i = k_1^{(2)} = S_0,\ \alpha_1^{(2)} = -\phi'(S_0),$$

$$\begin{aligned} i = 1\colon \quad & Z_{i+1} = Z_i - \Delta/2, \\ i \geq 2\colon \quad & k_{i+1}^{(2)} = k_i^{(2)} - \Delta, \quad Z_{i+1} = Z_i - \Delta, \\ & \alpha_{i+1}^{(2)} = -\phi'(Z_{i+1}) - \sum_{j=1}^{i} \alpha_j^2. \end{aligned}$$

This successive definition verifies that at the ith step the function $x \mapsto \sum_{j=1}^{i+1} \alpha_j^{(.)}(x - k^{(.)}j)^+$ should be tangent to $\phi(x)$ at Z_{i+1}.

Note that the weights depend only on the function f and S_0, but not of the maturity of the option T. This remark can be used to gain time in pricing.

5.5. Vanilla Interest Rate Options

5.5.1. *Interest rate caps and floors*

An interest rate cap[9] is an agreement that allows a borrower to seek the maximum interest rate payable over a set period. In other words it is a contract that guarantees a maximum level of a floating interest rate. A cap can be a guarantee for one particular date, in this case it is called a caplet. Each period is known as the tenor. On each of the rollover dates for the life of the cap, the purchaser can make

[9]Cap is also known as ceiling rate agreement CRA.

claims under the guarantee should floating interest rate above the level agreed on the cap. If the floating interest rate is greater than maximum level (strike), the second party must pay the difference between the rates. In the reverse, the holder is free to take advantage of the lower rate and effectively abandons the portion of premium only. Let us emphasize that borrower pays a premium; however, the lower the maximum level sought by a purchaser the higher premium. At settlement a caplet has a profit profile as follows:

Figure 5.6 shows the investor's net profit or loss on caplet. When floating interest rate is below the strike 5%, the caplet has no value. Claims will only be made when floating interest rate is above 5%. Suppose that the principal US$10 million and the floating interest rate is a 6-month Euribor that at reset date is 7%. The caplet therefore provides a payoff as follows[10]:

$$€10.000.000 \times 0.5 \times \mathrm{Max}(7\% - 5\%, 0.0)$$

whereas the holder of a cap is hoping that the floating interest rate will increase, the purchaser of a floor is hoping that it will decrease. An interest rate floorlet is an agreement that guarantees a minimum level of floating interest rate for one particular date. A floor is a sum of continuous series for floorlets. Like cap, in return for making this guarantee, the buyer pays a premium. Figure 5.7 shows the investor's

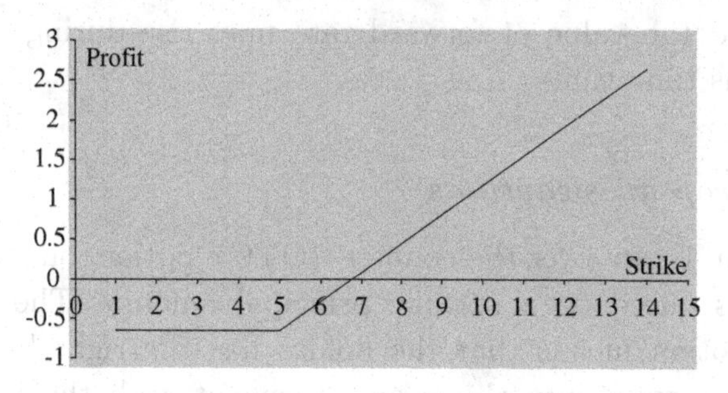

Figure 5.6: Profit from buying a caplet; premium = 0.65% strike or cap rate = 5%.

[10]Interest rate period must take account of the exact number of days and using the market convention.

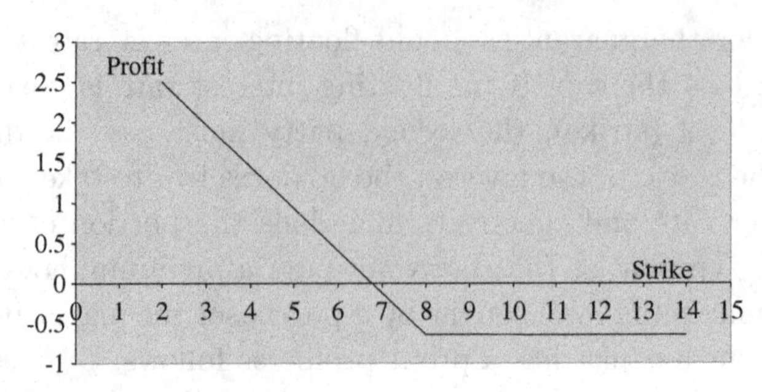

Figure 5.7: Profit from buying a Floorlet; premium = 0.65% strike or floor rate = 8%.

net profit or loss on floorlet. A comparison is made between the floor and floating rate. If the floating rate is less than the floor rate, the second party will pay the difference between the rates. However, if the floating interest rate is greater than floor rate the option is abandoned. With the notation already introduced, the floorlet provides a payoff of:

$$€10.000.000 \times 0.5 \times \text{Max}(8\% - 7\%, 0.0)$$

Example: Consider a cap that is strike at 7% and will expire in 6 months. If the current value of forward rate is 6.5%, is cap worthless? The cap has no intrinsic value, but 6 months remain until it expires, and the value of forward rate may rise during that time. The cap has time value.

5.5.2. *European swaptions*

A swaption is an over-the-counter (OTC) option on a specified IRS that is based on a specific principal amount. The flexibility of the swaption means that the holder has the right but not the obligation to enter into the swap agreement with the issuer on a specified future date. In return for this, the holder pays a premium or seller. The premium depends of the duration of the option period, the term and strike of the swap and the volatility of interest rates.

There are two types of swaptions, the *payer's* and the *receiver's swaptions*. By paying a premium, the seller enters into the obligation to pay floating interest rate and receive fixed rate if the buyer exercises his right to purchase a payer's swap. The holder of a receiver's swaption obtains the right to sell a receiver's swap.

Purpose of the product: A payer's swaption (receiver's) enables its holder to hedge against rising (falling) interest rates and speculate on rising (falling) interest rates. Swaptions grant all the benefits of an interest rate swap as well as being a perfect product to use where there is uncertainty of outcome.

5.5.3. *Spread options*

Literally, a spread option is an option on a difference of two indexes. But this definition has been widened to include all forms of options that are written as a linear combination of a finite set of indexes. In the fixed income market, spread options are based on two interest or swap rates, two yields, etc.

Example: Let us consider an option written on two CMS rates of different maturities. Its payoff is in the form $(CMS(1) - CMS(2) - K)_+$.

Pricing: There are several models for valuing this spread option: We can value the spread as normal, with good estimation of each CMS rate volatility. The spread option can be valuated as the difference of two lognormals. This can be done with numerical integration with Gauss Legendre polynomials. We value CMS rates with an SABR model. We use then copula to model the correlation between the different rates. For more details on spread options valuation, see Chapter 6. A panorama of mathematical tools can be found in Ref. [8].

5.6. Summary

The standard fixed income instruments are caps, floors, swaps and swaptions. A cap is insurance to hedges rising from interest rates. A floor is insurance to hedge from falling interest rates. An interest rate swap is an agreement to exchange two rates periodically for a specified period. The rates exchanged can be fixed or floating. They can be single currency or from different currencies. A swaption is a call option on a swap.

These instruments are the standard products. They are liquid and mainly used to valuate more complicated fixed income instruments.

Swaps are various. They range from plain vanilla swaps to more other complicated ones such as LIBOR-in-arrears and CMT–CMS swaps. The valuation of the latter uses a correction of forward volatilities commonly known as convexity adjustment.

Another important issue in the fixed income market is zero curves constructing. Money market rates, Euro futures, swap rates and bonds are used to construct the curve by bootstrapping and curve fitting. Bootstrapping is used for maturities for which we have market date. In order to fill gaps, we use curve fitting.

References

[1] Amblard G and Lebuchoux J (2000). BNP Paribas — models for CMS caps. *Euro Derivatives/Risk Magazine*, September.

[2] Benhamou E (2000a). A martingale result for convexity adjustment in the Black pricing model. Working Paper, London School of Economics.

[3] Benhamou E (2000b) Pricing convexity adjustment with Wiener Chaos. FMG Dp351.

[4] Carmona R and Durrleman V (2003). Pricing and hedging spread options. *SIAM Review*, **45**(4), 627–685.

[5] Hagan P (2003). Convexity conundrums: pricing CMS swaps, caps and floors. *Wilmott Magazine*, 38–44.

[6] Hull J (2002). *Options, Futures, and Other Derivative Securities*, 5th edn. New York: Prentice Hall.

[7] Selic N. An examination of the convexity adjustment technique in the pricing of constant maturity swaps. Advanced Mathematics of Finance Honours Project.

Chapter 6

SMILE MODELING

In this chapter, we will see first smile and digital models before reviewing some basic models. We will try to give the intuition behind the model and for complete proofs refer the reader to scientific papers.

6.1. Smile and Digital Models

According to BS model, if we take an option with different maturities and different strikes but written in the same stock, we should have the same volatility. That is not the case, at fixed maturity, if we plot the implied volatility of an option as a function of its strike, this is defined as a volatility smile. For equity options, where the graph is downward sloping, the term volatility skew is often used. In general, investors use volatility smile to price options in the foreign currency market and the equity option market.

Digital models are European products, whose value depends not only from the value of the implied volatility at their strike, but also from the slope of the smile at their strikes. We will discuss about these models in greater details in the following sections.

Also referred to as binary, cash-or-nothing or all-or-nothing options, digital option in its most vanilla version pays a predetermined amount Q if the underlying asset is above (respectively below) a certain strike for a digital call (respectively put). Common exotic versions of the digital are the following:

(1) The American digital paying a certain amount if the trigger condition is activated during the life of the option.
(2) The correlation digital paying something if another asset triggers the payment or non-payment.

6.1.1. *Target market*

Digital options are pure bet options. If the trader, investor is right, he will receive a known amount. The upside is even simpler than for standard call as one knows in advance what he should get. Because of this simplicity, digital options attract many market participants in the over-the-counter marketplace.

Besides this, structurers can use digitals to create a desired payoff profile. It is also worth noticing that some common structures are essentially series of digitals like, for instance, range accrual notes in fixed income market or mini-premium foreign exchange options.

European digital options are very easy to price under BS options as their pricing consists in simply computing the probability for a stock of being above or below a given threshold. For instance, for a binary call, the price is given by

$$\begin{aligned}\mathrm{Bin}\, C &= \mathrm{e}^{-rT} E\left[Q 1_{\{S_T > K\}}\right] \\ &= Q\mathrm{e}^{-rT} P(S_T > K) \\ &= Q\mathrm{e}^{-rT} N(d_2)\end{aligned} \tag{6.1}$$

with

$$d_2 = \frac{\ln(S_0/K) + (r - q - \sigma^2/2)T}{\sigma\sqrt{T}}, \tag{6.2}$$

where $N(x)$ is the cumulative normal density function, S_0 is the spot stock price, K the strike price, r the risk free rate, q the continuous yield dividend, T the option maturity and σ the BS implied volatility.

However, life is not as simple as the BS model assumes. Pricing with BS model would be quite misleading for very skewed market. Binary options are well known to be very sensitive to the product of the slope of the smile (at the barrier level) and its vega. An intuitive explanation is to approximate the binary option by a call spread. Obviously, the payoff of a binary option is just the limiting case of a call spread:

$$1_{\{S_T > K\}} = \lim_{\varepsilon \to 0} \frac{(S_T - K + \varepsilon)^+ - (S_T - K - \varepsilon)^+}{2\varepsilon}. \tag{6.3}$$

Therefore, writing that call option prices are function of the strike, the maturity, and the implied volatility itself function of the maturity and strike

$$C(T, K, \sigma(T, K)), \tag{6.4}$$

we have that the price of a binary should be equal to the limit of the call spread

$$\text{Bin}\, C = \lim_{\varepsilon \to 0} - \frac{C(T, K + \varepsilon, \sigma(T, K + \varepsilon)) - C(T, K - \varepsilon, \sigma(T, K - \varepsilon))}{2\varepsilon} \tag{6.5}$$

or using the chain rule

$$\begin{aligned} \text{Bin}\, C = - \bigg(& \frac{\partial}{\partial K} C(T, K, \sigma(T, K)) \\ & + \frac{\partial}{\partial \sigma} C(T, K, \sigma(T, K)) \frac{\partial \sigma(T, K)}{\partial K} \bigg) \end{aligned} \tag{6.6}$$

Proper modeling of the volatility smile is therefore essential. However, for European digital option like this case, there is no need to use a complicated modeling of the volatility as one can use static replication to price the European option. Indeed, as shown by Breeden and Litzenberger (1978) one can extract density information from European option.

6.2. Smile Models (Basic European Models)

In this section, we will review some of the most common models to handle smile. In general, there are plenty of models. We will concentrate on the following ones:

(1) jump models and in particular the Merton model or mixture of distributions (see Brigo, Mercurio and Rapisarda, 2004).
(2) local volatility model à la Dupire;
(3) stochastic volatility models, in particular the SABR and Heston model.

6.2.1. *The Merton model*

The first idea coming to the mind when thinking about the economical origin of the Smile is that it translates a fear of sudden moves of the market. Therefore, the difference with a deterministic volatility BS Model should be analyzable in term of jumps.

Let us model the underlying as:

$$\frac{dS_t}{S_t} = \mu dt + \sigma dW_t + (J_t - 1)dq_t, \tag{6.7}$$

where dq_t is a Poisson process with intensity λ

$$dq_t = \begin{cases} 0, & \text{with probability } 1 - \lambda dt, \\ 1, & \text{with probability } \lambda dt, \end{cases}$$

where J_t is a Gaussian variable with a mean conditional to the jump equal to

$$E[J_t - 1] = \varepsilon \tag{6.8}$$

and a variance conditional to the jump equal to

$$E\left[(J_t - 1)^2\right] = \delta^2. \tag{6.9}$$

Then by applying the Ito lemma (see Appendix A) to $\mathrm{Log}(S_t)$, we get:

$$S_t = S_0 \mathrm{e}^{\int_0^t \mu dt + \sigma dW_t + \mathrm{Log}(J_t) dq_t} \tag{6.10}$$

Merton (1977) showed that the value of a call can be priced as:

$$\mathrm{Call} = \mathrm{e}^{-\lambda(1+\varepsilon)T} \sum_n \frac{(\lambda(1+\varepsilon)T)^n}{n!} \left(S_0 \mathrm{e}^{b_n T} N[d_{1,n}] - K N[d_{2,n}]\right), \tag{6.11}$$

where

$$b_n = -\lambda + n\,\mathrm{Log}[1+\varepsilon] \tag{6.12}$$

and

$$d_{1,n} = \frac{\mathrm{Log}[F/K] + b_n T + 1/2(\sigma^2 T + n\delta^2)}{\sqrt{\sigma^2 T + n\delta^2}} \tag{6.13}$$

with

$$d_{2,n} = d_{1,n} - \sqrt{\sigma^2 T + n\delta^2} \tag{6.14}$$

In order to implement this formula, we have to truncate the sum on a level, but since exponential function grow more than any other, we have to express this formula as shown below to avoid numerical difficulties:

$$\mathrm{e}^{-\lambda(1+\varepsilon)T}\frac{(\lambda(1+\varepsilon)T)^n}{n!} = \exp\left(-\lambda(1+\varepsilon)T + n\ln(\lambda(1+\varepsilon)T) - \sum_{i=1}^{n}\ln(i)\right).$$

6.2.1.1. *Influence of the jumps (a la Merton) on the smile*

Negative jumps risk makes the put out of the money act as protection in case of market crashes. Therefore, the value of such put should be higher than predicted by a strait application of the BS formula. The same thing can be said about positive jumps risks, and out of the money calls. Therefore, jumps induce a smile effect as it can be seen in Fig. 6.1.

We see that when the maturity becomes large, the smile tends to disappear. This is a general theorem about jumps with stationary intensity. At large maturity the number of jump event makes them indistinguishable from the effect of ordinary volatility and ordinary drift.

6.2.2. *Local volatility models (a la Dupire)*

Local volatility models extend BS model to make it compatible with market options prices. Local volatility models are represented by a

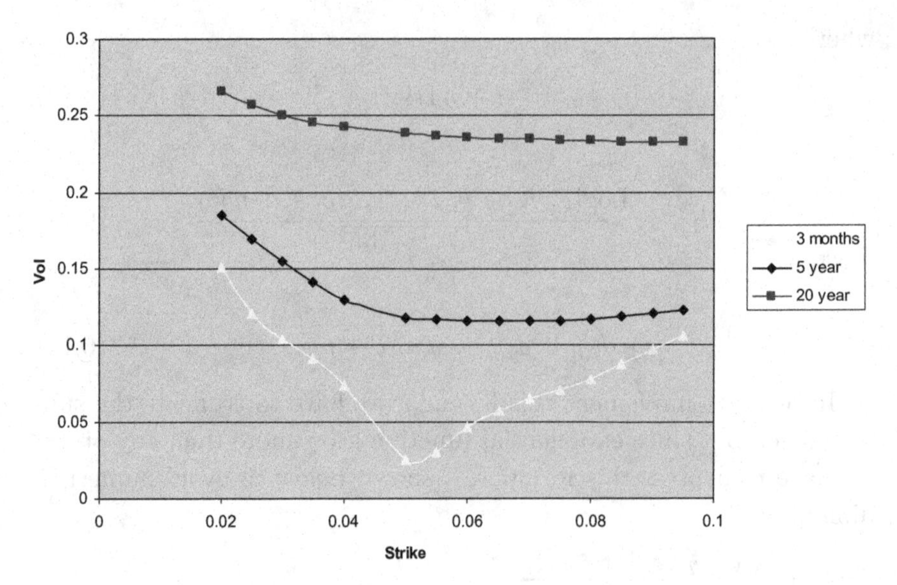

Figure 6.1: Merton implicit smile. Data for the preceding graph: forward: 0.05, Vol 0.05, jump intensity 0.02, average size of the jump: −10%; standard deviation of the jump: 10%.

dynamic equation such as

$$dS_t = \mu(S)dt + \sigma(S)dW_t. \tag{6.15}$$

These models were greatly appreciated in the late 1990s when Dupire and Derman released their methodologies. In essence, Dupire created the interest by showing that the local volatility was almost an observable in the market. Its formula is:

$$\sigma^2(K,T) = 2\frac{(\partial C_{K,T}/\partial T) + (r-d)K(\partial C_{K,T}/\partial K) + dC_{K,T}}{K^2(\partial^2 C_{K,T}/\partial K^2)}, \tag{6.16}$$

where r is the effective interest rate and d is the dividend rate associated with the underlying. When both can be put to zero, for example, in the case of options on forward price without dividends, then the local volatility is just the ratio of a calendar spread on the price of a butterfly strategy.

By exploiting the degrees of freedom of binomial trees, Derman *et al.* (1994, 1999) showed that it is possible to calibrate a binomial tree to replicate option prices observed in the market. But the lack of flexibility of binomial trees leading to negative probability problems in most of the cases led the authors to amend their methodology. Trinomial trees are more flexible because they allow us to control where to branch. By controlling the branching, the authors (see Derman and Kani, 1996) were able to build implicit trees re-pricing any given smile, without negative probabilities, as long as the discretization is small enough to accommodate for the variations in the smile.

6.2.3. *The CEV model*

Some paramerization of the volatility function are famous. The CEV process is probably the best known of them. The equation of the constant volatility process is:

$$dS_t = \mu S_t dt + \sigma S^{\beta+1} dW_t, \tag{6.17}$$

where the drift part is usually equal to the difference between the risk free and the dividend rate

$$\mu = r - q. \tag{6.18}$$

Obviously, the CEV model (6.17) nests the lognormal model of BS and Merton, the square-root model of Cox and Ross.

In addition, depending from the value of the elasticity parameter β, the process has different properties. For $\beta < 0$, infinity is a natural boundary. For $-(1/2) < \beta < 0$, the origin is an absorbing boundary. For $\beta < -(1/2)$, the origin is a killing boundary point. For $\beta = 0$, both zero and infinity are natural boundaries. For $\beta > 0$, infinity is an absorbing boundary.

The traditional way for finding the price of the call and put option is to notice that using the change of variable $z_t = S_t^{-\beta}/\sigma|\beta|$, the CEV process reduces to a Bessel process of order $(1/2\beta)$ in the case of no drift, while the case of drift is obtained via an extra time and scale change given by (Borodin and Salminen (1996), Goldenberg (1991) and Revuz *et al.* (2004).

Using the result of Godenberg (1991, p. 28), the CEV process S_t^u with the drift $\mu \neq 0$ is obtained from the process without drift via a scale and time change.

$$S_t^u = e^{\mu t} S^0_{\tau(t)}, \quad \tau(t) = \frac{1}{2\mu\beta}(e^{2\mu\beta t} - 1).$$

The corresponding pricing formula for the call option is given by:

$$C_t = S_0 e^{-qT} Q(\xi, n-2, y_0) - e^{-rT} K(1 - Q(y_0, n, \xi)). \tag{6.19}$$

For the case $\beta > 0$ and

$$C_t = S_0 e^{-qT} Q(y_0, n, \xi) - e^{-rT} K(1 - Q(\xi, n-2, y_0)). \tag{6.20}$$

For the case $\beta < 0$, and the usual BS formula for $\beta = 0$, where

$$n = 2 + \frac{1}{|\beta|}, \tag{6.21}$$

$$\xi = \frac{2\mu S^{-2\beta}}{\sigma^2\beta(e^{2\mu\beta T} - 1)}, \tag{6.22}$$

$$y_0 = \frac{2\mu K^{-2\beta}}{\sigma^2\beta(1 - e^{-2\mu\beta T})} \tag{6.23}$$

and $Q(x, u, v)$ is the complementary non-central χ^2-distribution function with the degree of freedom u and the non-centrality parameter v. A good algorithm to compute it can be find in Schroder (1989).

Example of smiles produced by such processes is shown in Fig. 6.2, where we see that contrary to the jump diffusion process, the skewness does not disappear with large maturity. We also notice that the kurtosis do exist for short maturities, but it has a completely different shape from the preceding Merton model. In particular, there is no singularity of the smile when the maturity tends toward zero, as it is the case for Jump models.

Dynamical behavior: Local volatility models have the wrong dynamic behavior. The underlying asset moves in the opposite direction of the smile: when the underlying asset decreases, the smile tends to move up and when the underlying asset increases, the smile moves down. This comes from the fact that the smile is created by the functional form of the volatility which does not change when the level of the underlying security moves.

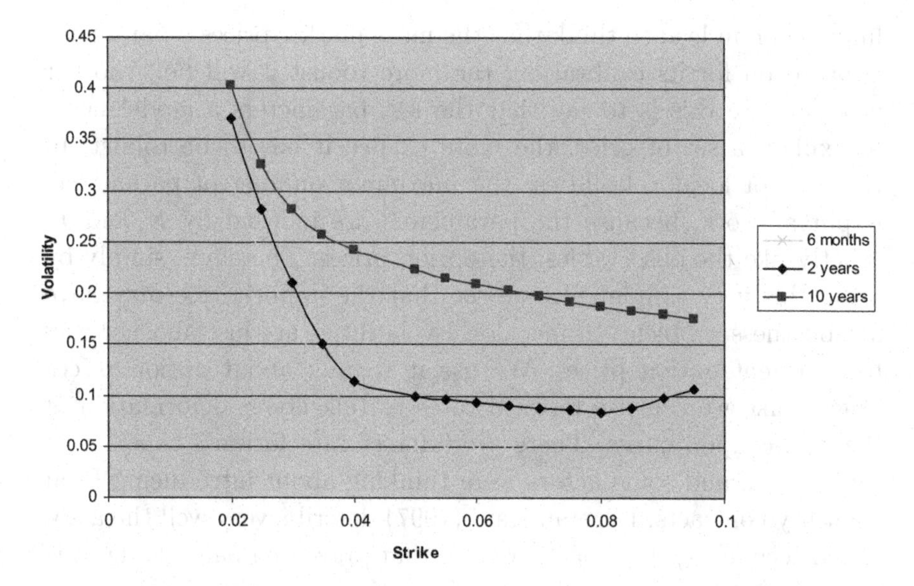

Figure 6.2: CEV implicit volatility.

In reality we observe on most of the markets that prices and smiles tend to move in the same direction (sticky strike rule; see Derman and Kani (1999)). Therefore the hedges computed with local volatility models tend to be worse than hedges computed with a constant volatility model.

6.2.3.1. *Extrapolation issues*

The practical way to use local volatility models is to build an implicit tree from market price. It implies that we need to have in advance a complete set of option prices that we will replicate with the implicit tree. It means that we will not be able to deduce the price of a very far out of the money or very deep in the money price of an option, because the tree will not be able to price it. So the local volatility models have very poor price extrapolation ability.

6.2.3.2. *Robustness issues*

Financial models are built to compute hedges. Hedges work only if model parameters are stable enough. Statistical results like central limit theorem lead to think that the more market prices a parameter is based on for its calibration, the more robust it will be. Another way to state this is to say that the less parameters a model need, to explain a set of price, the more chance it has to be robust. In the case of local volatilities, the maximum number of parameters is put at work, because the parameters are indexed by K and T, exactly like the observables: the option prices. Therefore, simply by using this very simple rule, we see that the implicit tree may have a robustness problem. In fact, local volatilities are just another way to represent option prices. We use it to talk about option prices exactly like we can use forward rates to talk about deformation of the zero-coupon curve. There are interest rate forward contract in the market, and some actors were thinking about introducing local volatility contracts. Derman Kani (1997) describe very well the associated technology. One of the reason that prevented bank doing so is the lack of robustness of the locked volatilities for these contract.

Despite their drawbacks, local volatility models have been extensively used by banks but mainly because of the easiness of associated technology. But why to produce price figures if you cannot rely on them to trade?

6.2.4. *Stochastic volatility models*

Stochastic volatility models are the answer to this problem. They have the required dynamic behavior and the necessary robustness. A general representation of their dynamic is:

$$\begin{aligned} \frac{dS_t}{S_t} &= \mu(S,\alpha)dt + \sigma(S,y_t,\alpha)dW_t, \\ dy_t &= \theta(y,\alpha)dt + \nu(y_t,\alpha)dW_t, \end{aligned} \tag{6.24}$$

where y_t represents the set of variables that drive the volatility of S_t and $\nu(y_t,t)$ is the volatility of the volatility (technically only an instantaneous standard deviation).

Here we parameterized all the functions by α, which stand for a vector of parameters. The calibration process aims to determine this vector α. The most important characteristic of such a model is its option formula. This formula can be used as an interpolating device to price options at the most likely level, according to the set of available prices, or to price exotic options by replication. But above all it can be used to calibrate the model. We will discuss three types of situations in the following sections.

6.2.4.1. *An analytical formula*

The only practical case known from the authors is the case where the volatility process is independent from the underlying process. Hull and White (1987) showed that if we know a formula for the option price conditioned by the volatility process, it is easy to get a formula including the stochastic volatility process by using an expectation of the realized variance of the process. For example, if we consider the underlying process as

$$\frac{dS_t}{S_t} = \sigma_t dW_t^S, \quad d\sigma_t = \frac{(\nu\sigma_t)^2}{2} + \nu\sigma_t dW_t^\sigma \tag{6.25}$$

where the two Brownian motions are independent $\langle dW_t^S, dW_t^\sigma \rangle = 0$.

Then, the pricing for a Call(K, T) is:

$$\begin{aligned} \text{Call} &= SN[d_1] - KN[d_2], \\ d_1 &= \frac{\text{Log}[S/K] + V/2}{\sqrt{V}}, \\ d_2 &= d_1 - V, \end{aligned} \tag{6.26}$$

where V is

$$V = \mathrm{e}^{\nu^2 t}\sigma_0^2 T. \tag{6.27}$$

The problem associated with such a model is the impossibility to have a skew, that is a slope for the smile. The only way to have a skew is to have a correlation between the volatility and the underlying process different from 0. This is obviously not the case for this model.

A possibility that has been investigated by Brigo *et al.* (2004) is to build an option formula with a mixture of option formula. For example, take n volatilities $\sigma_1, \sigma_2, \ldots, \sigma_n$ and find the coefficients $\lambda_1, \lambda_2, \ldots, \lambda_n$. Such that the resulting mixture matches the observed smile. It is extremely easy to calibrate maturity by using maturity. The main problem is that there is no associated process. Therefore, the hedge ratios computed by this model are extremely delicate to use due to the absence of any dynamic justification.

6.2.4.2. *Heston model and Fourier transform*

Very often at the level of the Fourier transform with respect to the underlying S, the integration of the differential equation in V can be done analytically. The Heston model is an example of such model:

$$\begin{aligned} \frac{dS_t}{S_t} &= rdt + \sqrt{V_t}dW_t, \\ dV_t &= \kappa(\theta - V_t)dt + \nu\sqrt{V_t}dB_t \quad d\langle W, B\rangle_t = \rho dt. \end{aligned} \tag{6.28}$$

The price of the call is then:

$$\text{Call} = S - Ke^{-rT}\frac{1}{\pi}\int_0^{\infty} \text{Re}\left[e^{-iK\frac{(k+i/2)}{k^2+1/4}}h(-k-i/2)\right], \tag{6.29}$$

where the function h is defined by:

$$h(z) = e^{A+BV_0} \tag{6.30}$$

with

$$\zeta = \sqrt{(\kappa - iz\rho\nu)^2 + \nu^2(iz + z^2)}, \tag{6.31}$$

$$\Psi^+ = -(\kappa - iz\rho\nu) + \zeta, \tag{6.32}$$

$$\Psi^- = (\kappa - iz\rho\nu) + \zeta, \tag{6.33}$$

$$A = -\frac{\theta\kappa}{\nu^2}(\Psi^+ T + 2\text{Log}((\Psi^- + \Psi^+ e^{-\zeta T})/(2\zeta)), \tag{6.34}$$

$$B = \frac{-(iz + z^2)(1 - e^{-\zeta T})}{\Psi^- + \Psi^+ e^{-\zeta T}}. \tag{6.35}$$

A complex analysis is needed to perform the calculations and we can use the fast Fourier transform to compute the integral (see, e.g. Sepp, 2002).

6.2.4.3. *Lewis model and the preservative approach*

When analytical solution is available, it is advisable to look for an analytical approximation. A lot of progress has been done in this area lately. The main idea is to consider the stochasticity of the volatility to be small. In that case we can make a Taylor development. A good example is the Lewis model (2000) as described by:

$$\frac{dS_t}{S_t} = rdt + \sqrt{V_t}dW_t,$$
$$dV_t = a(V_t)dt + b(V_t)dB_t, \quad d\langle W, B\rangle_t = \rho(V_t)dt, \tag{6.36}$$

where $a(V_t) = \kappa(\theta - V_t)$, $b(V_t) = V_t^{\varphi}$ and $\rho(V_t) = \rho$. The first step is to transform the pricing of the call into:

$$\text{Call} = S - K\mathrm{e}^{-rT}\frac{1}{2\pi}\int_{ik_i-\infty}^{ik_i+\infty}\frac{\mathrm{e}^{-ixK}}{x^2 - ix}H(x, V, T)dx,$$

where H is the fundamental transform of the process. It follows the partial differential equation:

$$\frac{1}{2}\xi^2\eta^2\frac{\partial^2 H}{\partial V^2} + (b + \xi d\chi)\frac{\partial H}{\partial V} - cVH - \frac{\partial H}{\partial T} = 0 \tag{6.37}$$

with

$$c = \frac{k^2 - ik}{2}, \quad d = -ik, \quad \chi = \rho(V)\eta(V)V^{1/2}, \quad a(V) = \xi\eta(V). \tag{6.38}$$

We now do the following development:

$$H = H_0(1 + \xi h_1 + \xi^2 h_2 + \cdot) \tag{6.39}$$

We get series of differential equations that we can solve to get the analytical approximation up to the desired order. The details are discussed in Lewis (2000).

An example of smile generated by such model is:

This is the smile observed and calibrated on the YEN/USD Market. The pure Heston (no jumps) parameters where calibrated maturity by maturity with some constraints enforcing some smoothness between maturities (Fig. 6.3).

If we include a jump sector in the model, then the short term part of the smile is deeply changed, allowing us to calibrate markets where the likelihood of credit event affects the smile. Such a smile looks like the model shown in Fig. 6.4.

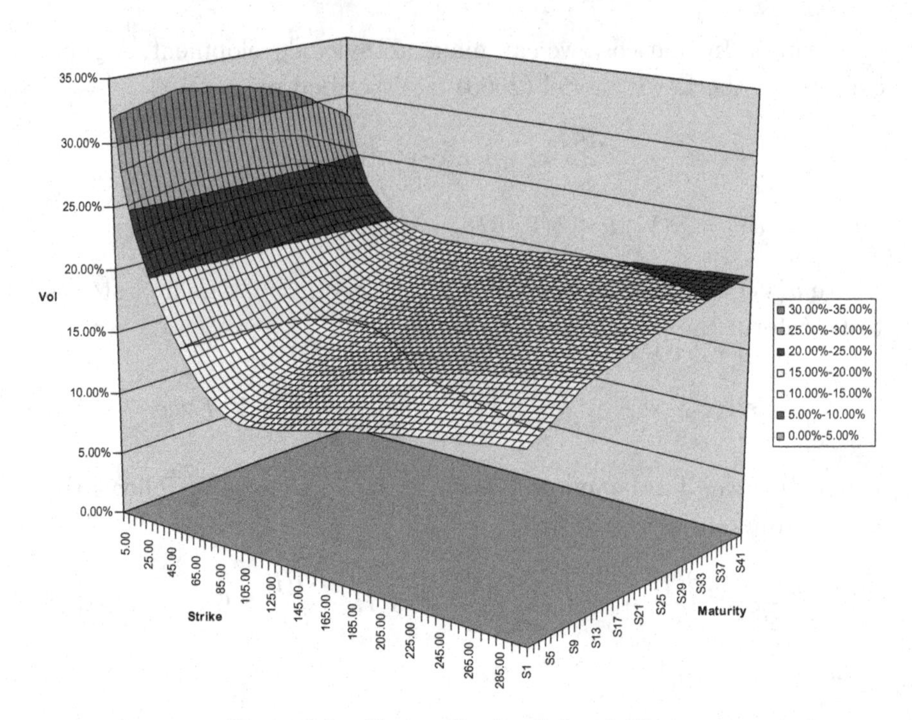

Figure 6.3: Heston Yen implicit volatility.

Contrarily to Merton model, stochastic volatility (Heston) induce smiles and skews that increases at maturiry increases for the interest. Then if we want to generate stochastic volatility and skews/smiles on both short and long maturity, we can use Bates model that combined the Merton and the Heston theorical modeling.

6.2.4.4. *SABR model and link to Riemannian geometry*

This is a major insight, whose precursors began with the paper from Minakshisundaram (1949), Varadhan (1967) and later Chavel (1984) in a clear synthetic book and Pleijel who showed that any pricing kernel G associated with the heat equation

$$\frac{\partial G}{\partial t} = \sum_{i,j} g^{i,j} \frac{\partial^2 G}{\partial S^i \partial S^j} + \sum \_ih_i \frac{\partial G}{\partial S^i}, \tag{6.40}$$

$$G(0) = \delta(S) \tag{6.41}$$

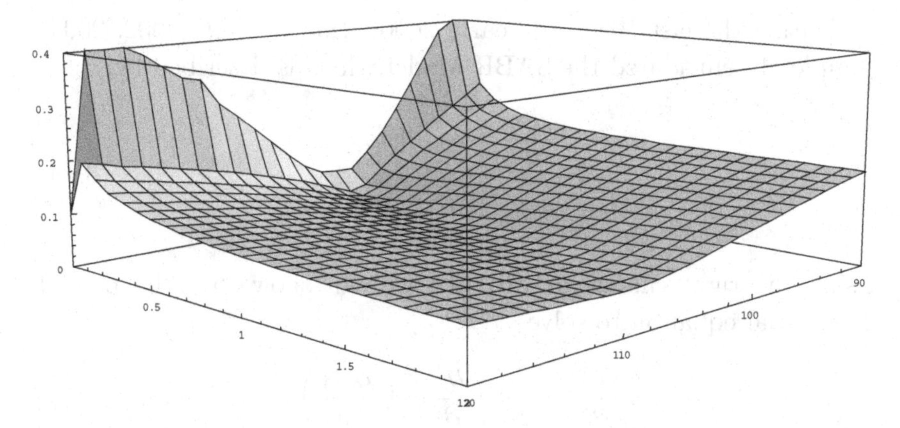

Figure 6.4: Heston and jumps models.

has the short-term development:

$$G(t, S_0, S) = \mathrm{e}^{-\frac{d(S_0,S)}{2t}}(R_0(S_0, S) + tR_1(S_0, S) + t^2R_2(S_0, S) + \cdots), \quad (6.42)$$

where $d(S_0, S)$ is the geodesic distance associated with the Riemannian space defined by the metric distance: $ds^2 = \sum_{i,j} g_{i,j}dx^i dx^j$ and $g_{i,j}$ being the inverse matrix of $g^{i,j}$. Therefore the short-time behavior of stochastic volatility models are perfectly well described by its Riemannian geometry. The shape of the leading term is reminiscent of the Gaussian transition probability, where the Euclidian distance is replaced by the distance inducted by the diffusion coefficient matrix. It is not by chance. In fact the existence of a change of variables, where the diffusion matrix is reduced to identity, is linked to the existence of geodesics. The use of these geodesics to reparametrize the equation lead to a behavior which essentially Gaussian in these new variables (called the normal coordinates). But first-order terms are still present most of the time and cannot got rid of. These first-order terms in the normal coordinates are linked to what is called the curvature of the Riemannian space. They are the origin of the terms $R_1(S_0, S)$, $R_2(S_0, S)$ and so on.

During the last 1990s and early 2000s, Hagan *et al.* (2002, 2004) completely elucidated the SABR Model which is described by:

$$\frac{dS_t}{S_t^\beta} = \alpha y_t dW_t,$$
$$\frac{dy_t}{y_t} = \nu dB_t, \quad d\langle B, W\rangle_t = \rho dt. \tag{6.43}$$

Using the right change of variable, we can show that the partial differential equation to solve

$$\frac{\partial G}{\partial t} = \frac{1}{2} y^2 \left(\frac{\partial^2 G}{\partial x^2} + \frac{\partial^2 G}{\partial y^2} \right)$$

with x being the "decorrelated underlying" and y being the "decorrelated volatility". Following our presentation, the metric distance of the problem is, therefore,

$$g = \begin{pmatrix} \frac{1}{y^2} & 0 \\ 0 & \frac{1}{y^2} \end{pmatrix}. \tag{6.44}$$

This exactly the metric of the Poincarré half plan (see Beardon, 1991). Therefore, it makes the connection with a very classical part of analysis. In particular, the geodesic distance is known analytically. It is given by

$$d(x_0, y_0, x, y) = \text{ArgCosh}\left(1 + \frac{(x - x_0)^2 + (y - y_0)^2}{2 y y_0}\right).$$

It also happens that the pricing kernel (Green function) of the Poincarré differential equation can be represented by a single integral: the McKean (1970) formula (see also Jost, 2002):

$$p(x_0, y_0, x, y, t) = \frac{e^{-t/8}\sqrt{2}}{(2\pi t)^{3/2}} \int_{d(x_0,y_0,x,y)}^{\infty} \frac{r e^{-r^2/2t} dr}{\sqrt{\cosh(r) - \cosh d(x_0, y_0, x, y)}}, \tag{6.45}$$

where $p(x_0, y_0, x, y, t)$ is the transition probability between the state (x_0, y_0) and (x, y).

A powerful pull-back technique given by the intrinsic differential geometric formulation of the concepts permit in theory to get formulas for cases with correlations, mean reverting and a lot of modified drift cases (see e.g. Jost, 2002).

Following intuitively first these ideas, then creating this powerful synthesis, Hagan *et al.* got an analytical approximate value of a call for the SABR model:

$$\begin{aligned}&\mathrm{SABR\_Call}(f,k,T,\alpha,\beta,\rho,v)\\&\quad=(f-k)^{+}+\frac{f-k}{2x\sqrt{2\pi}}\mathrm{e}^{\theta}\int_0^T\frac{\mathrm{e}^{-(x^2/2u)+\kappa u}}{\sqrt{u}}du\end{aligned}\tag{6.46}$$

with the definition

$$\begin{aligned}\theta&=\frac{1}{4}\frac{\rho v\alpha\beta z^2}{k^{1-\beta}}+\mathrm{Log}\left(\frac{\alpha(fk)^{\beta/2}z}{f-k}\right)\\&\quad+\mathrm{Log}\left(\frac{x}{z}(1-2v\rho z+v^2z^2)^{1/4}\right),\end{aligned}\tag{6.47}$$

$$\kappa=\frac{1}{8}(\alpha^2(\beta-2)\beta k^{2\beta-2}+6\alpha\beta k^{\beta-1}\nu\rho+\nu^2(2-3\rho^2)),\tag{6.48}$$

$$x=\frac{1}{\nu}\mathrm{Log}\left(\frac{-\rho+\nu z+\sqrt{1-2\nu\rho z+\nu^2z^2}}{1-\rho}\right),\tag{6.49}$$

$$z=\frac{f^{1-\beta}-k^{1-\beta}}{\alpha(1-\beta)}.\tag{6.50}$$

The integral happens to be analytically computable. Its value is given by:

$$\begin{aligned}\int_0^T\frac{\mathrm{e}^{-(x^2/2u)+\kappa u}}{\sqrt{u}}du&=\frac{\sqrt{\pi}}{2i\sqrt{\kappa}}\left[\mathrm{e}^{i\sqrt{2}x\sqrt{\kappa}}\left(\mathrm{erf}\left(\frac{x}{\sqrt{2T}}+i\sqrt{\kappa T}\right)-1\right)\right.\\&\quad\left.+\mathrm{e}^{-i\sqrt{2}x\sqrt{\kappa}}\left(\mathrm{erf}\left(\frac{-x}{\sqrt{2T}}+i\sqrt{\kappa T}\right)+1\right)\right],\end{aligned}\tag{6.51}$$

where for usual parameters, the erf function has just to be feed with a real parameter. In some circumstance, we need the extension of the erf function to complex numbers; in that case the following formula is useful (see Abramowitz and Stegun, 1964):

$$\begin{aligned}\mathrm{erf}(x+iy) &= \mathrm{erf}(x) + \frac{\mathrm{e}^{-x^2}}{2\pi x}[1-\cos(2xy)+i\sin(2xy)] \\ &\quad + \frac{2}{\pi}\mathrm{e}^{-x^2}\sum_{n=1}^{\infty}\frac{\mathrm{e}^{-(1/4)n^2}}{n^2+4x^2}\{2x-2x\cosh(ny)\cos(2xy) \\ &\quad + n\sinh(ny)\sin(2xy) + i(2x\cosh(ny)\sin(2xy) \\ &\quad + n\sinh(ny)\cos(2xy))\}.\end{aligned}$$

6.2.5. *Connection between the SABR and the "complex" BS models*

It is well known that the cost of a call and put option is equal to its intrinsic value plus the cost of a stop loss strategy. This stop loss strategy can be re-expressed in terms of the local time. It provides easily closed forms solution for model like BS model (Omberg, 1988; Carr and Jarrow, 1990). This section examines the theory of local time for stochastic volatility models and in particular the SABR model (Hagan *et al.*, 2002). It gives an approximated formula for the local time in SABR and shows that this model can be valued using a BS formula but where all the terms are complex number. This formula turns out to be more robust for low and high strikes. This solves in particular the problem of valuing the whole smile in SABR as required in the replication method for CMS and the copula integration for CMS spread options.

6.2.5.1. *Expected local time and vanilla option prices*

The strength of the local time theory lies in its general framework. Let the forward asset F_t follow a stochastic volatility diffusion:

$$dF_t = \sigma(t, F_t)C(F_t)dW_t^F, \quad F_0 = f \tag{6.52}$$

and

$$d\sigma(t, F_t) = \sigma_t M(t,\sigma)dt + \sigma_t S(t,\sigma)dW_t^\sigma, \quad \sigma_0 = \alpha, \tag{6.53}$$

where W_t^F and W_t^σ are two standard Brownian motions potentially correlated $\langle dW_t^F, dW_t^\sigma\rangle = \rho_t dt$ and $C(F_t)$ is a mapping function (usually a CEV $C(F_t) = F_t^\beta$ or a displaced diffusion function $C(F_t) = F_t + \mu_t$) and $M(t,\sigma)$, $S(t,\sigma)$ are some smooth functions. The Meyer–Tanaka formula, extension of the Ito formula to convex payoff (Meyer, 1976; Yor, 1978) provides an interesting framework to compute the price of a call with strike k given by (Carr and Jarrow, 1990):

$$\begin{aligned}(F_T - k)^+ = (f-k)^+ &+ \int_0^T 1_{F_u^- > k} dF_u \\ &+ \frac{1}{2}\int_0^T 1_{F_u = k}\sigma^2(F_u,u)C^2(F_u)du \end{aligned} \tag{6.54}$$

Taking the expected value of the above equation leads to a forward price of a call option given by

$$\begin{aligned} &E[(F_T-k)^+] \\ &\quad = (f-k)^+ E\left[\frac{1}{2}\int_0^T 1_{F_u=k}\,\sigma^2(F_u,u)C^2(F_u)du\right] \end{aligned} \tag{6.55}$$

The above equation summarizes that the forward price of a call is the sum of its intrinsic and its time value represented by its expected continuous[1] local time at strike k for a maturity time T. But more than a new formulation for the value of a vanilla option, it shows that this value lies in the computation of the expected local time for any model. In particular, in a BS model with volatility α whose diffusion equation is given by $dF_t = \alpha F_t dW_t^F$, the expected local time at strike k for a maturity time T is easy to compute and given by (Carr and Jarrow, 1990 or Dybvig, 1988):

$$E\left[\frac{1}{2}\int_0^T 1_{F_u=k}\alpha^2 du\right] = \frac{1}{2}\alpha^2 k^2 \int_0^T \frac{1}{\alpha k\sqrt{u}} \frac{\mathrm{e}^{-\frac{1}{2}\left(\frac{\mathrm{Log}(f/k)}{\alpha\sqrt{u}} - \frac{1}{2}\alpha\sqrt{u}\right)^2}}{\sqrt{2\pi}} du, \tag{6.56}$$

[1] We call it continuous in opposition to discrete local time. Discrete local time terms would appear if we had a jump diffusion. In the rest of the chapter, we will drop continuous as we only deal with continuous diffusion.

which implies a closed form solution given by:

$$\text{BS\_Call}(f, k, \alpha, T) = (f - k)^+ + \frac{1}{2}\alpha^2 k^2 \int_0^T \frac{1}{\alpha k\sqrt{u}} \frac{e^{-\frac{1}{2}\left(\frac{Log(f/k)}{\alpha\sqrt{u}} - \frac{1}{2}\alpha\sqrt{u}\right)^2}}{\sqrt{2\pi}} du \quad (6.57)$$

The above result can be derived either from a direct computation of the local time for geometric Brownian motion or by relating the vega to the local time (see Benhamou and Croissant, 2004, Section 7.1).

6.2.5.2. *Expected local time and probability density*

General framework: Define the probability density $p(t, f, \alpha; T, F, A)$ by:

$$p(t, f, \alpha; T, F, A)dFdA = \text{Prob}\{F < F_T < F + dF, A < \sigma_T < A + dA | F_t = f, \sigma_t = \alpha\}, \quad (6.58)$$

we then have the following connection between the local time and the probability density:

Proposition 6.1. The expected local time at strike is related to the density probability as follows:

$$E\left[\frac{1}{2}\int_0^T 1_{F_u = k}\, \sigma^2(F_u, u) C^2(F_u) du\right] = \frac{1}{2}C(k)^2 \int_{u=0}^T \int_{\sigma=0}^{+\infty} \sigma^2 p(t, f, \alpha, u, k, \sigma) d\sigma du, \quad (6.59)$$

which implies that the forward price of a call option $V(f, k, \alpha, T)$ with forward worth f today, with strike k, with volatility with boundary condition equal to α and with maturity T is equal to:

$$V(f, k, \alpha, \mathrm{T}) = (f - k)^+ + \frac{1}{2}C(k)^2 \int_{u=0}^T \int_{\sigma=0}^{+\infty} \sigma^2 p(t, f, \alpha, u, k, \sigma) d\sigma du. \quad (6.60)$$

Proof. See (Benhamou and Croissant, 2004, Section 7.2). □

Equation (6.59) shows that any expected local time problem can be reformulated as a probability density problem.

6.2.5.3. *Explicit computation of the local time for stochastic models*

Computing the local time in stochastic volatility models is not an easy task in general. The above theory assumes either a closed form for the expected local time or for the probability density. But this is far from being the case for general stochastic volatility models. However, we will see that we can derive very good approximation of the solution using perturbation theory for a stochastic volatility model given by Eqs. (6.53) and (6.54). Clearly, Eqs. (6.53) and (6.54) encompasses most of the common models, namely; The seminal SABR model [23],

$$dF_t = \sigma_t F_t^{\beta} dW_t^F, \quad d\sigma_t = \nu\sigma_t dW_t \tag{6.61}$$

with $C(F_t) = F_t^{\beta}$, $M(\sigma_t) = 0$, $S(\sigma_t) = \nu$ (CEV process on the underlying with a lognormal stochastic volatility). The Heston model:

$$dF_t = \sigma_t F_t dW_t^F, \quad d\sigma_t^2 = (\kappa - \lambda\sigma_t^2)dt + \nu\sigma_t dW_t), \tag{6.62}$$

which can be reformulated into the framework of (6.54) by

$$d\sigma_t = \left(\frac{\kappa - v^2/4}{2\sigma_t} - \lambda\sigma_t\right) dt + \frac{\nu}{2} dW_t \tag{6.63}$$

with $C(F_t) = F_t$, $M(\sigma_t) = (\frac{\kappa - v^2/4}{2\sigma_t^2} - \lambda)$, $E(\sigma_t) = \nu/2\sigma_t$ (lognormal diffusion on the underlying with a mean reverting square root stochastic volatility). The main result of the section is the following. We can obtain a very good approximation of the expected local time of our general stochastic volatility model. In deed, The expected local time of the stochastic volatility model (6.61) can be approximated at the second order (for details see Benhamou and Croissant, 2004, Section 7.3). In the particular case of a SABR model, this leads to the same approximation found by Hagan (see Eq. (6.46)).

Compared to the final result of Hagan *et al.* (2002, Eq. (2.17a)), the solution is more accurate as it does not involve the last approximation on the integral term $\int_0^T \frac{e^{-(x^2/2u)+\kappa u}}{\sqrt{u}} du$ as in (B.48c) where it is approximated as $\int_0^T \frac{e^{-\frac{x^2}{2u}}}{\sqrt{u}(1-\frac{2}{3}\kappa u)^{3/2}} du$. This approximation is clearly very off in general. We will see now how to efficiently compute this integral.

6.2.5.4. *Fast computation of the SABR stochastic integral and connection with BS model*

The SABR standard integral given by $\int_0^T \frac{e^{-(x^2/2u)+\kappa u}}{\sqrt{u}} du$ can be computed as follows:

$$\int_0^T \frac{e^{-(x^2/2u)+\kappa u}}{\sqrt{u}} du = \frac{\sqrt{\pi}}{2i\sqrt{\kappa}} \left[e^{i\sqrt{2}x\sqrt{\kappa}} \left(\operatorname{erf}\left(\frac{x}{\sqrt{2T}} + i\sqrt{\kappa T} \right) - 1 \right) \right. \\ \left. + e^{-i\sqrt{2}x\sqrt{\kappa}} \left(\operatorname{erf}\left(\frac{-x}{\sqrt{2T}} + i\sqrt{\kappa T} \right) + 1 \right) \right] \tag{6.64}$$

with the additional formula for the complex error function:

$$\operatorname{erf}(x+iy) = \operatorname{erf}(x) + \frac{e^{-x^2}}{2\pi x}[1 - \cos(2xy) + i\sin(2xy)] \\ + \frac{2}{\pi} e^{-x^2} \sum_{n=1}^{\infty} \frac{e^{-\frac{1}{4}n^2}}{n^2 + 4x^2} \{2x - 2x\cosh(ny)\cos(2xy) \\ + n\sinh(ny)\sin(2xy) + i(2x\cosh(ny)\sin(2xy) \\ + n\sinh(ny)\cos(2xy))\} \tag{6.65}$$

Proof. See Benhamou and Croissant, 2004, Section 7.5. □

The last series converges very quickly (five terms is in general enough for a good precision). See Press *et al.* (2002).

Connection with BS formula: Identifying the local time component both in BS and SABR formulae, we can find a relationship between these two. Namely, we can prove that the SABR model can be seen as an extension of the BS formula in the complex filed as proven by Proposition 6.2.

Proposition 6.2. The SABR model is equivalent to a BS models with the following parameters (complex numbers potentially but the price remains real:

$$\tilde{f} = \frac{x\sqrt{-2\kappa}e^{\pm x(\sqrt{-\frac{\kappa}{2}}+\sqrt{-2\kappa})-\theta}}{(e^{\pm x\sqrt{-2\kappa}}-1)} x\sqrt{-2\kappa}e^{x\sqrt{-\frac{\kappa}{2}}-\theta}, \tag{6.66}$$

$$\tilde{k} = \frac{x\sqrt{-2\kappa}e^{\pm x\sqrt{-\frac{\kappa}{2}}-\theta}}{(e^{\pm x\sqrt{-2\kappa}}-1)} x\sqrt{-2\kappa}e^{x\sqrt{-\frac{\kappa}{2}}-\theta}, \tag{6.67}$$

$$\tilde{\alpha} = \sqrt{-2\kappa}, \tag{6.68}$$

$$\text{SABR\_Call}(f,k,T,\alpha,\beta,\rho,v) = \text{BS\_Call}(\tilde{f},\tilde{k},\tilde{\alpha},T). \tag{6.69}$$

Proof. See Benhamou and Croissant, 2004, Section 7.6. □

Compared to the Hagan *et al.* (2002) formula, the above computation of the SABR model requires to compute the BS formula with the complex error function $\text{erf}(x+iy)$ using Remarks (4.1) and (4.2).

Quality of our approximation and numerical examples: The problem we meet with the Hagan approximation is that for long maturity (20 years for LIBOR markets) the low strike are mishandled. Our formula has a better behaviour. In the following example, the spot value of the LIBOR is 0.05, other parameters are: beta$=0.7$, rho$=-0.5$, nu$=0.2$.

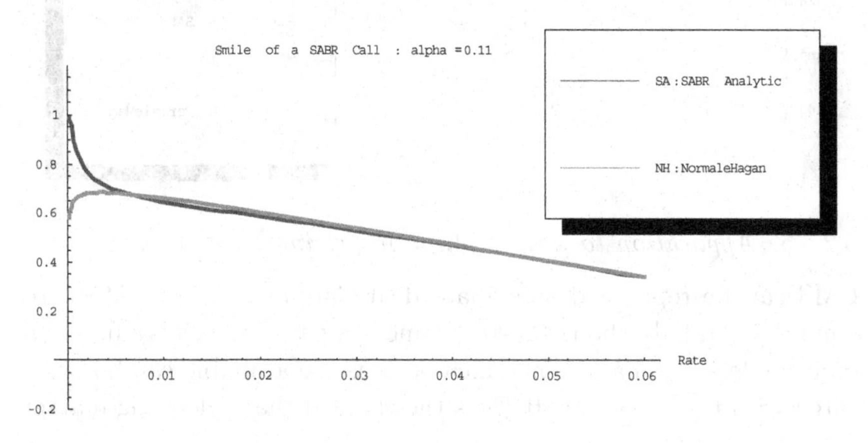

Of course, a higher at-the-money volatility makes the things worse:

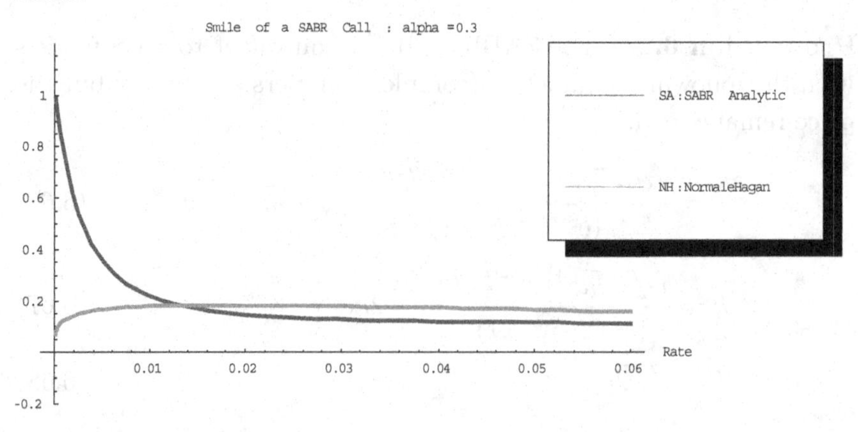

An even better insight is given when we look at the implicit probability density function computed as minus the derivative of the digital price. With the preceding example's parameters the density associated with normal Hagan's formula become negative below 0.5%.

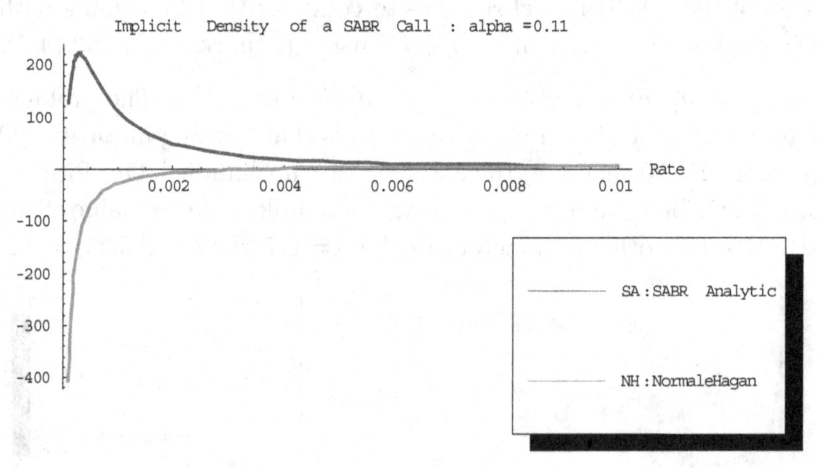

6.2.5.5. *Application to CMS replication pricing*

CMS can be replicated as explained in Carr and Jarrow (1990) or Amblard and Lebuchoux (2000). Namely, a payoff which is a function of an underlying market on which options are available can be replicated with these options. If F_T is the value of the underlying market at maturity T, and if $f(F_T)$ is the payoff we want to replicate, the replication formula is:

$$\begin{aligned} f(F_T) &= f(\kappa) + f'(\kappa)((F_T - \kappa)^+ - (\kappa - F_T)^+) \\ &\quad + \int_0^\kappa f''(K)(\kappa - F_T)^+ dK + \int_\kappa^\infty f''(K)(F_T - \kappa)^+ dK \end{aligned} \tag{6.70}$$

Therefore, when we select a value of κ such that $f'(\kappa) = 0$ and take the expectation of both members of this stochastic equation we get:

$$\begin{aligned} E[f(F_T)] &= f(\kappa) + \int_0^\kappa f''(K)\mathrm{Put}(K,T)dK \\ &\quad + \int_\kappa^\infty f''(K)\mathrm{Call}(K,T)dK \end{aligned} \tag{6.71}$$

This formula shows that knowing the price of vanilla options should be enough to compute the arbitrage-free price of the payoff that delivers $f(F_T)$.

We want to apply to compute the price of a security that pays a swap rate of maturity M at the date T. So we use the change of numeraire formula to determine that the forward price of such contract can be computed using the measure associated with the annuity as a numeraire:

The annuity is defined to be:

$$A_T(T) = \sum_{i=0}^{n-1} \delta_i B_{T_i}(T). \tag{6.72}$$

So the forward price is:

$$E^{Q_T}[s_{\mathrm{CMS}}] = E^{Q_A}\left[s_{\mathrm{CMS}}\frac{dQ_T}{dQ_A}\right] = E^{Q_A}\left[s_{\mathrm{CMS}}\frac{B_T(t)/B_T(0)}{A_T(t)/A_T(0)}\right]$$

and if we designate by $r_T(t)$ the zero-coupon rate of maturity T at the date t, we can reformulate the expected price as:

$$E^{Q_T}[s_{\mathrm{CMS}}] = \frac{A_T(0)}{B_T(0)} E^{Q_A}\left[\frac{s_{\mathrm{CMS}}(t)}{\sum_{i=0}^{n-1}\delta_i(1 + r_{T-T_i}(t))^{T_i - T}}\right]. \tag{6.73}$$

Obviously, we need a model able to address all forward rates and the swap rate. But, as a first cut we want to appreciate the influence of the volatility on the pricing of such non-linear functional, and with a good approximation, all rates have a correlation very close to 1. So we introduce the drastic approximation:

$$E^{Q_T}[s_{\text{CMS}}] \approx \frac{A_T(0)}{B_T(0)} E^{Q_A}\left[\frac{s_{\text{CMS}}(t)}{\sum_{i=0}^{n-1} \delta_i (1+s_{\text{CMS}}(t))^{T_i - T}}\right]. \tag{6.74}$$

We see now how we are going to apply the replication algorithm described in the beginning of this section.

We replicate the function

$$f(s_{\text{CMS}}) = \frac{s_{\text{CMS}}(t)}{\sum_{i=0}^{n-1} \delta_i (1+s_{\text{CMS}}(t))^{T_i - T}}$$

which has a positive convexity as we can see it on the following graph:

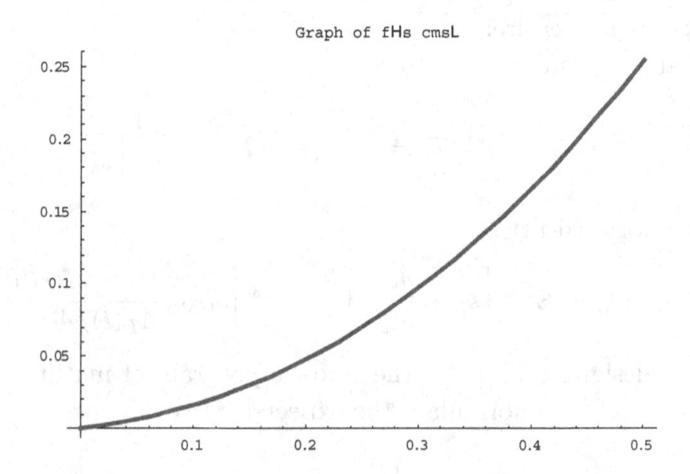

We want to compare the result of such replication for the SABR model and the BS model. It is obvious that the parameters of the

SABR model do depend on the maturity. This is checked by practitioners and comes from the fact that the SABR models assume a lognormal process for the volatility and the evolution of the market prices for the smile is more consistent with a stationary character of the volatility process. Therefore we use a black and sholes curve for the BS model calibrated at the money with the SABR model.

With such assumptions in the case of a small volatility of volatility the implied volatility looks like

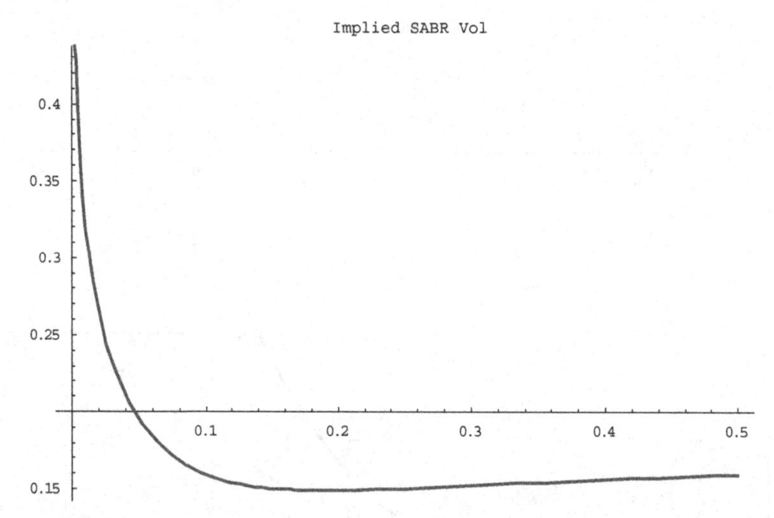

the prices look like:

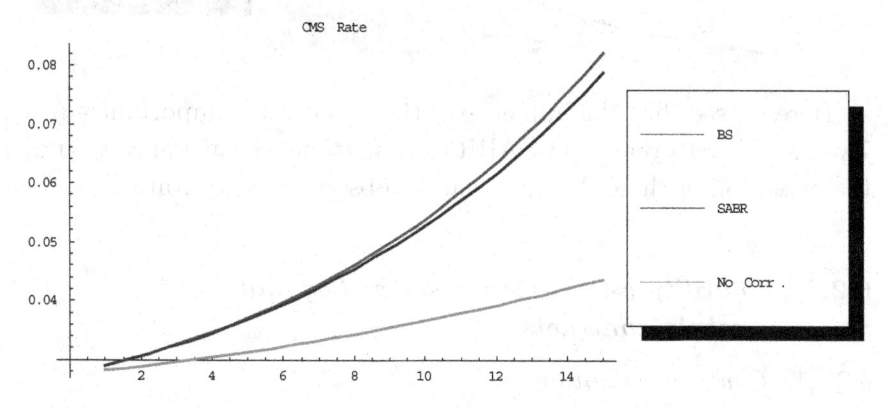

Here we observe that the influence of the smile on the CMS price is weak. In the case of a stronger vol of vol, the implied vol looks like

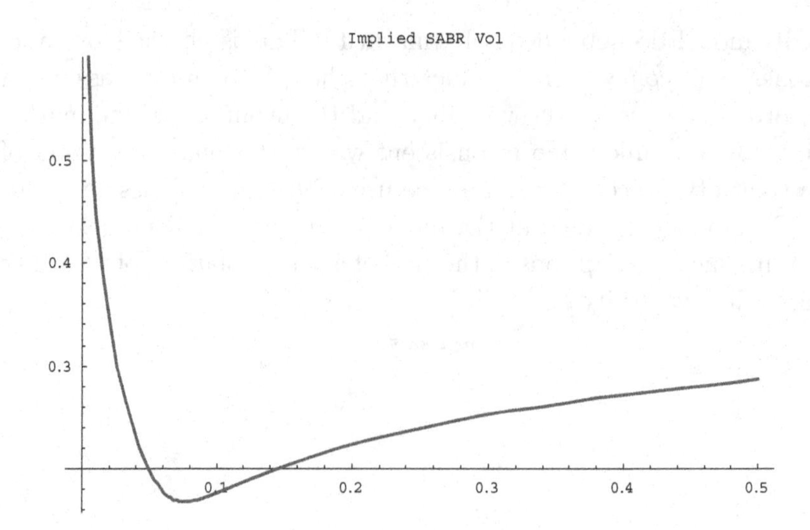

the prices look like:

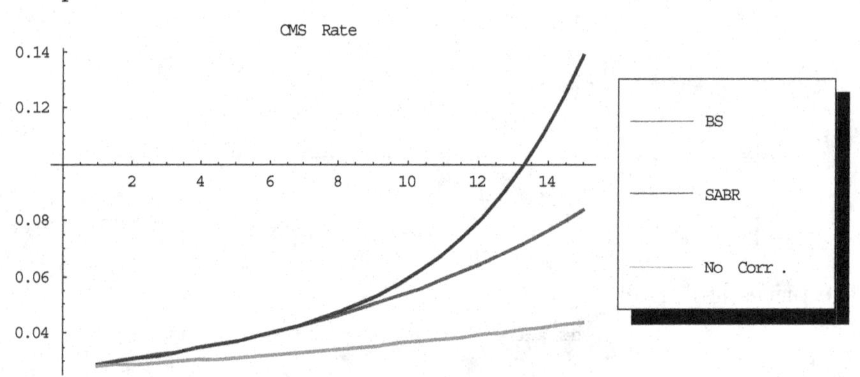

Here we see that the influence of the vol of vol is important after 7 years. After 15 years, the SABR computations become inexact and the explosion of the difference that we observe is inaccurate.

6.2.6. *Volatility and variance swap: Log and parabola contracts*

6.2.6.1. *Contract definition*

Let us begin with the definition of a vanilla variance swap contract: Counterparty A agrees to pay a fixed notional at the settlement date to Counterparty B. Counterparty B agrees to pay an amount proportional to the sum of daily squared returns. The proportionality constant is the relative price of such contract.

6.2.6.2. *Replication strategy and pricing*

We can define a replication strategy in a model independent way: this uses the method described in Breeden and Litzenberger (1978): the implicit probability density can be recovered from European option market prices by deriving either the call options prices either the put option prices with respect to the strike:

$$p(S_T, T, S_t, t) = \left.\frac{\partial^2 C(S_t, K, t, T)}{\partial K^2}\right|_{K=S_T} = \left.\frac{\partial^2 P(S_t, K, t, T)}{\partial K^2}\right|_{K=S_T}. \tag{6.75}$$

Then, by integrating twice, taking into account the boundary conditions, we get:

$$E[g(S_T)|S_t] = g(F) + \int_0^F P(S_t, K, t, T) g''(K) dK + \int_F^\infty C(S_t, K, t, T) g''(K) dK. \tag{6.76}$$

For any function g, F is here the expected value $F = E[S_T|S_t]$. For example if we apply the preceding formula to a call pay-off under the forward neutral world, then

$$g''(k) = \delta(K - k),$$
$$E[(S_T - K)^+|S_t] = (F - K) + P(K), \quad \text{if } K < F$$
$$E[(S_T - K)^+|S_t] = C(K), \quad \text{if } K > F.$$

Now if we apply the preceding formula to a log contract we get:

$$E[\text{Log}(S_T)|S_t] = \log(F) - \int_0^F \frac{P(S_t, K, t, T)}{K^2} dK - \int_F^\infty \frac{C(S_t, K, t, T)}{K^2} dK. \tag{6.77}$$

Now we can also express this expectation by using Ito Lemma and assuming that the process followed by S is an Ito Process:

$$\operatorname{Log}(S_T) - \operatorname{Log}(S_t) = \int_t^T \frac{dS_t}{S_t} - \frac{1}{2}\int_t^T \sigma_t^2 dt. \tag{6.78}$$

So we see now the relationship between a contract paying the realized volatility and a log contract. It gives us also the replication strategy and the pricing:

$$E\left[\int_t^T \sigma_t^2 dt | S_t\right] = 2\left(\int_0^F \frac{P(S_t, K, t, T)}{K^2} dK + \int_F^\infty \frac{C(S_t, K, t, T)}{K^2} dK - \log\left(\frac{F}{S_t}\right)\right) \tag{6.79}$$

6.2.6.3. *Variance swap in the Heston model*

A model which is very enlightening for studying the realized volatility is the Heston model. In this model, the variance follows a square root process, so its expectation can be shown as:

$$E\left[\int_0^T \sigma_t^2 dt\right] = \frac{1 - e^{-\kappa T}}{\kappa T}(V_0^2 - \theta^2) + \theta^2. \tag{6.80}$$

We see that in this model, it does not depends on the volatility of the volatility ν.

6.2.6.4. *Variance swap in the mean reverting lognormal model*

If the underlying follows:

$$\begin{gathered} \frac{dS_t}{S_t} = r dt + \sqrt{V_t} dW_t, \\ dV_t = \kappa(\theta - V_t)dt + \nu V_t dB_t \quad d\langle B, V\rangle_t = \rho(V_t)dt. \end{gathered} \tag{6.81}$$

Then the expected total variance can be computed as:

$$E\left[\int_0^T \sigma_t^2 dt\right] = \frac{2\kappa\theta^2}{(2\kappa - \nu^2)T}\left(T - \frac{1 - e^{-(2\kappa - \nu^2)T}}{(2\kappa - \nu^2)T}\right) + \frac{2\kappa\theta(V_0 - \theta)}{(2\kappa - \nu^2)T}$$

$$\times\left(\frac{1-\mathrm{e}^{-\kappa T}}{\kappa}-\frac{1-\mathrm{e}^{-(2\kappa-\nu^2)T}}{(2\kappa-\nu^2)T}\right)$$
$$+\frac{2\kappa\theta^2}{(2\kappa-\nu^2)T}(1-\mathrm{e}^{-(2\kappa-\nu^2)T}), \tag{6.82}$$

where we see an influence of the volatility of volatility ν. More generally, if we have a smile, there is a neat formula that can be used to compute the expectation of realized variance:

$$E\left[\int_0^T \sigma_t^2 dt\right] = \int_{-\infty}^{\infty} N'(d_2(K))\frac{\partial d_2}{\partial K}\sigma_{\text{imp}}(K)^2\,TdK, \tag{6.83}$$

where $\sigma_{\text{imp}}(K)$ is the implied volatility. We can use this formula to explore the consequences of the smile on the expected realized variance, for example, you can see that skewness has very little impact on realized variance, versus kurtosis which increase it almost linearly (see Gatheral, 2004).

If we accept the additional hypothesis that moves in the underlying is independent from moves in the volatility and that the underlying follows a martingale diffusion, then Carr and Lee established a very powerful formula that allow us to compute the expectation of any function of the quadratic variation. The formula computes its Laplace transform:

$$E[\mathrm{e}^{\lambda\int_0^T\sigma_t^2 dt}] = E[\mathrm{e}^{p(\lambda)x_T}], \tag{6.84}$$

where

$$p(\lambda) = \frac{1}{2} \pm \sqrt{\frac{1}{4}+2\lambda}. \tag{6.85}$$

6.2.6.5. *Application 1: Computation of expected volatility*

Using $\sqrt{y} = \frac{1}{2\sqrt{\pi}}\int_0^\infty \frac{1-\mathrm{e}^{-\lambda y}}{\lambda^{3/2}}d\lambda$ we can deduce that:

$$E\left[\sqrt{\int_0^T \sigma_t^2 dt}\right] = \frac{1}{2\sqrt{\pi}}\int_0^\infty \frac{1-E\left|\mathrm{e}^{-\lambda\int_0^T\sigma_t^2 dt}\right|}{\lambda^{3/2}}d\lambda$$
$$= \int_0^\infty c(k)w_{\text{vol}}(k)dk \tag{6.86}$$

where

$$k = \mathrm{Log}[F/K] \quad \text{and} \quad w_{\mathrm{vol}}(k) = \sqrt{\frac{\pi}{2}} e^{k/2} I_1\left(\frac{k}{2}\right) + \sqrt{2\pi}\delta(k). \tag{6.87}$$

And I_1 is a modified Bessel function of the first kind (see Gatheral and Friz, 2004).

6.2.6.6. *Application 2: Computation of the variance call*

By applying the general result to a call pay off we get:

$$E\left[\left(\int_0^T \sigma_t^2 dt - K\right)^+\right] = \int_0^\infty dk c(k) w_{\mathrm{call}}(k), \tag{6.88}$$

where

$$w_{\mathrm{call}}(k) = 4\frac{e^{k/2}}{2i\pi}\int_{a-i\infty}^{a+i\infty} \frac{e^{-\lambda K}}{\lambda} \cosh\left(k\sqrt{\frac{1}{4}+2\lambda}\right) d\lambda. \tag{6.89}$$

6.3. Summary

In this chapter, we studied the smile generated for the different model. In the first way, we have seen that Merton model induce a short-term smiles and skews that disappears quickly with increasing maturity because of the central limit theorem.

Second, we review the local volatility models which extends BS models in making the volatility depending on the underlying and are more compatible for the market, but they had the wrong dynamic behavior (the underlying asset moves in the opposite direction of the smile).

Finally, we introduce another correlated source of noise, and make the volatility dependant on it; here we defined stochastic models that allow the Markovian behavior of implied volatility:

(1) SABR that comes here in order to solve the problem that meet the local volatility for the moves of the smile, and we showed that closed formula of call in this model are an extension for the BS model with complex parameters.

(2) Heston which create a smile/skew that increase as the maturity increase, contrarily to Merton model.
(3) Bates that combined Merton and Heston models so as to generate smile/skew in both short and long maturity.

We conclude that adding jumps to stochastic models helps, but it make the model more difficult to calibrate and to hedge. Then, the solution is to find a compromise between a practical criterion and a theoretical one. Experience is then very appreciated.

References

[1] Abramowitz M and Stegun IA (1964). *Handbook of Mathematical Functions.* New York: Dover.
[2] Amblard G and Lebuchoux J (2000). Models for CMS caps, in *Euro Derivatives.* London: RISK publication.
[3] Beardon AF (1991). *The Geometry of Discrete Groups. Graduate Texts in Mathematics*, Vol. 91. Berlin: Springer.
[4] Benhamou E and Croissant O (2004). Local time for the SABR model connection with the "complex" Black Scholes and application to CMS and Spread Options.
[5] Brigo D, Mercurio F and Rapisarda F (2004). Smile at the uncertainty. *RISK Magazine*, May.
[6] Breeden and Litzenberger R (1978). Price of state-contingent claims implicit in option prices, *Journal of Business*, **51**, 621–651.
[7] Carr P and Jarrow RA (1990). The stop-loss start-gain paradox and option valuation — a new decomposition into intrinsic and time value. *Review of Financial Studies*, **3**, 469–492.
[8] Chavel I (1984). *Eigenvalues in Riemannian Geometry.* New York: Academic Press.
[9] Derman E and Kani I (1994). The volatility smile and its implied tree. *RISK*, **7**(2), 139–145, 32–39.
[10] Derman E and Kani I (1996). Implied trinomial trees of the volatility smile. *The Journal of Derivatives*, **3**(4), 7–22.
[11] Derman E and Kani I (1997). Trading & hedging local volatility. *The Journal of Financial Engineering*, **6**(3), 233–268.
[12] Derman E and Kani I (1999). Regimes of volatility. *RISK*, April.
[13] Derman E, Demeterfi K, Kamal M and Zou J (1999). More than you ever wanted to know about volatility swaps. *The Journal of Derivatives*, **6**(4), 9–32. (http://www.ederman.com/new/docs/gs-volatility_swaps.pdf).

[14] Dybvig P (1988). Inefficient dynamic portfolio strategies or how to throw away a million dollars in the stock market. *Review of Financial Studies*, **1**(1).

[15] Gatheral J (2004). Case studies in financial modeling course notes. Courant Institute of Mathematical Sciences, Fall Term.

[16] Gatheral J and Friz P (2004). Valuation of volatility derivatives as an inverse problem, Working Paper, Currant Institute of Mathematical Sciences and Merrill Lynch.

[17] Hagan P, Kumar D, Lesniewski A and Woodward D (2002). Managing smile risk. *Wilmott Magazine*, **7**, 84–108.

[18] Hagan P, Lesniewski A and Woodward D (2004). Probability distribution in the SABR model of stochastic volatility. Working Paper, Draft, of June.

[19] Hull J, John C and White AD (1987). The pricing of options on assets with stochastic volatilities. *Journal of Finance*, **42**(2).

[20] Jost J (2002). *Riemannian Geometry and Geometric Analysis*, 3rd edn. Berlin: Universitext, Springer.

[21] Lewis A (2000). *Option Valuation Under Stochastic Volatility.* Newport Beach, California: Finance Press.

[22] McKean HP (1970). An upper bound to the spectrum of Δ on a manifold of negative curvature. *Journal of Diff. Geom.* **4**, 359–366.

[23] Meyer P (1976). *Un Cours sur les Intégrales stochastiques, Séminaire de Probabilités X, Lecture Notes in Mathematics*, Vol. 511. Berlin: Springer-Verlag.

[24] Minakshisundaram S and Pleijel A (1949). Some properties of the eigenfunctions of the laplace operator on Riemannian manifolds. *Canadian Journal of Mathematics*, **1**, 242–256.

[25] Omberg E (1988). Binary Trading Strategies, Brownian Local Time, and the Tanaka Formulas, Working Paper, Santa Clara University.

[26] Press WH, Teukolsky SA, Vetterling WT and Flannery BP (2002). *Numerical Recipes in C++*, 2nd edn. Cambridge: Cambridge University Press.

[27] Schroder M (1989). Computing the constant elasticity of variance option pricing formula. *Journal of Finance*, **44**, 211–219.

[28] Sepp A (2002). Fourier inversion method for option pricing under jump-diffusion stochastic volatility and Levy processes. Master Thesis, University of Tartu.

[29] Varadhan RS (1967). Diffusion processes in small time intervals. *Communication Pure Applied Mathematics*, **20**, 659–685.

[30] Yor M (1978). Rappels et Préliminaires Généraux, in Temps Locaux, Société Mathématique de France, Astérisque, 52–53, 17–22.

Chapter 7

YIELD CURVE MODELING

In order to price securities with rate-dependent cash flows, it is important to predict the future evolution of rates. Interest rate models appear then fundamental in fixed income market. This chapter aims to give an overview of these different interest models ranging from the earliest 1-factor short-rate models to the more complicated ones such as market models (like in Mercurio, 2001 or Rebonato, 2002).

7.1. Model Typology

7.1.1. *Short rates*

The first approach was introduced by Vasicek (1997). In this approach, we suppose that the short rate r follows a general stochastic process:

$$dr_t = a(b - r_t)dt + \sigma dW_t, \tag{7.1}$$

where a, b and σ are constants. This approach is consistent with some economic arguments, since interest rates appear to be pulled back to some long-term average. Parameters have econometrics interpretation as a is the speed of reversion and b is the long-term average. See Figure 7.1.

Many extensions to the diffusion of the short rate have then been developed, but most of them can be written in the following form:

$$dr_t = \mu(t, r_t)dt + \sigma(t, r_t)dW_t. \tag{7.2}$$

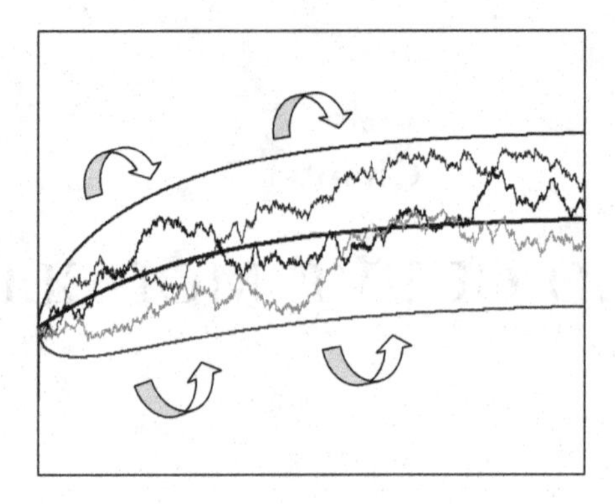

Figure 7.1: Mean-revearting process.

Standard models are obtained for simple forms of the volatility function:

Examples: $\sigma(t, r_t) = \sigma(t)$ leads to the Gaussian (or normal) model, $\sigma(t, r_t) = \sigma(t)r_t$ to Black–Derman–Toy and Black–Karasinski (BK) models (Black *et al.*, 1990) $\sigma(t, r_t) = \sigma(t)\sqrt{r_t}$ to Cox–Ingersoll–Ross (CIR) or square-root model (Cox *et al.*, 1985), $\sigma(t, r_t) = \sigma(t)r_t^\beta$ (constant elasticity of variance, CEV, extension) and $\sigma(t, r_t) = \sigma(t)(\mu + r_t)$ to shifted log or Q-model.

Under this framework, we assume that the evolution of the whole yield curve is completely determined by the dynamic of the short rate. Thus the price of a zero-coupon bond at time t, maturing at T can be calculated through the arbitrage-free relation:

$$B(t, T) = E^Q(\mathrm{e}^{-\int_t^T r_s ds}|F_t), \tag{7.3}$$

where E^Q is the expectation under the risk neutral measure. Then, from bond prices, we determine all kind of rates.

7.1.2. *Market (LIBOR and swap rate) models*

Short-rate models used to be the mainstream approach for modeling the yield curve dynamics. The principal feature of these models is their simplicity as well as the easiness to implement them efficiently.

However, neither the instantaneous short rate nor the instantaneous forward rate can be observed in the market.

A different approach, known as market models, has emerged over the years. In this approach, actively traded securities such as LIBOR or swap rates are used as the underlying variables. As a result, these models are now known as market models. There are two types of market models.

The first one is based on the description of the forward LIBOR and is referred to as the LIBOR market model, first introduced by Brace, Gatarek and Musiela (1997). The model is commonly referred to as the BGM model in an eponym fashion. In this approach, forward LIBOR rates are assumed to follow a log-normal diffusion. Then the model prices caps using the Black's formula which is the standard formula in the market.

The second type is the swap rate model. This model was first introduced by Miltersen *et al.* (1997) and then extended by Jamshidian. Here, the swap rate is assumed to be log-normal and swaptions are priced using Black's formula.

These two kinds of models are incompatible since forward and swap rates cannot be simultaneously log-normal. Once a forward- or swap-rate based model is chosen, we need some approximations or numerical methods to fit the complementary plain vanilla option (see the relationship between the volatility of the FRA and the swap discussed later in this chapter).

Furthermore, market models assume that the forward LIBOR (swap) rate is a martingale under its associated (natural) measure. This assumption, although very practical for the pricing of European Vanilla options, makes the model difficult to use for American-style products. This is because the model cannot be characterized by low-dimensional Markov process. We need to diffuse as many rates as the number of relevant rates in the deal.

7.1.3. *Markov functional models*

Markov functional models are a class of interest rate models in which the value of the zero-coupon risk free bonds can be expressed as a functional of some Markov process. The idea introduced by Pelsser, Kennedy and Hunt is to take a low-dimensional Markov Process (x_t) (in practice x will be one-or at most two-dimensional process) and to define its relationship to market prices (Pelsser, 2004). This allows a perfect fit for forward rates distributions. At each time t, the state of the economy is summarized via x_t. Therefore, natural subclasses of these models are classical short-rate models, where $x_t = r_t$.

In this framework, the zero-coupon bond prices are functions of the Markov process x:

$$B(t,T) = f(t,T,x_t) \quad \text{for } 0 \leq t \leq \partial_T \leq T,$$

where ∂_T is the "boundary curve," function of the maturities.

For example, $\partial_T = T$ if $T \leq \overline{T}$, or $\partial_T = \overline{T}$ if $\overline{T} \leq T$, where $\overline{T}$ is some terminal maturity. Besides, x is assumed to be Markovian under a numeraire N, which is in turn a function of $x : N_t = N(t, x_t)$.

Thus, given a process (x_t), the model becomes fully specified by choosing the functional form f of the discount factors on the boundary ∂_T for $0 \leq T \leq \overline{T}$, and the functional form of the numéraire $N(t, x_t)$ for $0 \leq t \leq \partial_T$. In fact, all the others bonds are given by:

$$B(t,T,x_t) = N(t,x_t)E^N\left[\frac{B(\partial_T,T,x_{\partial_T})}{N(\partial_T,x_{\partial_T})}\right], \tag{7.4}$$

where E^N is the expectation associated to the numeraire N. For more details on change of numeraire see appendices. More theorical details can be found in Karlin. The functional form f can then be chosen so as to calibrate the model to relevant market prices. The remaining degrees of freedom can be specified so as to achieve some desirable properties (mean reversion, time homogeneity, etc.). For more details, see Pessler (2004).

To sum up, Markov models allow us to fit the distributions of the relevant market rates (as in the market models) while keeping low dimensionality. Besides, these models can easily be implemented in a recombining lattice.

However, their calibration can require the numerical resolution of a non-linear integral equation. In addition, this class of models is not suited for some products, which clearly depend on more than one rate per fixing time (such as spread options), or products with clear dependencies on more than one strip of volatilities.

One of the most efficient Markov models is the Cheyette model. This model will be described later as a smile extension of short-rate models.

7.2. Heath–Jarrow–Morton Framework

7.2.1. *Philosophy*

Heath, Jarrow and Morton (HJM) developed a quite general framework for the modeling of interest-rate dynamics. In this approach, the t instantaneous forward rate, given by:

$$f(t,T) = -\partial_T \ln(B(t,T)) \tag{7.5}$$

is the fundamental driving quantity for the modeling of the yield curve term structure. The specification of its diffusion allows us to determine completely the yield curve dynamics.

As we have seen in Chapter 7, the "no arbitrage" assumption leads to some constraints on the expected return of financial assets.

In the interest rate market, zero-coupon bonds are the core of financial assets. Therefore, the assumption of "no arbitrage" leads to the following diffusion of zero-coupon bonds under the risk-neutral measure[1]:

$$\frac{dB(t,T)}{B(t,T)} = r_t dt + \Gamma(t,T)dW_t \tag{7.6}$$

We basically say that the drift term of zero-coupon bonds should be equal to the risk free rate r_t. The price at time t of a zero-coupon bond maturing at T can then be written as:

$$B(t,T) = B(0,T)\exp\left(\int_0^t r_s ds + \int_0^t \Gamma(s,T)dW_s - \frac{1}{2}\int_0^t |\Gamma(s,T)|^2 ds\right) \tag{7.7}$$

[1] W_t denotes a n-dimensional Wiener process.

As the zero-coupon bond is equal to 1 at maturity, $B(t,t) = 1$ at each time t, we can eliminate the short-rate term Eq. (7.7) to get:

$$B(t,T) = \frac{B(0,T)}{B(0,t)} \exp\left(\int_0^t (\Gamma(s,T) - \Gamma(s,t))dW_s - \frac{1}{2} \int_0^t (|\Gamma(s,T)|^2 - |\Gamma(s,t)|^2)ds \right). \quad (7.8)$$

7.2.2. *Forward bond volatility and drift*

Prices of zero-coupon bonds are then directly linked to the actual yield curve as well as to the structure of local volatilities.

We are now going to express these constraints in terms of interest rates. Let $\gamma(t,T)$ denote the derivative of $\Gamma(t,T)$ with respect to the second variable T.[2]

$$\gamma(t,T) = \partial_T \Gamma(t,T) \quad (7.9)$$

Using Eq. (7.9) we can derive the following formula for the diffusion of $f(t,T)$:

$$f(t,T) = f(0,T) - \int_0^t \gamma(s,T)dW_s + \int_0^t \gamma(s,T)\Gamma(s,T)^* ds. \quad (7.10)$$

The resulting short-rate dynamics is then given by:

$$r_t = f(t,t) = f(0,t) - \int_0^t \gamma(s,t)dW_s + \int_0^t \gamma(s,t)\Gamma(s,t)^* ds. \quad (7.11)$$

Equations (7.9) and (7.10) impose that the drift and the volatility of the instantaneous forward rate depend only on forward rates volatilities and their correlations. Hence, the specification of the form of the volatility imposes the model.

Note that the HJM approach specifies general relations coming from no-arbitrage relationship. The HJM framework is not a model

[2]We assume that the vector $\Gamma(t,T)$ of local volatility is differentiable with a bounded derivative.

but rather conditions, as sometimes said. All short-rate based models can be regarded as special cases of the HJM approach.

For instance, here below some local volatility functions for different models:

$$\begin{aligned}
&\text{Vasicek: } \Gamma(t, t+\theta) = \sigma \tfrac{1-e^{-a\theta}}{a} \quad \text{with } \theta \geq 0, \\
&\text{Ho and Lee: } \Gamma(t, t+\theta) = \sigma\theta, \\
&\text{CIR: } \Gamma(t, t+\theta) = \sigma\sqrt{r_t}B(\theta), \\
&\quad \text{where } B(\theta) \text{ is a deterministic function,} \\
&\quad \text{solution of a Riccati differential equation.} \\
&\text{Hull and White (HW): } \Gamma(t, T) = \int_t^T \sigma_s \exp\left(-\int_s^T \lambda_h dh\right) ds, \\
&\quad \text{where } \lambda \text{ is the mean reversion.}
\end{aligned}$$

7.3. Short Rate Models

7.3.1. *HW 1,2,...,nF*

7.3.1.1. *Diffusion equation*

In the Vasicek framework, the diffusion of short rates is given by:

$$dr_t = a(b - r_t)dt + \sigma dW_t. \tag{7.12}$$

A natural extension of this model was suggested by Hull and White (1993), in order to fit an arbitrary exogenous set of bond prices. The modification consists in making the reversion level in the real world dynamics, time-dependent. This extension allows us to fit exactly the initial term structure.

In this model, the diffusion of short rates becomes then:

$$r_t = f(0, t) + Y_t, \tag{7.13}$$

$$dY_t = -\lambda_t Y_t dt + \theta_t dt + \sigma_t dW_t, \tag{7.14}$$

where $f(0, t)$ denotes the instantaneous forward rate described previously. We can easily generalize the model for an n-dimensional process.[3]

$$r_t = f(0,t) + \sum_{1\leq k\leq N} Y_t^k, \tag{7.15}$$

$$dY_t^k = -\lambda_t Y_t^k dt + \theta_t^k dt + \sigma_t^k dW_t^k. \tag{7.16}$$

Let ρ_{ij} denote the correlation between W_i and W_j.

Starting from the diffusion of short rates, we calculate zero-coupon prices through the "no arbitrage" relation by integrating the process r_t.

$$B(t,T) = E^Q\left(\mathrm{e}^{-\int_t^T r_s ds}\,|F_t\right) \tag{7.17}$$

Since $\int_t^T r_s ds$ is still Gaussian, we have

$$B(t,T) = \mathrm{e}^{-E_t\left(\int_t^T r_s ds\right)+\frac{1}{2}(-1)^2\mathrm{Var}_t\left(\int_t^T r_s ds\right)}, \tag{7.18}$$

$$B(t,T) = \frac{B(0,T)}{B(0,t)}\mathrm{e}^{-c(t,T)-\sum_{1\leq k\leq N}\beta_k(t,T)Y_t^k}, \tag{7.19}$$

where $c(t,T)$ and $\beta_k(t,T)$ are deterministic functions of t and T.

$$\beta_k(t,T) = T-t \quad \text{if } \lambda_k = 0, \quad \frac{1-\mathrm{e}^{-\lambda_k(T-t)}}{\lambda_k} \quad \text{otherwise.}$$

$$c(t,T) = \int_0^t \sum_{1\leq k,j\leq N} (\beta_k(s,T)\beta_j(s,T) - \beta_k(s,t)\beta_j(s,t))\sigma_k(s)\sigma_j(s)\rho_{ij}ds \tag{7.20}$$

In this expression, we impose θ_t so as to fit exactly current zero-coupon bond prices and have $B(t,t) = 1$ for all t. Each factor depends then on two parameters: the volatility and the mean reversion.

We have now to determine the diffusion of zero-coupon bonds. Under these assumptions, a straightforward application of Itô's lemma shows that zero-coupon bond prices are log-normal with the

[3]Hereafter we will treat only the n-dimensional case. We have then the 1-factor case by setting N at 1.

following diffusion equation:

$$\frac{dB(t,T)}{B(t,T)} = r_t dt + \Gamma(t,T)dW_t \quad \text{where } \Gamma(t,T) = \sigma_t \beta(t,T) \quad (7.21)$$

This is consistent with the HJM framework.

7.3.1.2. *Model's calibration*

As we have already pointed out, the model's diffusion depends on two different parameters: the volatility and the mean reversion. Calibrating the model means finding value for these two parameters consistent with some market prices. These market prices should obviously be actively traded options. These instruments will then be used by the trader to effectively hedge his portfolio (see Chapter 12 on trading and hedging). It is therefore crucial to have the model to match these instruments. Caps and swaptions are the two main markets in the interest rate derivatives world. In the following, we treat the case of 1-factor model. The analysis can be easily generalized to the n-factor case.

7.3.1.3. *Pricing of caplets*

In the HW framework, the price of a caplet resetting at T_{R}, starting at T_1, ending at T_2, paying at T_{p} and with strike K is given by:

$$\text{Price}_{\text{Caplet}}(t) = E_t\left[\mathrm{e}^{-\int_t^{T_{\mathrm{p}}} r(s)ds}\delta(F_1(T_{\mathrm{R}}) - K)^+\right], \quad (7.22)$$

where $F_1(T_{\mathrm{R}})$ is the LIBOR forward price, as seen at time, T_{R} for the interest period $[T_1, T_2]$.

$$\text{Price}_{\text{Caplet}}(t) = E_t\left[\mathrm{e}^{-\int_t^{T_{\mathrm{p}}} r(s)ds}\left(\frac{B(T_{\mathrm{R}}, T_1)}{B(T_{\mathrm{R}}, T_2)} - 1 - \delta K\right)^+\right] \quad (7.23)$$

where δ is the interest period. Applying the Girsanov lemma, we shift to the forward neutral probability associated to the payment date.

Therefore, we get:

$$\text{Price}_{\text{Caplet}}(t) = E_t^{Q_{T_\text{P}}}\left[\left(\mathrm{e}^{\Phi(Y_{T_\text{R}})} - (1+\delta K)\right)^+\right], \tag{7.24}$$

where $\Phi(Y_{T_\text{R}})$ is an affine function of the Y_{T_R}. Then, we determine the exercise boundary and integrate with respect to the law of Y_{T_R} under Q_{T_P}, using the cumulative normal distribution, obtaining a trivial Black–Scholes formula.

7.3.1.4. *Pricing of swaptions*

Let us take the example of receiving a swaption resetting at T_R, with payment schedule $(T_i)_{0\le i\le n}$ and strike K. The price of the swaption is given by:

$$\text{Price}_{\text{Swaption}}(t)$$
$$= E_t\left[\mathrm{e}^{-\int_t^{T_\text{P}} r(s)ds}\left(\sum_{1\le i\le n} C_i B(T_\text{R}, T_i) - B(T_\text{R}, T_0)\right)^+\right], \tag{7.25}$$

where $C_i = K\delta_i$ for $1 \le i \le n-1$ and $C_n = 1 + K\delta_n$. Similarly to caplets, the price of the swaption can be written as:

$$\text{Price}_{\text{Swaption}}(t) = E_t^{T_0}\left[\left(\sum_{1\le i\le n} C_i \mathrm{e}^{\phi_i(Y_{T_\text{R}})} - 1\right)^+\right], \tag{7.26}$$

where Φ_i is a monotonous affine function of x. Then, we can find y_0 so that $\left(\sum_{1\le i\le n} C_i \mathrm{e}^{\phi_i(y_0)} - 1\right) = 0$.[4] Therefore,

$$\text{Price}_{\text{Swaption}}(t) = E_t^{T_0}\left[\sum_{1\le i\le n} C_i\left(\mathrm{e}^{\phi_i(Y_{T_\text{R}})} - \mathrm{e}^{\phi_i(y_0)}\right)^+\right]. \tag{7.27}$$

This can be easily calculated, knowing the law of Y_{T_R} under Q_{T_0}. Numerical integration is done with Gauss–Legendre polynomial points.

[4]This argument is similar to Jamshidian's (1989) one for pricing coupon bearing bonds.

7.3.1.5. *Example of calibration of the model*

One very important feature of the model is that, given a level of mean reversion λ, the price of a caplet (or swaption) expiring at t will only depend on the volatility up to time t. $\{\sigma_s\}_{s<t}$. As a result, an easy way to calibrate a strip of different expiry options is to take a piece-wise volatility function and then bootstrap the different values as follows: we first determine the volatility segment $\{\sigma_s\}_{s<t_1}$ for the model to fit the option with expiry t_1. Then, we move on to calibrate the second segment $\{\sigma_s\}_{t_1<s<t_2}$ to fit the second option with expiry t_2, and so on...

Once the model is calibrated in volatility, we can do a global optimisation on the mean reversion to reduce pricing errors on a second set of options.

Then, the determination of the model is directly linked to the choice of the calibration portfolio. Consequently, it is very important to choose relevant options for the calibration, depending on the characteristics of the deal to price.

For the 2-factor case, we have three additional parameters to calibrate: the second mean reversion λ_2, the second volatility σ_2 and the correlation between the two factors ρ. Generally, it is useful to fix mean reversion spread and volatility ratio to impose stationary covariance structure:

$$\mathrm{var}(df(t,T)) = \sigma_1(t)^2 \mathrm{e}^{-2\int_t^T \lambda_1(s)ds} b(t,T,T)dt, \tag{7.28}$$

$$\mathrm{corr}(df(t,T_1), df(t,T_2)) = \frac{b(t,T_1,T_2)}{\sqrt{b(t,T_1,T_1)b(t,T_2,T_2)}}, \tag{7.29}$$

$$\begin{aligned} b(t,T_i,T_j) = 1 &+ \rho\frac{\sigma_2(t)}{\sigma_1(t)}\left(\mathrm{e}^{-\int_t^{T_i}(\lambda_2-\lambda_1)ds} + \mathrm{e}^{-\int_t^{T_j}(\lambda_2-\lambda_1)ds}\right) \\ &+ \frac{\sigma_2^2(t)}{\sigma_1^2(t)}\mathrm{e}^{-\int_t^{T_i}(\lambda_2-\lambda_1)ds-\int_t^{T_j}(\lambda_2-\lambda_1)ds} \end{aligned} \tag{7.30}$$

Finally, we can calibrate the correlation using some products of correlation such as spread options.

7.3.2. *Cox–Ingersoll–Ross model*

7.3.2.1. *Diffusion equation*

The models developed by Vasicek and Hull and White allow short rates to be negative. In order to raise this objection, Cox, Ingersoll and Ross introduced the "square-root" model (or CIR). A square-root term was added in the diffusion of short rates. The formulation of the diffusion under the risk-neutral measure is:

$$r_t = f(0,t) + Y_t, \tag{7.31}$$
$$dY_t = (\theta_t - k_t Y_t)dt + \sigma_t \sqrt{\alpha_t + Y_t} dW_t. \tag{7.32}$$

Analogously to the HW model, the coefficients of the diffusion are assumed to be time dependent to fit exactly the initial yield and volatility curves.

Since the square root function is not a Lipchitz function, it is more difficult to prove the existence of a solution to the general stochastic differential equation. A proof of the existence and uniqueness of the solution can be found in the paper of Karlin. In addition, we can prove that if $2\theta_t > \sigma^2$, the solution could not reach 0 in a finite time. Therefore 0 is an absorbing boundary.

7.3.2.2. *Reconstruction formulae*

The model has the advantage to have closed formula for zero-coupon bonds and vanilla options. In fact, as for the Vasicek case, zero-coupon bond can be written as:

$$B(t,T) = \frac{B(0,T)}{B(0,t)} \mathrm{e}^{-b(t,T)Y_t - c(t,T)}, \tag{7.33}$$

b_t and c_t solve the following differential equations[5]:

$$\partial_t b = k_t b + \frac{1}{2}\sigma_t^2 b_t^2 - 1, \tag{7.34}$$

[5]The Riccati differential equation $y_t' = f_t y_t^2 + g_t y_t + h_t$ can be solved if it is possible to find constants c, d such that $c^2 f_t + cd g_t + d^2 h_t = 0$ and $|c| + |d| > 0$. If $d = 0$, then $f \equiv 0$, so Riccati's equation is a linear equation. If $d \neq 0$, then it could be simplified into a Bernoulli equation by substituting.

$$\partial_t c = \theta_t b_t + \frac{1}{2}\sigma_t^2 \alpha_t b_t^2. \tag{7.35}$$

These equations are obtained from an application of the Feynman Kac theorem (see Chapter 7 for more technical details). They are subjects to the boundary conditions: $b(T,T) = 0$; $c(T,T) = 0$; so as to ensure that $B(T,T) = 1$.

Equations (7.34) and (7.35) are Riccati equations. In general case, Riccati's differential equation cannot be solved analytically, if the coefficients are not constants or we do not know at least one solution. Such model is not then analytically tractable.

A simple tractable version of the diffusion was proposed by Jamshidian (1995). He assumed that the ratio k_t/σ_t^2 is equal to a positive constant, leading to closed solutions for the above differential equations.

Another method could be to assume that coefficients are step-wise functions of t. We find then the solutions to Eqs. (7.34) and (7.35) in each interval where the coefficients are constants. We link then the different parts of the curve by continuity. This allows us then to fit exactly prices for liquid securities. More details are discussed in Section 7.5.2.

7.3.2.3. *Remark*

We saw that HW and CIR models propose similar form for zero-coupon bond prices. More generally, the formulation of the price of a zero-coupon bond as exponential of an affine function characterize affine class models. Duffie and Kan (1993) show that a necessary and sufficient condition for the affine class is that the short rate dynamics under the risk-neutral measure is of the form:

$$dr_t = b(t, r_t)dt + \sigma(t, r_t)dW_t \tag{7.36}$$

with $b(t, r_t)$ and $\sigma(t, r_t)^2$ are given respectively by:

$$b(t, r_t) = \lambda_t r_t + \theta_t \quad \text{and} \quad \sigma(t, r_t)^2 = \alpha_t r_t + \varphi_t \tag{7.37}$$

λ, θ, α and ϕ are deterministic functions of t. We have then:

$$B(t,T) = A(t,T)\mathrm{e}^{-\beta(t,T)r_t}. \tag{7.38}$$

Nonetheless, the coefficients of zero-coupon bond prices are not, in general, analytically tractable. One can do some approximation to obtain closed form. Alternatively, one can use numerical methods to get better accuracy. To our experience, it is advisable to find a closed form approximation and benchmark it to real pricing with numerical method and refine the approximation up to the desired level. Hence, the calibration can still be done in a very efficient way and the only problem lies in the accuracy of the closed form approximation.

7.3.3. *Black–Karasinski model*

7.3.3.1. *Diffusion equation*

Along the same lines of positive interest rates, Black and Karasinski (1991) suggested a model where the diffusion of the instantaneous spot rate is of the form:

$$d\ln(r_t) = [\alpha_t - a_t \ln(r_t)]dt + \sigma_t dW_t, \tag{7.39}$$

where a and α are deterministic functions of time that can be chosen so as to fit precisely market prices of instruments that trade actively in the market. Under this framework, short rates can be written as:

$$r_t = \mathrm{e}^{\theta(t)+x_t}, \tag{7.40}$$

where

$$\theta(t) = \ln(r_0)\mathrm{e}^{-\int_0^t a_s ds} + \int_0^t \mathrm{e}^{-\int_s^t a_h dh} \alpha_s ds \tag{7.41}$$

and x_t follows a mean reverting process:

$$dx_t = -a_t x_t dt + \sigma_t dW_t. \tag{7.42}$$

7.3.3.2. *Remarks*

However, the BK model is not analytically tractable. Thus, we do not have analytical formulas even for zero-coupon bonds. This makes the model harder to calibrate than the HW model. The calibration of the model is then performed through numerical procedures or once again through closed-form approximation. It is indeed not too hard to find an extension of the HW reconstruction formula that matches the BK distribution up to the second order.

For example, the calibration of a BK tree goes as follows: We first construct a trinomial tree, and then shift each slice by a deterministic drift so as to fit exactly the initial zero-coupon bond curve. For more details on the procedure, see Chapter 9.

Admittedly, the model is more flexible than other short-rates models nevertheless its difficulty to be calibrated makes it less interesting than other mapping models.

7.4. Market Model (BGM)

7.4.1. *Motivations*

Nowadays, the LIBOR market model is among the most popular interest-rate models. Its success is mainly due to its ability to reproduce market prices for caps or swaptions using Black's formula. Besides, the model is based on directly observable variables (LIBOR rates). It, then, turns out to be the most convenient in many cases because of the nature of the payoffs commonly traded and its better tractability. Nevertheless, the use of this model is not exempt from difficulties.

In fact, the non-compatibility between swaptions and caps and the non-Markovian feature of the model in terms of the driving Brownian motions imposes either to make approximations or to degenerate the model. In this section, we present the model and how it is used in practice. Later, we will present a simple extension to smile of the model (shifted forward rate model).

7.4.2. *Diffusion*

As we have already pointed out, in the HJM framework, zero-coupon bond dynamics under the risk-neutral probability (see Geman *et al.*, 1995), are of the form:

$$\frac{dB(t,T)}{B(t,T)} = r_t dt + \Gamma(t,T)dW_t. \tag{7.43}$$

Recall that the forward LIBOR rate between T_k and T_{k+1} can be written as:

$$F_k(t) = \frac{1}{\delta_k}\left(\frac{B(t,T_k)}{B(t,T_{k+1})} - 1\right), \tag{7.44}$$

where δ_k denotes the interest period between T_k and T_{k+1}, $B(t,T_k)$ the zero-coupon bond with maturity T_k. A straightforward application of the Itô lemma leads to the following diffusion for $F_k(t)$ under Q:

$$\frac{\delta_k dF_k(t)}{1+F_k(t)} = (\overrightarrow{\Gamma(t,T_k)} - \overrightarrow{\Gamma(t,T_{k+1})})(\overrightarrow{dW_t^Q} - \overrightarrow{\Gamma(t,T_{k+1})}\,dt). \tag{7.45}$$

We define the new probability $Q_{T_{k+1}}$, through its density

$$\frac{dQ_{T_{k+1}}}{dQ} = \mathrm{e}^{-\frac{1}{2}\int_0^t \Gamma_{k+1}^2(s)ds + \int_0^t \overrightarrow{\Gamma_{k+1}(s)}\overrightarrow{dW_s^Q}}. \tag{7.46}$$

The Girsanow theorem shows that the new Wiener process under the new probability $Q_{T_{i+1}}$ can be written as:

$$W_t^{Q_{T_{i+1}}} = W_t^Q - \int_0^t \overrightarrow{\Gamma(s,T_{k+1})}ds. \tag{7.47}$$

So that the diffusion of $F_k(t)$ becomes:

$$\frac{\delta_k dF_k(t)}{1+F_k(t)} = (\overrightarrow{\Gamma(t,T_k)} - \overrightarrow{\Gamma(t,T_{k+1})})\overrightarrow{dW_t^{Q_{T_{i+1}}}}, \tag{7.48}$$

$$\frac{dF_k(t)}{F_k(t)} = \vec{\sigma}_k(t)\overrightarrow{dW_t^{Q_{T_{i+1}}}} \tag{7.49}$$

with a relative volatility of

$$\vec{\sigma}_k(t) = \frac{1+\delta_k F_k(t)}{\delta_k F_k(t)}(\overrightarrow{\Gamma(t,T_k)} - \overrightarrow{\Gamma(t,T_{k+1})}). \tag{7.50}$$

Thus, the dynamics of the forward rate are driftless (it is indeed a martingale) under its own probability. Note that this log-normal diffusion is exactly what we need to fit exactly in Black's formula, without approximation. In the BGM framework, it is the volatility of the forward LIBOR $\vec{\sigma}_k(t)$ that is a parameter of the model.

As a result, contrary to short rates models, zero-coupon bonds volatilities as shown in Eq. (7.50) are stochastic.

Each forward has then a simple diffusion under its own probability. However, to use the model, we need to express this diffusion under a unique probability for all the forwards. Generally, two probabilities are used: terminal and spot probabilities.

Let consider a structure with $N+1$ dates $T_1, \ldots, T_{N+1}$. We define the terminal probability as the forward neutral probability $Q^{T_{N+1}}$ associated to the last forward. Again, we can apply recursively Girsanov to find the diffusion of the forward rate $F_k(t)$ under the terminal probability:

$$\frac{dF_k(t)}{F_k(t)} = -\sum_{i=k+1}^{N} \frac{\delta_i F_i(t)}{1+\delta_i F_i(t)} \vec{\sigma}_i(t).\vec{\sigma}_k(t)dt + \vec{\sigma}_k(t)\overrightarrow{dW_t^{Q_{T_{N+1}}}} \tag{7.51}$$

Analogously, the diffusion of the forward rate under the spot probability is

$$\frac{dF_k(t)}{F_k(t)} = \sum_{i=n(t)}^{k} \frac{\delta_i F_i(t)}{1+\delta_i F_i(t)} \vec{\sigma}_i(t).\vec{\sigma}_k(t)dt + \vec{\sigma}_k(t)\overrightarrow{dW_t^{Q_{\mathrm{Spot}}}} \tag{7.52}$$

where $n(t) = k$ if $t \in]T_k, T_{k+1}]$. Notice that we have a positive drift for the forward diffusion under spot probability. We have then a risk of explosion of the solutions. That is why we prefer, in general, working under the terminal probability, which induces a negative drift on the diffusion. It can also be interesting to use a rolling numeraire defined as at each date T_k as the $Q_{T_{k+1}}$ forward measure. The advantage of this measure is in a sense to have the best adapted measure to a given time.

7.4.3. *Interpolation*

In the previous section we saw that the BGM model is based on a scheduler given by a set of dates $T_1, \ldots, T_{N+1}$. Then, to determine precisely the dynamics of the whole yield curve, we need to compute the forward prices (or discount prices) resetting at any date T_{reset}

and ending at any date T_{end}. T_{rese} and T_{end} dates do not necessarily belong to the previous set of dates. The art of using the BGM model lies in our opinion in correctly interpolating the model's forward rate to compute other rates. In a sense, the forward LIBOR market model provides us with only the rates corresponding to the period of the forward LIBOR model, let us say 6 months for instance. Hence, we need to have some way of interpolating these 6 month rates to get 3 month rates as our product depends not only on 6 months rates but also on 3 month rates. Whenever, we think that the liquidity given by the 6 month rate is more reliable than the one of the shorter maturity. This should be obviously depends on the liquidity of the market. The principle consists of interpolating the value of the forward using the values of the adjacent regular forwards. An example of interpolation could be the following:

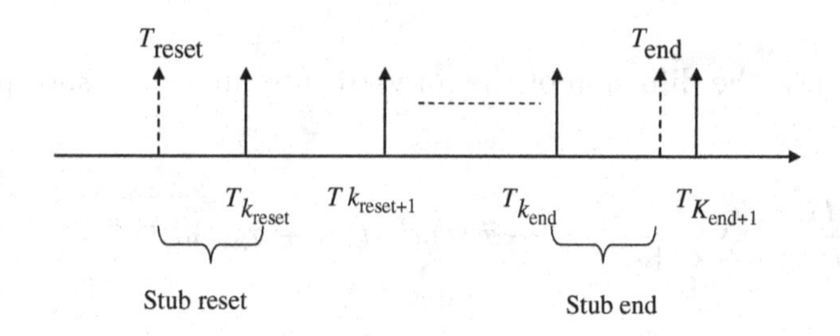

$$F_{\text{stub\_reset}}(T_{\text{reset}}, T_{k_{\text{reset}}}) = (F_{\text{stub\_reset}}(0))e^{(\text{StochasticTerm } F_{Tk_{\text{reset}}+1}(T_{\text{reset}}))} \tag{7.53}$$

$$\begin{aligned}(1 + F(T_{\text{reset}}, T_{\text{end}})) &= (1 + F_{\text{stub\_reset}}(T_{\text{reset}}, T_{k_{\text{reset}}}))\left[\prod_{i=k_{\text{reset}}}^{k_{\text{end}}} (1 + F(T_i, T_{i+1}))\right] \\ &\quad \times (1 + F_{\text{stub\_end}}(T_{k_{\text{end}}}, T_{\text{end}}))\end{aligned} \tag{7.54}$$

Other more complex methods of interpolation can be used, but it is important to have the same price as the forward seen at time 0 when the volatility is null (for "no arbitrage" purposes).

The method can be generalized to determine continuous values from a discrete diffusion.

7.4.4. *Handling drift*

In this section, we suppose that we work under a terminal probability associated to the last forward. Similar results can be found if we work under any other probability.

As we have already pointed out, the diffusion of forward rates under $Q_{T_{N+1}}$ can be written as:

$$\frac{dF_k(t)}{F_k(t)} = \mu_k(t)dt + \vec{\sigma}_k(t)\overrightarrow{dW_t^{Q_{T_{N+1}}}}, \tag{7.55}$$

where

$$\mu_k(t) = -\vec{\sigma}_k(t) \sum_{k+1 \leq i \leq N} \frac{\delta_i F_i(t)}{1 + \delta_i F_i(t)} \vec{\sigma}_i(t). \tag{7.56}$$

Recall that $\mu_N = 0$ so that F_N be a martingale under $Q_{T_{N+1}}$. The resulting drifts of the other forwards are stochastic. This characteristic poses some problems in the implementation of some numerical methods with this model. We will take the case of the tree method.

To construct a recombining tree (see Chapter 9 for details on tree or PDE construction), we suppose the drift either deterministic or just Markovian (depending only on forwards evaluated at the pricing time). However, with the exception of the last forward, all the other drifts are stochastic and depend on previous values of the forwards:

Two main approaches have been developed in the literature. The first approach consists in calculating a deterministic approximation of the drift. This method is equivalent to a projection of the drift on the space of constants:

$$\int_0^t \mu_k(s)ds = \frac{\exp\left(\frac{1}{2}\int_0^t (\sigma_k(s))^2 ds\right) \frac{B(0,T_{k+1})}{B(0,T_{N+1})}}{E_s^{N+1}\left(\exp\left(\int_0^t \sigma_k(s)dW_s^{k+1}\right) \frac{B(t,T_{k+1})}{B(t,T_{N+1})}\right)}, \tag{7.57}$$

since $\mu_N = 0$ and $B(t,T_{k+1})/B(t,T_{N+1}) = \prod_{k+1 \leq i \leq N} (1 + \delta_i F_i(t))$ (from the arbitrage-free relation, we can calibrate all the drifts by a backward induction on the reset dates indexes).

However, this deterministic drift approach is no longer efficient when the local BGM volatility is high or the maturity is long. This is due to the variance of diffused processes.

The second approach consists of a Markovian approximation of the drift. Then, we write the drift as a function of a limited number of stochastic processes at time t. We use the processes already diffused in the tree.

The method was first detailed in the paper "Fast drift approximated pricing in the BGM model" by Pietersz *et al.* (2004).

Similar methods can also be performed for other numerical methods.

Another important reason to use the forward LIBOR market model is to avoid paying the price for the Euler scheme in Monte Carlo. Rigorously, the drift is path dependent. Its path dependency should be captured by the Euler scheme. But this can slow down the Monte Carlo. In most common cases, this refinement is not so much required and one can just diffuse the model for each forward LIBOR date. Whenever refinement is required, which comes to our experience for long maturities product, it is better to use predictor correction and avoid the Euler scheme price. More details on this can be found in Pietersz *et al.* (2004).

7.4.5. *Calibration: Relation vol Swap vol Fra*

Before using the model, we need to calibrate the parameters on markets prices of caplets and swaptions. As the diffusion of forward rates is log-normal, caplets can easily be priced using the Black's formula. Let C_t denotes the price of a caplet on the forward F_i with a strike K. Then we have

$$C_{t_0} = F(t_0, T_i, T_{i+1})N(d_1) - KN(d_2), \tag{7.58}$$

where $N(.)$ denotes the cumulative normal function and

$$d_{1,2} = \frac{\text{Log}\left(\frac{(F(t_0,T_i,T_{i+1})}{K}\right)}{\sigma_{F_i}\sqrt{T_i - t_0}} \mp \frac{1}{2}\sigma_{F_i}(T_i - t_0) \tag{7.59}$$

By taking a piecewise-constant instantaneous volatility structure we can then calibrate a portfolio of caplets by a Boostrap algorithm.

As for swaptions, we show previously that a log-normal diffusion for forward rates is not compatible with a log-normal diffusion for forward swap rates. This makes the pricing of swaptions more complex than the pricing of caplets. In the following, we try to overcome these difficulties by making some approximations on swap rate diffusion. We have seen in Chapter 5 that a swap rate is simply given as the value of the floating leg divided by the annuity of the fixed leg:

$$S_t = \frac{B(t,T_0) - B(t,T_N)}{\sum_{1\le k\le N} B(t,T_k)}. \tag{7.60}$$

The price of a swaption with a maturity T_0 and a strike K can then be written as:

$$S_{t_0} = B(t_0,T_0)E^{Q_{T_0}}\left[\sum_{1\le k\le N} B(T_0,T_k)(S_t - K)^+\right]. \tag{7.61}$$

Recall that the market is used to price swaptions using a Black–Scholes volatility type. Then, a natural probability that could lead to Black's formula for swaptions is the probability associated to the numeraire $A(T_0) = \sum_{1\le k\le N} B(T_0,T_k)$ known as the annuity.[6]

The resulting measure is called the annuity measure, or more frequently, the forward swap measure.

Applying Itô's and then Girsanov's theorems, we can prove that the diffusion of the swap rate under this new probability is of the form:

$$\frac{dS_t}{S_t} = \Sigma_t dW^{Q_{01}} \tag{7.62}$$

where Σ_t is stochastic and depends on the values of the forwards F_t and their volatilities. Besides, we can write Σ_t as

$$\Sigma_t = \sum_{1\le k\le N} \mu_t^k \sigma_t^k \tag{7.63}$$

[6]We note that in general the annuity $A(t) = \sum_{1\le k\le N} \delta_k B(t,T_k)$ where δ_k is the day count fraction for the time interval $[T_{k-1}, T_k]$. Here we have $\forall k, \delta_k = \delta$.

One can find an approximation to get a deterministic formulation for Σ_t by taking for the weight μ_i^s their forward values at time 0. After this approximation, swap rate has a log-normal diffusion under the probability measure. Swaptions can then be priced easily using the Black's formula.

7.5. Extension to Smile

7.5.1. *Short-rate model extension and Cheyette*

7.5.1.1. *Quadratic Gaussian model*

An extension to general affine models has been introduced by El Karoui *et al.* (1992). In this approach, spot rate is a quadratic function of an Ornstein Uhlenbeck process X_t.

$$\begin{aligned} r_t &= f(0,t) + a_r(t)X_t^2 + X_t + c_r(t), \\ dX_t &= -\lambda(t)X_t dt + \sigma(t)dW_t. \end{aligned} \tag{7.64}$$

Notice that if we set $a_r(t)$ at 0, we recover a classical affine short-rate model such as HM model. The model has the advantage to have a reconstruction formula since zero-coupon bonds can be written as:

$$B(t,T) = \mathrm{e}^{-\int_t^T f(0,s)ds}\mathrm{e}^{-\left[A(t,T)X_t^2 + B(t,T)X_t + C(t,T)\right]}, \tag{7.65}$$

where the coefficient $A(t,T)$, $B(t,T)$ and $C(t,T)$ are linked to the parameters of diffusion of short rate through the following differential equations[7]:

$$\begin{aligned} \partial_t A(t,T) &= 2A(t,T)^2\sigma(t)^2 + 2\lambda(t)A(t,T) - a_r(t), \\ \partial_t B(t,T) &= 2\sigma(t)^2 A(t,T)B(t,T) + \lambda(t)B(t,T) - b_r(t), \\ \partial_t C(t,T) &= -\frac{1}{2}\sigma(t)^2\left[2A(t,T) - B(t,T)^2\right] - c_r(t). \end{aligned} \tag{7.66}$$

[7]These equations are due to an application of the Feynman Kac formula to the no-arbitrage relation $B(t,T) = E^Q(\mathrm{e}^{-\int_t^T r_s ds}|F_t)$.

Besides, the zero-coupon bond volatility is of the form: $\Gamma(t,T) = [2A(t,T)X_t + B(t,T)]\sigma(t)$ which keeps X_t Gaussian under any forward neutral probability. Then we can price caplets and swaption using the same approach as presented for the HM model.

In terms of implied volatility smiles, the model allows us different forms of skew. We present different shapes of smiles generated by the model for different values of a_r as shown in the following:

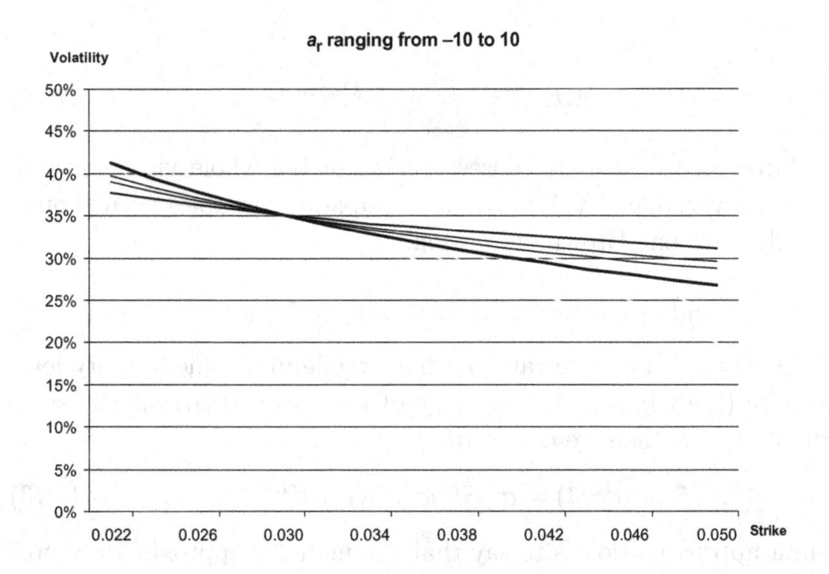

The model can be extended to multidimensional.

7.5.1.2. *Cheyette model*

This model has been simultaneously addressed by Ritchken and Sankarasubramanian (1995) and Cheyette. It incorporates volatility skew but not convexity smile. In the 1-factor case, it is given by:

$$r_t = f(0,t) + X_t, \tag{7.67}$$

$$dX_t = (Y_t - k(t)X_t)dt + \sigma(r_t,t)dW_t, \tag{7.68}$$

$$dY_t = [\sigma(t)^2 X_t^{2\beta} - 2k(t)Y_t]dt. \tag{7.69}$$

Notice that with $\sigma(r_t,t) = \theta(t)\sqrt{r_t}$ and $\sigma(r_t,t) = \theta(t)r_t$, we have the same diffusion as the one of the CIR model or the BK model.

Nevertheless, in each case, the drift and term structure of volatility are different. It is common to set the volatility function at:

$$\sigma(r_t, t) = \theta(t) r_t^{\beta}. \tag{7.70}$$

In this model, bonds are also priced with a closed formula:

$$B(t,T) = \frac{B(0,T)}{B(0,t)} \mathrm{e}^{-A(t,T)X_t - \frac{1}{2}A(t,T)^2 Y_t}, \tag{7.71}$$

where

$$A(t,T) = \int_t^T \mathrm{e}^{-\int_t^s k(u)du} ds. \tag{7.72}$$

We have then a Markov representation of the whole yield curve in two-state variables (X, Y). We have two-state variables even though we only have one Brownian motion.

7.5.1.3. *Swaption pricing approximation in Cheyette*

We know that the swap rate is a martingale under the annuity level measure (see Chapter 7 on change of measure). If $S(t)$ is the swap rate at time t, then we can write:

$$dS(t) = \partial_X S(t) \sigma(t, X_t) dW_t^{LVL}. \tag{7.73}$$

A first approximation is to say that the model is approximately normal and get a volatility given by:

$$\eta(t) = \sqrt{\frac{1}{t} \int_0^t ([\partial_X S(u) \sigma(u, X_u)]_{x=y=0})^2 du}. \tag{7.74}$$

This does not perform too bad, at least for ATM options.

A better approximation is to take a general shifted log-normal model for the diffusion of the swap rate. Let us consider:

$$\varphi(t) = \int_0^t [\partial_X S(u) \sigma(u, X_u)]^2 du, \tag{7.75}$$

then we get:

$$\begin{aligned} \partial_S \varphi(t)|_{x=y=0} = \int_0^t & 2[\partial_X^2 S(u) \sigma(u, X_u)^2 \\ & + \partial_X S(u) \sigma(u, X_u) \partial_X \sigma(u, X_u)]_{x=y=0} \, du. \end{aligned} \tag{7.76}$$

In a simple shifted log-normal model, we have:

$$dS(t) = \overline{\lambda}[\overline{m}S(t) + (1-\overline{m})S(0)]dW_t^{LVL}, \tag{7.77}$$

$$\kappa(t) = \int_0^t \overline{\lambda}^2[\overline{m}S(u) + (1-\overline{m})S(0)]^2, \tag{7.78}$$

$$\kappa(t)|_{S=S(0)} = \overline{\lambda}^2 S(0)^2 t, \tag{7.79}$$

$$\partial_S \kappa|_{S=S(0)} = 2\overline{m}\overline{\lambda}S(0)t. \tag{7.80}$$

The last step is to find the average mean reversion and mean $\overline{\lambda}$, $\overline{m}$, so that:

$$\kappa(t)|_{S=S(0)} = \varphi|_{x=y=0}, \tag{7.81}$$

$$\partial_S \kappa|_{S=S(0)} = \partial_S \varphi|_{x=y=0}. \tag{7.82}$$

7.5.2. *BGM extension: constant elasticity of variance (CEV) (Cox and Ross, 1976), SFRM*

The volatility shape generated by the LIBOR market model (BGM) as presented previously is flat. This is not convenient for the pricing of away-from-the-money options. A simple way to generalize the dynamics to capture the market skew is to add a shift to the diffusion of the forward rates. Then, we assume that the forward-rate diffusion is given by:

$$\frac{dF_k(t)}{F_k(t) + m_k} = \vec{\sigma}_k(t)\overrightarrow{dW_t^{Q_{T_{i+1}}}}. \tag{7.83}$$

Under the terminal forward probability $Q_{T_{N+1}}$, the diffusion becomes:

$$\frac{dF_k(t)}{F_k(t) + m_k} = \lambda_k(t)dt + \vec{\sigma}_k(t)\overrightarrow{dW_t^{Q_{T_{N+1}}}}, \tag{7.84}$$

where

$$\lambda_k(t) = -\vec{\sigma}_k(t) \sum_{k+1 \le j \le N} \frac{\delta_j(F_j(t) + m_j)}{1 + \delta_j F_j(t)} \vec{\sigma}_j(t). \tag{7.85}$$

We can then apply the same approach of the Markovian drift and swaption volatility approximations to calibrate the model and use it with numerical methods.

7.5.3. *Stochastic volatility with BGM models*

The classical forward LIBOR model has been established as the traditional tool to price and hedge interest rates derivatives because of its ability to calibrate the volatility surface of at the money swaption across all expiries and tenors. However, as interest rates vanilla desk have started to use more and more intensively stochastic volatility model such as the SABR model, extension of the forward LIBOR model to account for a slope of the smile either through a shifted log-normal or a CEV diffusion has turned out to be insufficient.

7.6. Summary

Earliest 1-factor short-rate models: The earliest model was Black (1976) and Rendleman and Batter (1980) one with log-normal short rate:

$$dr_t = \mu r_t dt + \sigma r_t dW_t.$$

But the log-normality assumption was immediately criticized as it cannot capture the mean reverting feature of interest rates (Vasicek, 1997):

$$dr_t = a(b - r_t)dt + \sigma dW_t$$

But the main drawback of the Vasicek model is that short rates can run negative.

The CIR model (1985) added square-root diffusion term to avoid such a problem:

$$dr_t = a(b - r_t)dt + \sigma\sqrt{r_t}\, dW_t.$$

No-arbitarge models: These models take as input the initial term (and volatility) structure.

Ho and Lee (1986) pionnered an arbitrage-free lattice approach for interest-rate models. They studied binomial version of $dr_t = \mu_t dt + \sigma dW_t$ taking initial term structure of interest rates as inputs.

Heath *et al.* (1990, 1992) extended the Ho and Lee model in three directions:

(i) they chose forward rates as basic building blocks,
(ii) incorporated continuous trading,
(iii) allowed multiple factor models.

Dybvig (1988) studied the Ho and Lee model in the HJM framework for the case of 2-factors.

Hull and White (1990) extended the Vasicek model by making the short-rate model parameters time dependent:

$$dr_t = a_t(b_t - r_t)dt + \sigma_t dW_t.$$

This model is widely used as it produces closed-form solutions for bond prices (Black *et al.*, 1990).

$$d\ln r_t = \left(\alpha_t + \frac{\sigma_t}{\sigma_t}\ln r_t\right)dt + \sigma_t dW_t.$$

This model combined the mean reverting behaviour of the short rate with log-normality. The major appeal of the model is that it is easy to calibrate. However, its main drawback is the dependency between mean-reversion and volatility terms.

Black and Karasinski (1991) rectified this shortcoming of the BDT model by decorrelating mean reversion and volatity.

$$d\ln(r_t) = [\alpha_t - a_t\ln(r_t)]dt + \sigma_t dW_t.$$

Sandmann and Sondermann (1993) studied a general arbitrage-free model dynamically incorporating properties of both normal and log-normal models:

$$R_t = \ln(1 + r_t)$$
$$dR_t = \mu_t R_t dt + \sigma_t dW_t$$

Thus interest rates cannot be negative and do not explode.

However, neither the instantaneous short rate nor the instantaneous forward rate can be observed in the market. A different approach has then emerged over the years referred to as market models. The most popular is the BGM or the LIBOR market model

(1997). Its success is mainly due to its ability to reproduce market prices for caps or swaptions. Another well-known model is Ritchken and Sankarasubramanian or Cheyette model, which under some regularity conditions on the volatility structure can be Markovian in a bi-dimensional diffusion system.

References

[1] Brace A, Gatarek G and Musiela M (1997). The market model of interest rate dynamics. *Mathematical Finance* **7**, 127–155.
[2] Black F, Derman E and Toy W (1990). A one factor model of interest rates and its application to treasury bond options. *Financial Analysts Journal*, **46**, 33–39.
[3] Cox J, Ingersoll J and Ross S (1985). An intertemporal general equilibrium model of asset prices. *Econometrica*, **53**(2), 384–386.
[4] Duffie D and Kan R (1996). A yield-factor model of interest rates. *Mathematical Finance*, **6**(4), 379–406.
[5] Geman H, El Karoui N and Rochet J (1995). Changes of numeraire, changes of probability measure and option pricing. *Journal of Applied Probabilitiy*, **32**, 443–458.
[6] Heath D, Jarrow A and Merton A (1992). Bond pricing and the term structure of interest rate: a new methodology. *Econometrica*, **60**(1), 77–105.
[7] Hull J and White A (1990). Pricing interest rate derivatives securities. *Review of Financial Studies*, **3**(4), 573–592.
[8] Mercurio B (2001). *Interest Rate Models, Theory and Practice*, Berlin: Springer Verlag.
[9] Pessler A. Markov-Functional Interest Rate Models. *Finance and Stochastics*, **4**(4), 391–408; Reprinted in *The New Interest Rate Models*, L. Hughston (ed.). London: RISK Publications (with Phil Hunt and Joanne Kennedy).
[10] Pietersz R, Pelsser A and Van Regenmortel M (2004). Fast drift approximated pricing in the BGM model. *Journal of Computational Finance*, **8**(1), 93–124.
[11] Rebonato R (2002). *Modern Pricing of Interest-Rate Derivatives.* Princeton, NJ: Princeton University Press.
[12] Ritchken P and Sankarasubramanian L (1995). Volatility structure of forward rates and the dynamics of the term structure. *Mathematical Finance*, **5**, 55–72.
[13] Taylor HM and Karlin S (1984). *A Second Course in Stochastic Processes.* New York: Academic Press.

Chapter 8

INFLATION

In the recent years, inflation derivatives market has experienced considerable growth with an outstanding notional volume over $100 bn and is expected to continue in the coming years. The pace of the growth has been swift.

In this chapter, we discuss the plain vanilla inflation derivatives products: inflation swaps in their different forms, zero-coupon, cap and floor swaps. In order to determine the price of vanilla product concerning inflation and to give more coherence to the forward CPI ratio values, we have to take a look at the convexity adjustment and the curve modeling: that is the aim of Section 8.2. As inflation exhibits seasonal patterns, we will see the different estimation of seasonality. Nowadays, more and more exotic products dealing with the inflation appear. So, to conclude, we will expose the hybrid models that enable us to price the hybrid inflation products.

8.1. Vanilla Products

8.1.1. *History of inflation markets*

The development of inflation products has occurred differently depending on the market: In Europe, products that developed first where year-on-year swaps (see later), whereas first products in the United States were inflation bonds. This comes from the fact that the US government developed very early a liquid market for inflation-indexed bonds. So, European market has mainly driven the growth of inflation-linked derivatives for several years.

8.1.2. *Swap: YoY, zero coupon and bond*

Basically all inflation-linked products are swaps: they are an exchange of an inflation leg against another leg. It is a bilateral contract that enables the inflation payer, also referred to as the inflation receiver, to pay a predetermined fixed rate and to receive from the inflation seller, also referred to as the inflation donor, inflation-linked payment. It is obvious that inflation swaps can be used to hedge out inflation risk.

We will write the consumer price index CPI(t) at time t. There are mainly three types of inflation swaps:

- year-on-year (YoY) swaps,
- bonds,
- zero-coupon swaps.

A YoY payoff is of the form $\mathrm{CPI}(t-3)/\mathrm{CPI}(t-15)$, which is paid at time t. Hence, a YoY swap is of the form:

$$\frac{\mathrm{CPI}(T_1 - 3\,\mathrm{months})}{\mathrm{CPI}(T_1 - 15\,\mathrm{months})} - 1\,\frac{\mathrm{CPI}(T_2 - 3\,\mathrm{months})}{\mathrm{CPI}(T_1 - 3\,\mathrm{months})}$$
$$\cdots\frac{\mathrm{CPI}(T_n - 3\,\mathrm{months})}{\mathrm{CPI}(T_{n-1} - 3\,\mathrm{months})} - 1$$

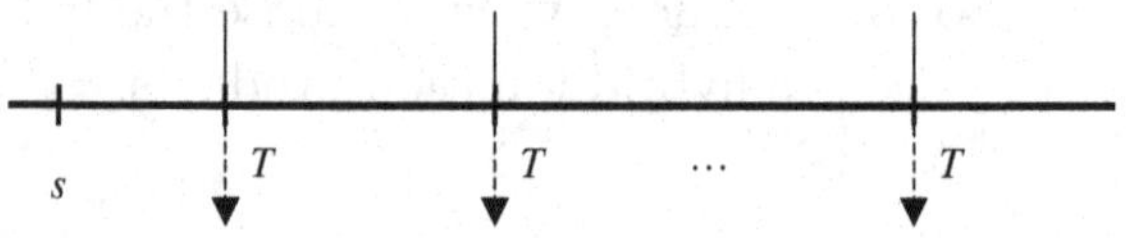

with $T_{i+1} - T_i = 1$ year.

A zero-coupon payoff is of the form $\mathrm{CPI}(t-3)/\mathrm{CPI}(s-3)$ where s is the start date of the swap.

An inflation bond security whose cash flows are of the forms:

$$\frac{\mathrm{CPI}(T_1 - 3\,\mathrm{months})}{\mathrm{CPI}(s - 3\,\mathrm{months})} - 1\,\frac{\mathrm{CPI}(T_2 - 3\,\mathrm{months})}{\mathrm{CPI}(s - 3\,\mathrm{months})} - 1$$
$$\cdots\frac{\mathrm{CPI}(T_n - 3\,\mathrm{months})}{\mathrm{CPI}(s - 3\,\mathrm{months})} - 1$$

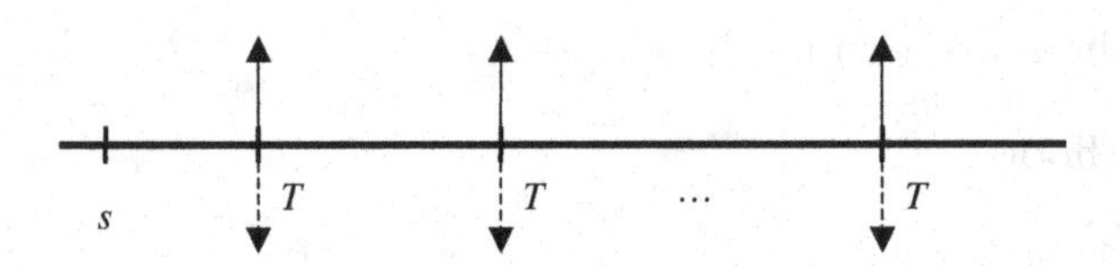

A bond that involves cash flows at only one date is named a zero-coupon swap. Its cash-flow schedule can be represented as:

$$\frac{\mathrm{CPI}(T - 3\,\mathrm{months})}{\mathrm{CPI}(s - 3\,\mathrm{months})} - 1$$

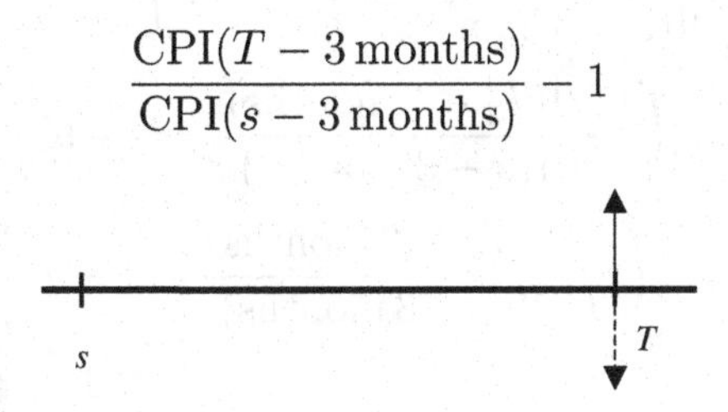

We can notice that all these payoffs involve cash flows of the form $\mathrm{CPI}(t - \delta)$ for payments at date t. This is due to the fact that at date t, only the index of date $(t - \delta)$ is known. There is a lag for indexation during which figures are processed to build the index. This indexation lag is for instance 3 months for HICP (European Index).

It appears that bonds are more liquid on exchanges than YoY; this can be explained by the fact that they are easier to understand and used to protect against inflation.

8.1.3. *Option: Cap, floor and swaption*

Options can also be traded on inflation indices. There are puts and calls on zero-coupon swaps. Puts are generally called floorlets and calls are referred to as caplets. They are usually spot starting and can be used to leverage a view on inflation. An inflation cap–floor parity relation exists that should be satisfied at any time by cap and floor, which have the same strike:

$$\text{Inflation Cap} - \text{Inflation Floor} \;=\; \text{Payer Inflation Swap}. \qquad (8.1)$$

An inflation collar is just the combination of an inflation floor and cap with the same strike.

Roughly, there are two kinds of options:

- cap and floors,
- swaptions.

Cap and floors exist both on YoY and zero-coupon payoffs.

$$\left(\frac{\mathrm{CPI}(T_1 - 3\,\mathrm{months})}{\mathrm{CPI}(s - 3\,\mathrm{months})} - 1 - K\right)^+ \\ \times \left(\frac{\mathrm{CPI}(T_2 - 3\,\mathrm{months})}{\mathrm{CPI}(s - 3\,\mathrm{months})} - 1 - K\right)^+ \\ \cdots \left(\frac{\mathrm{CPI}(T_n - 3\,\mathrm{months})}{\mathrm{CPI}(s - 3\,\mathrm{months})} - 1 - K\right)^+$$

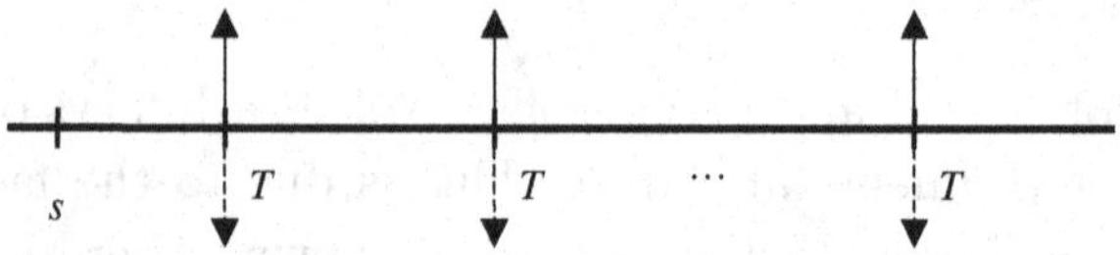

s T T ... T

$$\left(\frac{\mathrm{CPI}(T_2 - 3\,\mathrm{months})}{\mathrm{CPI}(s - 3\,\mathrm{months})} - 1 - K\right)^+ \\ \times \left(\frac{\mathrm{CPI}(T_2 - 3\,\mathrm{months})}{\mathrm{CPI}(T_1 - 3\,\mathrm{months})} - 1 - K\right)^+ \\ \cdots \left(\frac{\mathrm{CPI}(T_n - 3\,\mathrm{months})}{\mathrm{CPI}(T_{n-1} - 3\,\mathrm{months})} - 1 - K\right)^+$$

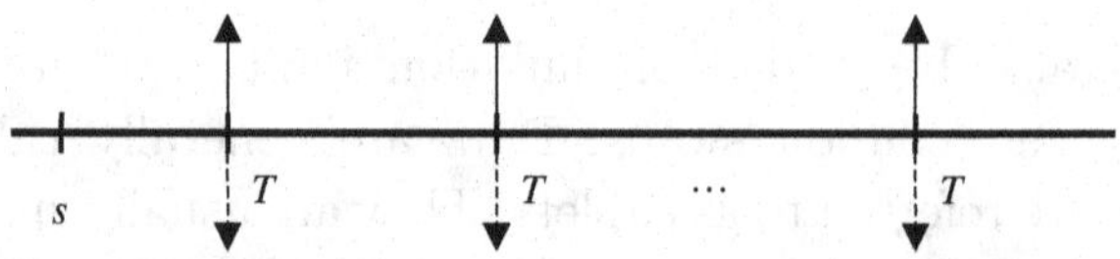

s T T ... T

Swaptions are options to enter into a swap. They actually exist for YoY swaps only. There are mainly three kinds of such options: options to enter into a YoY swap against a fixed rate, YoY against LIBOR or YoY against another YoY based on another inflation rate (for instance, inflation with tobacco against inflation without). Commonly, YoY swaptions, are more liquid than bonds swaptions. There exist more exotic product like breakeven inflation swaptions and real swaptions (for more information, see Belgrate, 2006).

8.1.4. *Pricing example*

This example is an extract from Belgrade *et al.* (2005). In this section, we present a Bermudan inflation swaption where one has the right to enter at the exercise dates into a swap where one pays inflation and receives LIBOR minus a spread capped and floored to certain targets.

For instance, let us look at the following deal:

- as of date: 1 September 2004,
- start date: 3 September 2004,
- end date: 3 September 2010.

At the various notification dates, the investor has the right to enter into a swap until year 2010:

(1) He receives semi-annually 6 m LIBOR minus 15 bps on Act/360 basis, capped to 6% and floored to 1.50%. We will refer to this leg as the pure interest rates leg.
(2) He pays quarterly the average between the 6 month LIBOR and the inflation return of the CPALEMU on a 30/360 basis. We will refer to this leg as the mixed leg.

Here the notification dates are given by:

1-Sep-04, 1-Mar-05, 1-Sep-05, 1-Mar-06, 31-Aug-06,
1-Mar-07, 30-Aug-07, 28-Feb-08, 1-Sep-08, 27-Feb-09,
1-Sep-09, 1-Mar-10.

The interest leg payoff can be easily decomposed into a spread of two cap at 1.50% and 6%:

$$\begin{aligned}&\mathrm{Max}(1.50\%, \mathrm{Min}(6\%, \mathrm{LIBOR}(\mathrm{Eur}, 6\,\mathrm{m}))\\&\quad = 1.50\% + \mathrm{Max}(\mathrm{LIBOR}(\mathrm{Eur}, 6\,\mathrm{m}) - 1.50\%, 0)\\&\qquad - \mathrm{Max}(\mathrm{LIBOR}(\mathrm{Eur}, 6\,\mathrm{m}) - 6\%, 0))\end{aligned}$$

The mixed leg has risk to inflation and LIBOR. But ignoring the cap and floor on the pure interest rates leg, we can roll the LIBOR part from the mixed to the pure interest rates leg and obtain a swap between inflation and LIBOR, given by:

- receives 0.5*LIBOR-15 bps,
- pays 0.5*the inflation return.

This leads to think that the calibration strike for the inflation model should be at the level of this swap. Inversely, the interest rates model should tackle correctly the smile at the level of the forward inflation return + 30 bps.

The Bermuda feature imposes us to find equivalent swaption strike to capture the "callability" feature through an equivalent strikes. This description shows a natural swaption strike, but this makes the simplification of cap and floor features on the pure interest rates leg. Ideally, the model should correctly "reprice" LIBOR cap at 1.50% and 6% level and account for swaption volatility at the equivalent strike level. This adjustment is in fact quite tiny and can be ignored at first sight.

8.2. Vanilla Product Pricing

8.2.1. *Presentation*

As we described in Section 8.1, the most liquid vanilla derivatives on inflation are some swaps and options. The purpose of inflation derivatives is to transfer the inflation risk. The market quotes daily a zero-coupon inflation swap rate as an equivalent of the variable leg of the swap, generating therefore a forward CPI curve. So, the market of the zero-coupon swaps guarantees a minimal value of inflation forward whereas the YoY swaps guarantee an office plurality of annual inflations forward. In order to evaluate the YoY inflation swap rate, we should determine before the forwards value of the annual inflation cash flows. This is the challenge of the curve modeling and the convexity adjustment.

8.2.2. *Curve modeling*

The challenge of the curve modeling is, first, to set a framework regrouping the CPI historical monthly data and "merge" it with the forward market CPI curve. The idea is to keep the same convention and the interpolation between the set guaranteeing the coherence and the continuity of the past and the future series.

Example of a curve for inflation is given in Figure 8.1.

2D	2.02
1W	2.02
1M	2.184
2M	2.184
3M	2.189
DEC04	97.8189
MAR05	97.7894
JUN05	97.7281
SEP05	97.63504
DEC05	97.52216
MAR06	97.42104
JUN06	97.28383
SEP06	97.14441
3Y	2.7325
4Y	2.953
5Y	3.149
6Y	3.326
7Y	3.486
8Y	3.623
9Y	3.741
10Y	3.84
11Y	3.922
12Y	3.993
13Y	4.056
14Y	4.113
15Y	4.166
20Y	4.356
25Y	4.443
30Y	4.479

Figure 8.1: Inflation curve example.

8.2.3. *Convexity adjustment*

On inflation markets, forward CPI ratios and forward CPIs do not have coherent values, in the sense that

$$E^{Q^T}\left[\frac{\mathrm{CPI}(T)}{\mathrm{CPI}(T-\delta)}\right] \neq \frac{E^{Q^T}[\mathrm{CPI}(T)]}{E^{Q^{T-\delta}}[\mathrm{CPI}(T-\delta)]}. \tag{8.2}$$

The difference between those two values is named convexity adjustment:

$$\lambda_a = E^{Q^T}\left[\frac{\mathrm{CPI}(T)}{\mathrm{CPI}(T-\delta)}\right] - \frac{E^{Q^T}[\mathrm{CPI}(T)]}{E^{Q^{T-\delta}}[CPI(T-\delta)]}, \tag{8.3}$$

$$\lambda_m = E^{Q^T}\left[\frac{\mathrm{CPI}(T)}{\mathrm{CPI}(T-\delta)}\right] \Big/ \frac{E^{Q^T}[\mathrm{CPI}(T)]}{E^{Q^{T-\delta}}[CPI(T-\delta)]}. \tag{8.4}$$

Convexity adjustments can be expressed in either an additive or a multiplicative way (see Eqs. (8.3) and (8.4)).

The name convexity adjustment comes from Jensen's inequality which states that for any real-valued and convex function and any random variable, the average of the function applied to the random variable is greater than the function applied to the average of the random variable:

$$E[f(X)] \geq f(E[X]). \tag{8.5}$$

8.3. Seasonality

8.3.1. *Motivations and static seasonality modeling*

Unlike interest rates, inflation forwards are not really smooth and exhibit repeatable seasonal patterns. The market provides only a little information on seasonal effects. These patterns take their roots in various recurrent economic phases like consumer spending increase during Christmas winter and/or summer sales, cyclic variations in energy and food consumption and many other periodic effects. Consequently, the assumption of a linear growth for month-on-month inflation changes is clearly misleading. This month-on month change may be far away from its monthly average for month with strong seasonality effect. This advocates energetically for the incorporation of seasonality in any inflation pricing model.

In fact, macro-economists have known for quite some time now that inflation markets should and do exhibit strong seasonality behavior Bryan-Cecchetti (1995). But when inflation derivatives were still in their infancy with only a few traded products, seasonality was not really an issue. This is not the case any more. The market has experienced tremendous growth both in terms of hedging instruments (inflation linkers issued by states, sovereigns or corporate firms) and OTC derivatives (inflation swap and other vanilla inflation derivatives). In Europe, for year 2004, an estimated €3 to 7 bn has been traded every month. This compares with only €500 million a year ago. And more is to come. In France, for instance, the total amount of regulated saving account Livret A, tied up to inflation rates should reach €400 bn at the end of year. Seasonality modeling is now not a choice but a necessity.

Incorporating seasonality in a stochastic model may at first sight look complex. One may think to include seasonality in the diffusion equation itself via its drift or volatility term or both. The model could be taken as the limit of seasonal discrete time model like SARIMA and periodic GARCH. This would come at the price of both confusing the model and making it harder to calibrate. In this paper, we suggest a simpler approach similar in the spirit to the one used for end-of-year effect in interest rate derivatives. Instead of modeling seasonality dynamically, we use a static pattern to reshape the forward curve of CPIs. Hence, it is only the forward curve that is modified while the inflation dynamics stays unchanged. In addition, we take yearly seasonality on a monthly basis. Yearly pattern is not very restrictive as most of the seasonality is on a year basis. We therefore use a vector of yearly seasonal up and down bumps $\{B(i)\}_{i=1,\dots,12}$ indexed by their corresponding months i with the convention that January equals 1.

The incorporation of seasonality in our model is then straightforward. When computing the spot value of the CPI forward value $\text{CPI}(0,T)$, maturing at time T with corresponding month m, we first look at the two adjacent liquid market points with corresponding times $T_{\rm d}$ and $T_{\rm u}$ that bound lower and upper the time T. We then compute the interpolated forward CPI value $\mathrm{CPI}(0,T)$ using the CPI market value $\mathrm{CPI}(0,T_{\rm down})$ and $\mathrm{CPI}(0,T_{\rm up})$. This interpolation may be assuming linear, stepwise constant or cubic spline interpolation on either inflation spot zero-coupon rates, forward CPI or forward zero coupon. Insights of the chapter are not on this interpolation method but rather on what follows. To the interpolated forward CPI value $\mathrm{CPI}^{\rm int}(0,T)$, we apply a seasonal bump, computed as the difference between the seasonal bump of month M and its interpolated seasonal bump using the same interpolation assumption as the one for the forward CPI. For the sake of simplicity, we will assume linear interpolation on forward CPIs. In this case, the seasonal bump for the time T and denoted by $\mathrm{SB}(T)$ is given by:

$$\mathrm{SB}(T) = B(m) - \left((m - m_{\rm d})\frac{B(m_{\rm u}) - B(m_{\rm d})}{m_{\rm u} - m_{\rm d}} + B(m_{\rm d})\right), \quad (8.6)$$

where $m_{\rm d}$ and $m_{\rm u}$ are the months corresponding to the times $T_{\rm d}$ and $T_{\rm u}$. We can notice that the (linear) interpolation reshapes our vector of yearly up and down bumps to guarantee market points to have zero-seasonality adjustment. This is because market points incorporate already seasonality in their price. In addition, the seasonality interpolation is done consistently with the one of CPI to create similar interpolation effect. Equation (8.6) assumes additive bumps, meaning that the corrected CPI is obtained as the interpolated CPI plus the seasonality bumps:

$$\mathrm{CPI}(0,T) = \mathrm{CPI}^{\rm int}(0,T) + \mathrm{SB}(T). \quad (8.7)$$

If we were to assume a multiplicative correction, we would correct the raw interpolated forward CPI as follows:

$$\mathrm{CPI}(0,T) = \mathrm{CPI}^{\rm int}(0,T)^{*}\mathrm{SB}(T) \quad (8.8)$$

with the seasonal bump computed as the monthly bump renormalized by the interpolated bump as given in Eq. (8.9). This equation is the simple translation of the interpolation Eq. (8.6) of the seasonal bump but for multiplicative bump:

$$\mathrm{SB}(T) = \frac{B(m)}{\left((m - m_{\mathrm{d}})\frac{B(m_{\mathrm{u}}) - B(m_{\mathrm{d}})}{m_{\mathrm{u}} - m_{\mathrm{d}}} + B(m_{\mathrm{d}})\right)}. \tag{8.9}$$

Once the spot value for forward CPI has been estimated, it is easy to adapt the stochastic model described in Belgrade *et al.* (1994, 2004). The forward CPI inflation index $\mathrm{CPI}(t,T)$ observed at time t that fixes at time T keeps its diffusion equation given by

$$\frac{d\mathrm{CPI}(t,T)}{\mathrm{CPI}(t,T)} = \mu(t,T)dt + \sigma_{\mathrm{Inf}}(t,T)dW(t) \tag{8.10}$$

but the boundary condition is modified to include the seasonality adjustment. This is given by Eq. (8.7) in an additive correction and by Eq. (8.8) in a multiplicative one. In this framework, it is then easy to price any inflation-linked derivatives.

We understand that the seasonality estimation comes to the finding of a vector of 12 up and down (additive or multiplicative) bumps from historical CPI data. Following standard econometric theory, we decompose our time series of CPI data into a variety of components, more or less observable and easy to discriminate. We assume that our time series of CPI data is summarized by a general trend, that may be stochastic or not and that represents the long-term evolution of the CPI, a seasonal effect corresponding to yearly fluctuations and a noise represented by a random variable. Note also that a multiplicative correction is as an additive one on the logarithm of the CPI data. We will therefore look at additive correction as the processing for multiplicative is then straightforward.

At this stage, we could decide to either one of the following:

(1) assume a parametric form for the data and do an OLS to estimate the seasonal component;
(2) determine sequentially the trend of the time series and on the de-trended data, the seasonal component.

The first method is a parametric estimation of the seasonal component that spreads the noise error on both the trend and the seasonal component. But this advantage is counterweighted by the parametric assumption used to do everything in one go.

The second method is a non-parametric method, straight application of the X11 method to our CPI data. Key advantage is to make no assumption on the trend and seasonal component. We let the filter recover the trend iteratively. However, because of this two-stage estimation, noise affects only the seasonal component and not the trend.

8.3.2. *Parametric estimation of seasonality*

The method first used by Buys-Ballot (1847) to see seasonality in astronomy assumes that our time series $(X_t)_{t=1,\ldots,T}$ (here the logarithm of CPI) has an additive decomposition schema "trend + seasonality + noise". The two first components have also parametric forms. For an annual seasonality, the model is written as follows, a time series given by:

$$X_t = T_t + S_t + \varepsilon_t, \quad t = 1, \ldots, T, \tag{8.11}$$

where the trend is modeled by a polynomial of degree p and given by:

$$T_t = \alpha_0 + \sum_{j=1}^{p} \alpha_j T_t^j = \sum_{j=0}^{p} \alpha_j t^j, \tag{8.12}$$

the annual seasonality seen as 12 average bumps and given by:

$$S_t = \sum_{i=1}^{12} B(m_i) S_t^i = \sum_{i=1}^{12} B(m_i) 1_{\{t \bmod i = 0\}}: \tag{8.13}$$

the white noise sequence is given by:

$$(\varepsilon_t)_{t=1,\ldots,T}. \tag{8.14}$$

The parameters $(\alpha_j)_{j=1,\ldots,p}$ and $(B(m_i))_{i=1,\ldots,12}$ are estimated by the OLS's best-linear unbiased estimators. The seasonal average bumps $(B(m_i))_{i=1,\ldots,12}$ constitute successively a regular cycle and they compensate themselves $\sum_{i=1}^{12} B(m_i) = 0$.

This approach presents four advantages:

(1) First, being parametric, it provides easily estimation's error for forecast.
(2) Second, it can detect not only yearly seasonality but also monthly pattern.
(3) Third, it spreads the error estimation not only on the seasonality but also on the trend estimation.
(4) Fourth, it uses the whole set of data computing in one go both the trend and the seasonality pattern.

These advantages are offset by the strong assumption on the parametric form.

Figures 8.2 and 8.3 give the estimation of the seasonality, while Figures 8.6 and 8.7 present the trend component of the European and the US CPI. We can notice less irregularity in the United States seasonality pattern than in the European one. This is consistent with empirical studies that confirm stronger seasonality of the US CPI data compared to European ones. This estimation is done with a fifth-order polynomial estimation.

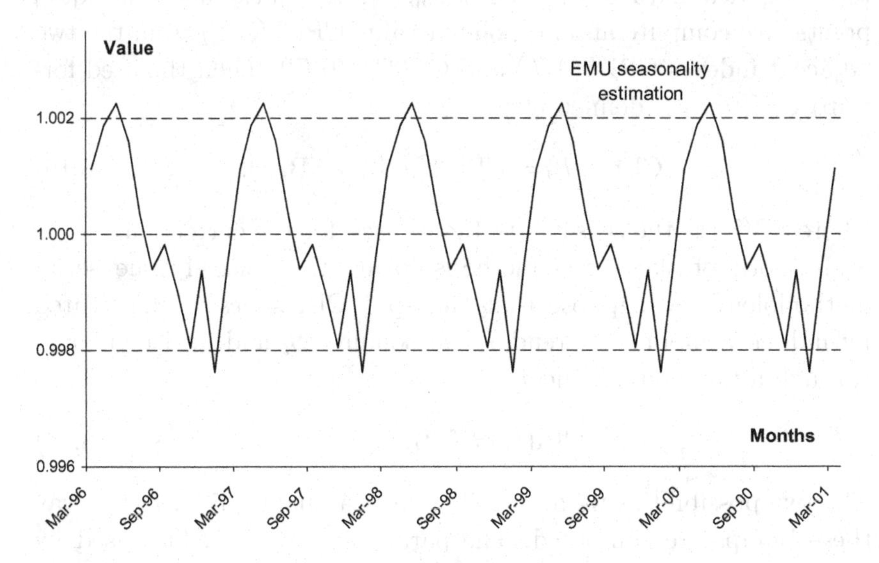

Figure 8.2: The estimation of the European and the CPI seasonality.

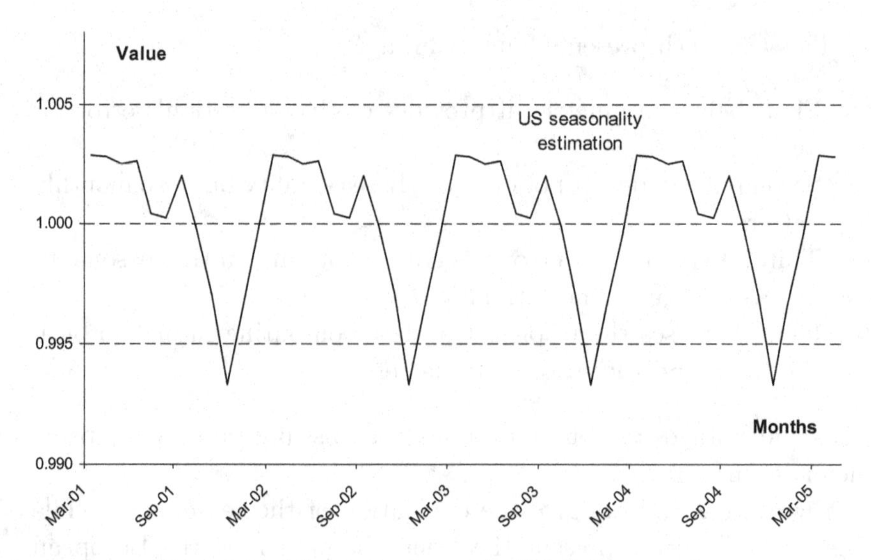

Figure 8.3: The estimation of the US CPI seasonality.

8.3.3. *Parametric vs. non-parametric*

Amongst possibilities to handle seasonality, there are two solutions: parametric and non-parametric. Both of them rely on the same pattern. For each maturity T, we find T_{up} and T_{down}, two adjacent liquid points. We compute an interpolated value $\mathrm{CPI}^{\mathrm{int}}(0,T)$ from the two adjacent indexes $\mathrm{CPI}^{\mathrm{up}}(0,T)$ and $\mathrm{CPI}^{\mathrm{down}}(0,T)$. Then the used forward $\mathrm{CPI}(0,T)$ is defined by:

$$\mathrm{CPI}(0,T) = \mathrm{CPI}^{\mathrm{int}}(0,T) + \mathrm{SB}(T), \tag{8.15}$$

where $\mathrm{SB}(T)$ is the seasonality bump used for the correction.

In both of these two methods, using the general time series methodology, we suppose that the spot CPI series is constituted by hidden components trend T_t, seasonality S_t and an irregular ε_t through a function scheme f:

$$\mathrm{CPI}(t) = f(T_t, S_t, \varepsilon_t). \tag{8.16}$$

The two possibilities we have to handle seasonality differ in the way these bumps are computed. The parametric method, first used by Buys-Ballot in astronomy, assumes that X_t, the logarithm of CPI has the following decomposition:

$$X_t = S_t + T_t + \varepsilon_t, \tag{8.17}$$

where the trend T_t is modeled by a polynomial in t, the annual seasonality S_t, seen as 12 average bumps is modeled as

$$S_t = \sum_{i=1}^{12} \mathrm{SB}(i) 1_{t \bmod i=0} \tag{8.18}$$

and ε_t is a white noise. These coefficients are computed in two steps using historical data. First T_ts coefficients are computed using least square regression, and then SBs values are computed using monthly CPI without the trend T_t.

The second method, namely the non-parametric method, works using time lag operator. The trend T_t is extracted using a moving average over of a period of thirteen data:

$$T_t = M(X_t), \tag{8.19}$$

where the operator M is given by:

$$M = \sum_{i=-6}^{6} \theta_i L^i \tag{8.20}$$

and L is the standard one month lag operator, and $\theta_i = 1/6$ for $i = -5, \ldots, 5$, $\theta_i = 1/12$ otherwise.

The seasonal component is then extracted from the series without trend:

$$S_t = (1 - M)M'(X_t - T_t) \tag{8.21}$$

with

$$M' = \sum_{i=-2}^{2} \theta_i L^i, \tag{8.22}$$

where $\theta_i = 1/9$ for $i \in \{-2, 2\}$, $\theta_i = 2/9$ for $i \in \{-1, 1\}$ and $\theta_i = 1/3$ for $i = 0$.

The seasonality component and the corresponding estimation of the trend of the European and the US CPI are given in Figs. (8.4) and (8.5).

In Figs. 8.4 and 8.5, we can remark that the "crude" US seasonal component is more regular than the European one: the amplitude being less variable. The difference of regularity between European and US inflation is even more striking than in the parametric method. The two CPIs' trend components in Figs. 8.6 and 8.7 are very close to the parametric results.

To obtain the 12 monthly bumps $(B(m_i))_{i=1,\ldots,12}$ for the seasonality adjustment, we have to regularize the seasonality component to an annual cycle. For this, we can simply apply to seasonality component the parametric model (2.1) without trend. We obtain for the European and the US CPI the best yearly approximations of the seasonality component as shown in Figs. 8.8 and 8.9.

For both the European and US the CPI, the seasonality estimation is here less important in magnitude than in the parametric estimation. This results from the fact that the approximation replicates at best the yearly effect but smoothes the pattern. Hence Figs. 8.7 and 8.1 are quite different.

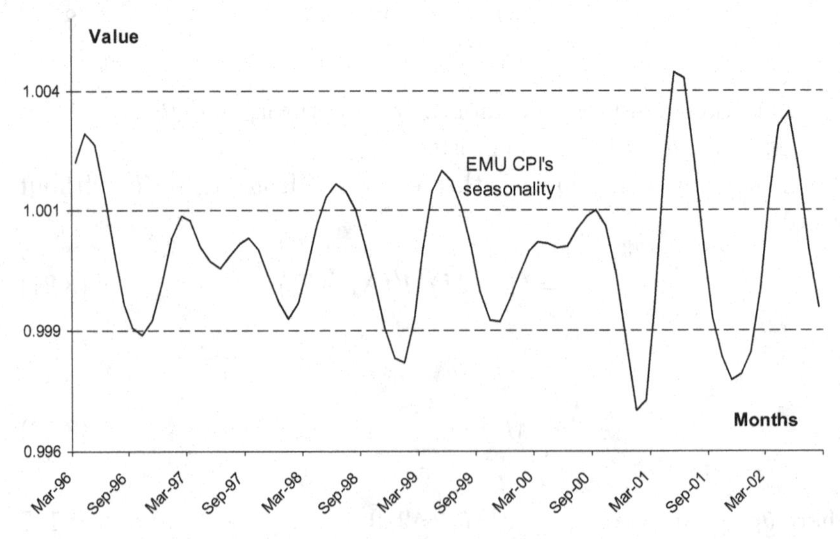

Figure 8.4: The seasonality component extracted from the European CPI data.

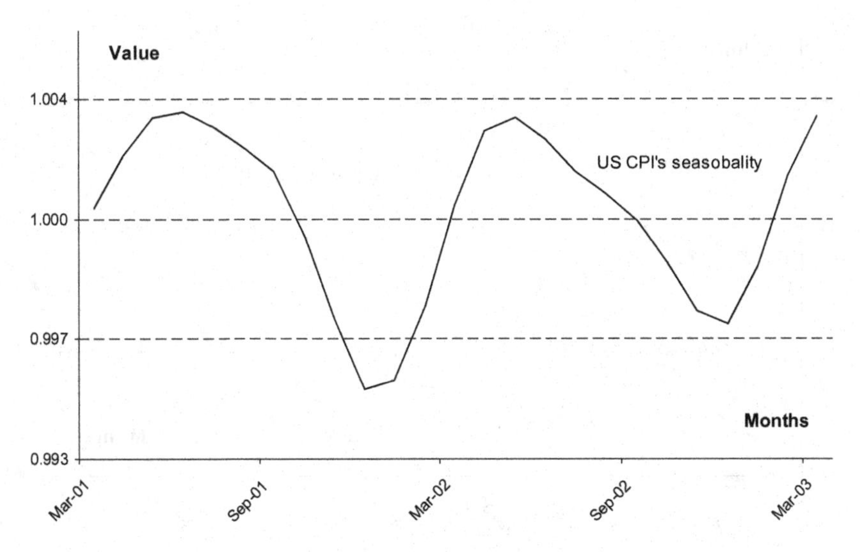

Figure 8.5: The seasonality component extracted from the US CPI data.

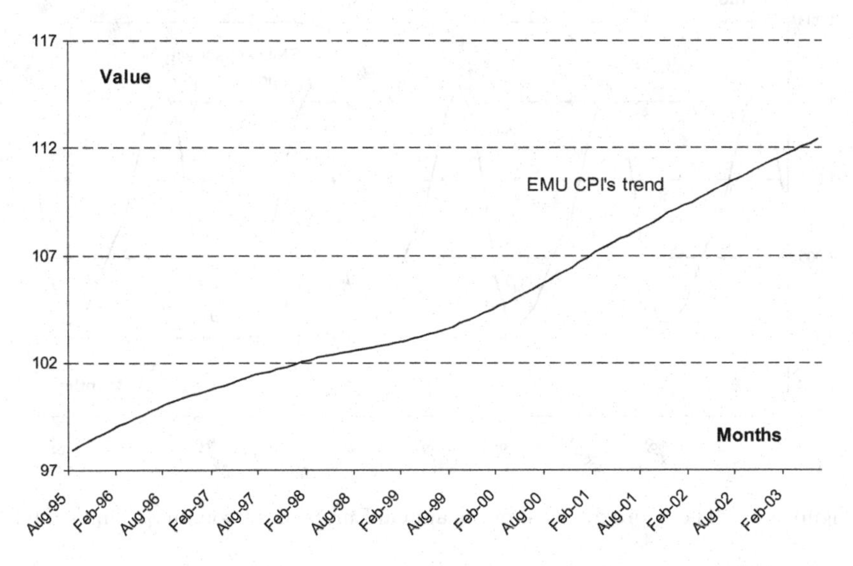

Figure 8.6: The trend component extracted from the European CPI data.

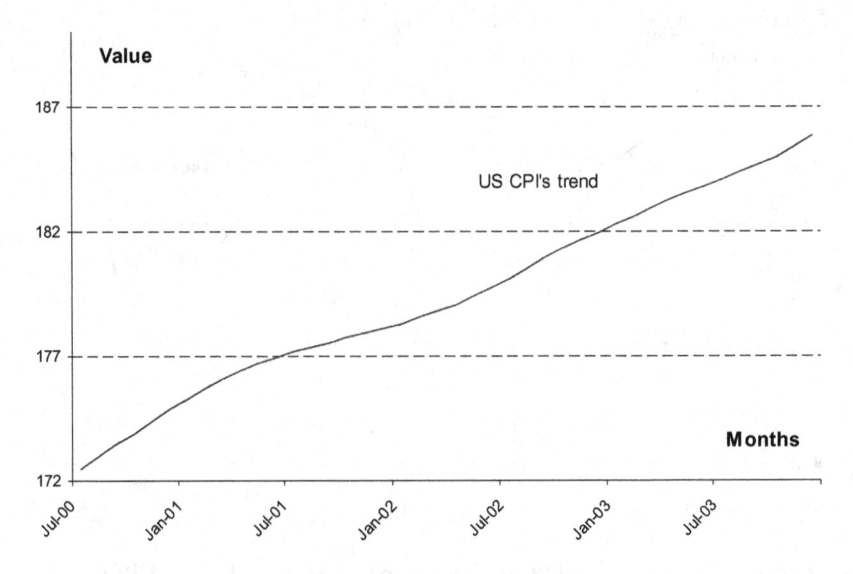

Figure 8.7: The trend component extracted from the US CPI data.

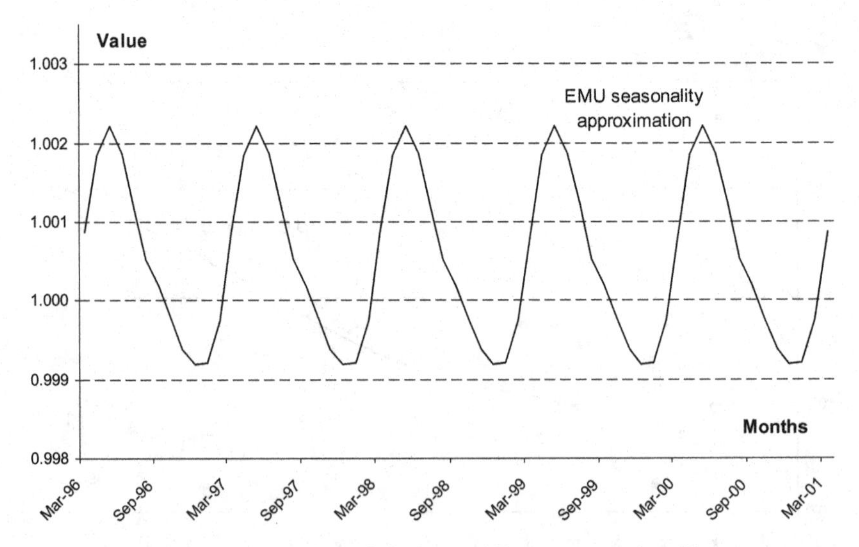

Figure 8.8: The approximation of the annual European seasonality components.

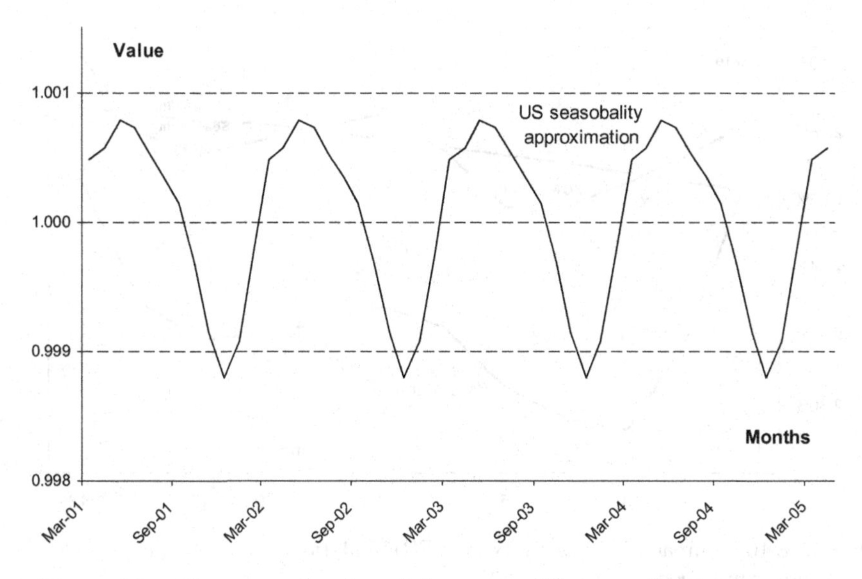

Figure 8.9: The approximation of the annual US seasonality components.

8.3.4. *Pricing impact on various inflation linked derivatives*

To measure the influence of seasonality, we price with and without various products: Zero-coupon and YoY inflation swap. Impact on more exotic structures such as options on real interest rates (and interest rates structured coupons floored with inflation-linked strikes) is similar but harder to analyze because of the increased complexity of these products. However, the same conclusions hold.

First of all, accounting for seasonality imposes a periodic cycle on the naïve price (price without seasonality). Impact varies according to the month of the fixing. For instance, the impact of seasonality on a 10-year IFRF inflation zero-coupon swap fluctuates between −1.5 and 2 bps. On a 10-year annual IFRF inflation YoY swap, this difference goes from −2 to 2 bps (Fig. 8.10).

Second, we should not ignore seasonality on options. Obviously, this changes along the lines of the options characteristics. For example, on 1% IFRF inflation floor, seasonality adjusts the price by −6 to 9 bps (Fig. 8.11).

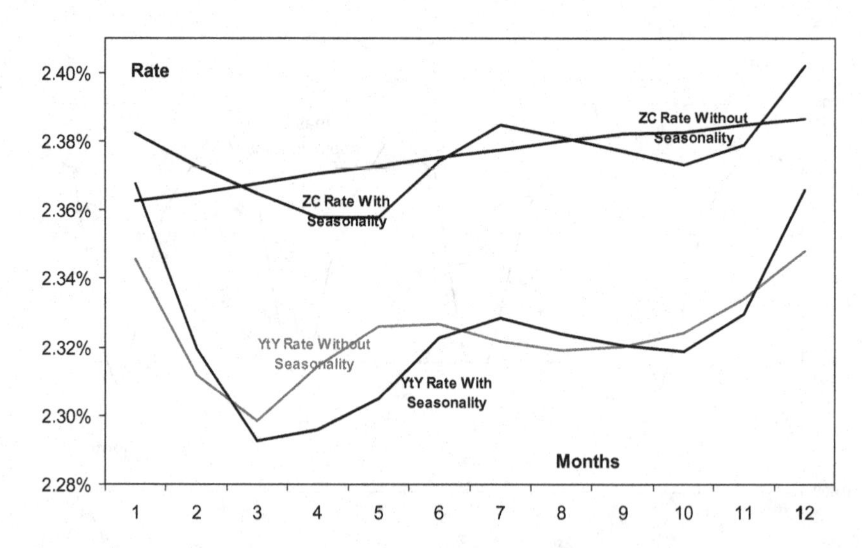

Figure 8.10: Impact of seasonality on IFRF inflation swap rate (10-year YoY and zero-coupon rates).

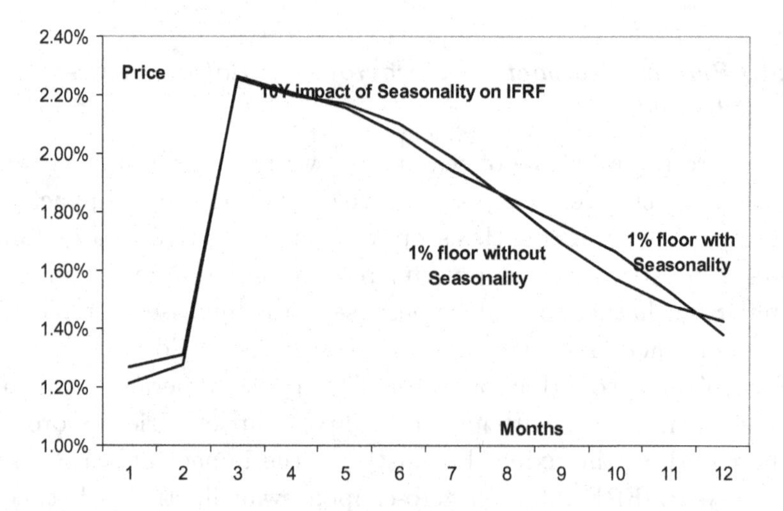

Figure 8.11: Impact of seasonality on 10-year 1% inflation floor on IFRF.

Because of the tightening of quotes on inflation-linked derivatives, model should incorporate seasonality for more accurate pricing. Using simple harmonic analysis tools, we can extract seasonality component of the inflation data and superimpose this effect to the bootstrapped inflation curve. Pricing shows that it can have a significant impact on seasonality sensitive trades.

8.4. Hybrid model

The actual tendency of the inflation market to include more exotic products than vanilla on inflation seems evident and natural. The future inflation being treated and "well" defined on the market, there is a growing demand of products on forward real yield which would be seen as the difference between the nominal forward yield and the forward inflation rate. The pricing of these new derivatives requires the definition of at least two-factor framework: the nominal rate factor and the inflation ones, from where the hybrid theory starts. In the following sections, we expose the classic and the eventual approaches models proposed to price the hybrid inflation vs. interest rate products. The various approaches taken to try to model the inflation as an exchange rate or as a complex market model.

8.4.1. *Jarrow–Yildirim model*

The Jarrow-Yildirim model was the first academic and publicly proposed model to price the inflation derivatives. The underlying idea is to consider the spot CPI as an exchange rate between the nominal and the real values of asset; the authors defines an HJM framework on both of the nominal and the real interest forward rates and deduces from non-arbitrage conditions, as for an exchange rate model, the drift of the spot CPI.

The hypotheses of the model are the followings:

$$\begin{cases} f_k(t,T) = -\dfrac{\partial B_k(t,T)}{\partial T} \\ r_k(t) = f_k(t,t) \end{cases}, \quad k \in \{n, r\}, \tag{8.23}$$

where we denote by

- n for "nominal" and r for "real",
- f_k the forward interest rate process of type k,
- B_k the zero-coupon bond process of type k,
- r_k the shorts interest rate process of type k.

We suppose also the following diffusions of each element under the historical probability and with its own factor:

$$\begin{cases} \dfrac{dB_k(t,T)}{B_k(t,T)} = r_k(t)dt + \Gamma_k(t,T)dW_k(t), \quad k \in \{n,r\}, \\ \dfrac{d\text{CPI}(t)}{\text{CPI}(t)} = \mu(t)dt + \sigma_{\text{CPI}}(t)dW_{\text{CPI}}(t). \end{cases} \tag{8.24}$$

Jarrow and Yildirim set as non-arbitrage conditions the following assets:

$$\frac{B_n(t,T)}{e^{\int_0^t r_n(s)ds}}, \quad \frac{\text{CPI}(t)B_r(t,T)}{e^{\int_0^t r_n(s)ds}}, \quad \frac{\text{CPI}(t)e^{\int_0^t r_r(s)ds}}{e^{\int_0^t r_n(s)ds}}$$

should be martingales under the risk neutral probability $\mathbf{Q}_n$ leading to the diffusions under the risk neutral probability:

$$\begin{cases} \dfrac{d\text{CPI}(t)}{\text{CPI}(t)} = (r_n - r_r)(t)dt + \sigma_{\text{CPI}}(t)dW_{\text{CPI}}(t), \\ \dfrac{dB_n(t,T)}{B_n(t,T)} = r_n(t)dt + \Gamma_n(t,T)dW_n(t), \\ \dfrac{dB_r(t,T)}{B_r(t,T)} = (r_r(t) + \rho_{r,\text{CPI}}\sigma_{\text{CPI}}(t)\Gamma_r(t,T))dt + \Gamma_r(t,T)dW_r(t), \end{cases} \tag{8.25}$$

where we denote by $d\langle W_k, W_j\rangle(t) = \rho_{k,j}dt$, $k,j \in \{n,r,\text{CPI}\}$ the correlations between the three factors. The volatilities of both the nominal and the real zero-coupon bonds, $\Gamma_k(.,.)$, $k,j \in \{n,r\}$, satisfy the usual HJM condition with those of their respective forward rates ones.

This approach allows us to have closed formula on inflation swap legs values and Black–Scholes formulas on inflation options values, due to the deterministic hypothesis on all the volatility coefficients of the model. This last being engendered by volatility terms, the closed formulas are essentially function of the volatility coefficients and the factors correlations. Under a Hull–White parametric form, Jarrow and Yildirim propose to estimate the parameters from the historical data on CPI, the nominal and the real zero-coupon bonds.

We can remark that this model represents a mixture of two economic and econometric approaches. First, remark that the instantaneous inflation is deduced empirically from the CPI data by the relation:

$$i(t) = \frac{\Delta \ln \mathrm{CPI}(t)}{\Delta t} \tag{8.26}$$

and therefore the CPI in the Jarrow–Yildirim model is defined through an instantaneous stochastic implicit inflation rate equal to the difference of the nominal and the real interest rate:

$$i(t) = (r_n - r_r)(t). \tag{8.27}$$

This equation is called the Fisher relationship used in macro-economy and concern the long-term expected dynamics, so the drift of the CPI.

And now, remark that the inflation rate process $(i(t))_t$ would follow a Hull–White model for little which the parameters a_n and a_k would be close. So, Jarrow and Yildirim defines a macro-economy dynamics for the long-term CPI dynamics and a local random as an own factor, and the whole structured in Hull–White framework.

The Jarrow–Yildirim model for inflation has several advantages: it guarantees non-arbitrages of the model through defining conditions; it is also very intuitive and easy to understand because it is very similar to interest rate models and it gives direct modeling of the interest rate yield. Finally, it represents a good way of recycling complex models developed for hybrid products such as FX vs. interest rates. However, despite all these advantages and its familiarity, it exhibits many severe drawbacks from its users. First, it does not give relation between zero-coupon and YoY swap rates. Second and most of all, it does not give a good modeling of market observable for the 3 various markets as only short-term rates are diffused.

The real interest rate part of the model seems abstract for traders and the market operators which is normal because that is not a real market. Using the exchange rate analogy, the model gives a good understanding but it is just a metaphor because the real assets seen as foreign from the nominal market do not exists!

In economy, we deduce the real interest rate from the nominal one and the inflation rate since they are observed whereas this model defines the inflation from the real interest rate.

Another important drawback is the estimation of the parameters historically, which makes the estimation instable. Indeed, the historical estimation gives approximations that depend on both the data interval and its length. Furthermore, this method of estimation is more difficult for the real factor.

For all these reasons, we should then set a framework with which we define the nominal interest rate and the inflation commonly and from which we could redefine the real yield to price the interest rate, the inflation and real yield derivatives. At first, and more important, the inflation market being not very liquid we should link the zero-coupons and the YoY products in the swap and the option market, which gives more consistency to the pricing model.

8.4.2. *Mercurio model*

In 2004, to perform the Jarrow–Yildirim framework, Mercurio kept the exchange rate analogy but supposed more a BGM for both of the nominal and the real factors. The BGM model being a special case of the HJM framework, he redefined the nominal and the real zero-coupon bonds through LIBORS following each the dynamics:

$$\begin{cases} \dfrac{dF_k(t,T_{i-1},T_i)}{F_k(t,T_{i-1},T_i)} = \sigma_{k,i} dW_i^k(t), \\ \dfrac{B_k(t,T_i)}{B_k(t,T_{i-1})} = \dfrac{1}{1+\tau_i F_k(t,T_{i-1},T_i)}, \end{cases} \quad k \in \{n,r\}, \tag{8.28}$$

where the volatilities $\sigma_{k,i}$ are deterministic and the forward period $\tau := (T_i - T_{i-1})$ Within this extended exchange rate analogy, the author introduced two different market models: the two allowing pricing swaps and the second gave more the options pricing. He argued that using these market-model approaches, would give a better understanding of the model parameters and of a more accurate calibration to market data, and this by reducing the number of the parameters. But this fact, the author avoids the possibility of negative rates and the historical estimating of the real rate parameters.

Mercurio based the valuation of the inflation swap on one result of Jarrow–Yildirim work from the non-arbitrage conditions "that the forward value of a ratio of future CPIs could be expressed only through nominal and real bonds."

$$\mathbf{E}^{\mathbf{Q}}\left[\mathrm{e}^{-\int_t^{T_i} r_n(s)ds}\,\frac{\mathrm{CPI}(T_i)}{\mathrm{CPI}(T_{i-1})}\middle|\,\mathrm{F}_t\right] = \frac{B_r(t,T_i)}{B_r(t,T_{i-1})}B_n(t,T_i). \tag{8.29}$$

Then, he defines two martingale processes to evaluate the forward value of an inflation ratio: one on the terminal probability of the nominator CPI and the second on the terminal probability of the denominator CPI. Both of the two are expressed through nominal, the real LIBORS and the correlations between them.

In the first model, the martingale process under its terminal nominal probability is the following forward CPI process:

$$\tilde{I}_i(t) := \mathrm{CPI}(t)\frac{B_r(t,T_i)}{B_n(t,T_i)}. \tag{8.30}$$

The author supposed furthermore that it is lognormal:

$$\frac{d\tilde{I}_i(t)}{\tilde{I}_i(t)} = \sigma_{I,i}dW_i^I(t). \tag{8.31}$$

Equation (8.29) becomes:

In order to compute this expectation, Mercurio used numerical integration in one dimension and volatility approximation at time $t = 0$ of both of the nominal and the real LIBOR rates because these lasts are stochastic as a BGM result. The computation contained also the instantaneous deterministic CPIs, the nominal and the real forward LIBOR's volatilities, the correlation between forward CPIs and the forward LIBORs themselves:

$$\begin{aligned}\sigma_{I,i-1} = \sigma_{I,i} + \sigma_{r,i}&\frac{\tau_i F_r(t,T_{i-1},T_i)}{1+\tau_i F_r(t,T_{i-1},T_i)}\\ -\sigma_{n,i}&\frac{\tau_i F_n(t,T_{i-1},T_i)}{1+\tau_i F_n(t,T_{i-1},T_i)}.\end{aligned} \tag{8.32}$$

The second model takes the forward CPI defined in Eq. (8.30), of the denominator but under the nominal terminal probability of the numerator. By the classical changing of the numeraire, the process is written as follows:

$$\frac{d\tilde{I}_i(t)}{\tilde{I}_i(t)} = -\sigma_{n,i}\sigma_{n,i-1}\rho_{n,I,i}\frac{\tau_i F_n(t,T_{i-1},T_i)}{1+\tau_i F_n(t,T_{i-1},T_i)}dt + \sigma_{I,i-1}dW_{i-1}^I(t). \tag{8.33}$$

The next step is straightforward, using the martingale propriety, we deduce that the forward value of the CPI ratio is equal to the ratio of the forward CPIs multiplied by the drift correction

$$\mathbf{E}^{\mathbf{Q}_{T_i}}\left[\frac{\tilde{I}_i(T_i)}{\tilde{I}_{i-1}(T_{i-1})}\middle|\mathbf{F}_t\right] = \frac{\tilde{I}_i(t)}{\tilde{I}_{i-1}(t)}\mathrm{e}^{D_i(t)} \tag{8.34}$$

with

$$D_i(t) := \sigma_{I,i-1}\left[\frac{\tau_i F_n(t,T_{i-1},T_i)}{1+\tau_i F_n(t,T_{i-1},T_i)}\rho_{I,n,i} - \rho_{I,i}\sigma_{I,i} + \sigma_{I,i-1}\right](T_{i-1}-t). \tag{8.35}$$

This convexity adjustment coefficient is deterministic at time $t = 0$ and provides close formula on the inflation swap rate or value. It includes the instantaneous volatilities of forward CPIs and their correlations, the instantaneous volatilities of nominal forward LIBOR rates, the instantaneous correlations between forward CPIs and the nominal forward LIBOR rates.

This second approach gave furthermore closed formula on YoY inflation options (and naturally for zero-coupons), where the initial conditions are respectively the forward CPIs or the forward ratio of CPIs and the implied volatility is done by:

$$V_i(t) := \sqrt{(\sigma_{I,i}^2 + \sigma_{I,i-1}^2 - 2\rho_{I,i}\sigma_{I,i}\sigma_{I,i-1})(T_{i-1}-t) + \sigma_{I,i}^2(T_i - T_{i-1})}, \tag{8.36}$$

for an option of maturity T_i on the underlying $\mathrm{CPI}(T_i)/\mathrm{CPI}(T_{i-1})$.

8.4.3. *Market model*

As opposed to the underlying modeling to describe a derivative market, we could model the most liquid of product generating this market. In fact, the inflation market trades a forward expected CPI value, seen from today, which is different then the statistical forecast used in econometrics. The inflation traded in the market is defined in a forward world specific to the market, affected but not the same as that of the market.

As for LIBORs in the BGM model described in Section 7, we can consider the processes of the forward CPI given by the zero-coupon swap market. Conceptually, we can see the CPI either as

(1) a n sampling of one single process observed at different times, or
(2) a single sampling of n different processes observed each at one time.

The main difference between the direct modeling of the forward and the spot underlying is to allow the correlations between the forwards processes. Econometrically, the taking of all forward processes correlations equal to 1 (different to 1) corresponds to a heteroscedastic process without (respectively with) autocorrelation on errors.

We can consider the CPI data fixing (bill book) at various dates $\{T_i, 1 \leq i \leq N\}$ and the family of the forward CPI: $\{\mathrm{CPI}(t, T_i), 0 \leq t \leq T_i, 1 \leq i \leq n\}$, where we denote by $\mathrm{CPI}(t, T_i)$ the forward CPI process applying for a period of one month seen at time t and fixing at time T_i. Being tradable in the market, these processes are martingales under their respective terminal probabilities and we suppose more that they are driven by a deterministic volatility structure and lognormal dynamics:

$$\frac{d\mathrm{CPI}(t, T_i)}{\mathrm{CPI}(t, T_i)} = \sigma(t, T_i) dW^i(t), \tag{8.37}$$

$(W^i(t))_{t \geq 0}$ being a standard Brownian motion under the probability Q^T. We set simultaneously a lognormal structure on the zero-coupon obligations under the Q risk neutral probability:

$$\frac{dB(t, T_i)}{B(t, T_i)} = r(t)dt + \Gamma(t, T_i) dZ(t), \tag{8.38}$$

associated to the nominal interest rate factor. The different processes are linked by the following static correlation structure:

$$
\begin{aligned}
d\langle W^i, W^j\rangle(t) &= \rho_{i,j}^{\mathrm{Inf}}\,dt,\\
d\langle W^i, Z\rangle(t) &= \rho_i^{\mathrm{Inf},B}\,dt.
\end{aligned} \tag{8.39}
$$

The framework is mainly generated by the volatility term structure $\{\sigma(t,T_i), 0 \le t \le T_i, 1 \le i \le N\}$. Many schemes of volatility can be chosen as the Hull–White or the integrated Hull–White[1] form but we use a homogeneous from

$$
\sigma(t,T_i) = f(T_i - t). \tag{8.40}
$$

This particular form allows us to reduce the degree of freedom because of temporal stationary and to propagate the volatility structure in the future keeping the forward structure.

Within this framework, the YoY swap value is calculated by closed formulas reducing the forward inflation cash flows as the ratio of forward CPIs corrected from convexity effect. A forward CPI ratio is equal for $0 \le t \le T_j \le T_i$:

$$
\mathbf{E}^{\mathbf{Q}_{T_i}}\left[\left.\frac{\mathrm{CPI}(T_i,T_i)}{\mathrm{CPI}(T_j,T_j)}\right|\mathbf{F}_t\right] = \frac{\mathrm{CPI}(t,T_i)}{\mathrm{CPI}(t,T_j)}\lambda(t,T_j,T_i), \tag{8.41}
$$

to the ratio of forward CPIs multiplied by $\lambda(t,T_j,T_i)$ the inflation convexity adjustment at time t between T_j and T_i equal to:

$$
\lambda(t,T_j,T_i) = \mathrm{e}^{\int_0^{T_j}\sigma(s,T_j)\left(\rho_{i,j}^{\mathrm{Inf}}\sigma(s,T_i)-\sigma(s,T_j)+\rho_j^{\mathrm{Inf},B}(\Gamma(s,T_i)-\Gamma(s,T_j))\right)ds}, \tag{8.42}
$$

including the covariance between the forward CPI, $\mathrm{CPI}(t,T_j)$, and the forward CPI ratio $\mathrm{CPI}(t,T_i)/\mathrm{CPI}(t,T_j)$ and the covariance the forward CPI, $\mathrm{CPI}(t,T_j)$, with the forward zero-coupon bond $B(t,T_j,T_i) = B(t,T_i)/B(t,T_j)$.

We link thus the zero-coupon and the YoY markets through the inflation-bond volatility-correlation structure. The estimation volatility of the zero-coupon bond $\Gamma(.,.)$ is model free.

[1]For example, where $\sigma(t,T_i)$ can be $\sigma\mathrm{e}^{-\lambda(T_i-t)}$ or $\sigma\frac{1-\mathrm{e}^{-\lambda(T_i-t)}}{\lambda}$.

Another important target of the market model was to "connect" the inflation zero-coupon and YoY option markets by filling up the volatility matrix. Indeed, assuming the log normality of the forward CPIs and the determinism of theirs volatilities, we can then implied volatilities of a YoY option and of two zero-coupon options by the correlations between theirs underlying. This is defined as the "volatility cube" by Belgrade *et al.* (2004).

Let us denote by $\psi_K(t, \delta_i, T_i)$ the implied Black–Scholes volatility seen at time t of the inflation option with exercise date T_i, of tenor δ_i for a given strike K. Typically, the YoY volatility would be denoted by $\psi_K(t, T_i - T_{i-1}, T_i)$ and the zero-coupon volatility by $\psi_K(t, T_i, T_i)$ and we will have always in the general case the relation:

$$\psi_K(t, \delta_i, T_i) = \sqrt{\frac{1}{\delta_i - t} \int_t^{T_i} \sigma^2(s, T_i) ds}. \tag{8.43}$$

Therefore, we can deduce easily that:

$$\begin{aligned}
&(T_i - T_{i-1} - t)\psi_K^2(t, T_i - T_{i-1}, T_i) \\
&\quad = (T_{i-1} - t)\psi_K^2(t, T_{i-1}, T_{i-1}) \\
&\qquad + (T_i - t)\psi_K^2(t, T_i, T_i) \\
&\qquad - 2\rho_{i-1,i}^{\text{Inf}} \chi(t, T_{i-1}, T_i),
\end{aligned} \tag{8.44}$$

where $\chi(t, T_{i-1}, T_i) := \int_t^{T_{i-1}} \sigma(s, T_{i-1})\sigma(s, T_{i-1}) ds$ is the integrated covariance between the two forward CPIs: $\text{CPI}(t, T_i)$ and $\text{CPI}(t, T_{i-1})$. The χ term is all dependent on the volatility from $\sigma(.,.)$. For the two difficulties remain, the reconstruction of the covariance χ and the choice of the implied CPIs correlations ρ^{Inf}, because the market is not very liquid for all strikes (only for some points of -2% and 0%). The traders construct the zero-coupon volatility or the YoY ones from the other by the relation (8.44) strike by strike but use the CPIs correlations implied for 0%.

The homogeneous hypothesis makes the computation of χ easier, reducing the degree of freedom; but more important, it can provide additional level information on the market data. First, we can provide a minimum level for the YoY volatility equal to the zero-coupon volatility maturing at the tenor of the YoY option:

$$\psi_K(t, T_{i-1}, T_i) \geq \psi_K(t, T_i - T_{i-1}, T_i - T_{i-1}). \tag{8.45}$$

Second, the relation (8.44) contains three market data namely the zero-coupon and YoY volatilities and implied correlation, the whole is consolidated by the χ term. Bounding this last between two bounds:

$$\chi(t, T_{i-1}, T_i) \in [\chi_1, \chi_2], \tag{8.46}$$

where

$$\begin{aligned}\chi_1 &= (T_i - t)\psi_K^2(t, T_i, T_i) \\ &\quad - (T_i - T_{i-1} - t)\psi_K^2(t, T_i - T_{i-1}, T_i - T_{i-1}), \end{aligned} \tag{8.47}$$

$$\chi_2 = (T_{i-1} - t)\psi_K^2(t, T_{i-1}, T_{i-1}), \tag{8.48}$$

we can provide using the homogenous property of the volatility function, confidence intervals for each of the zero-coupon and YoY volatilities and implied correlation by fixing the two others:

$$\text{for} \quad \kappa \in \{\psi_K(t, T_i - T_{i-1}, T_i - T_{i-1}), \psi_K(t, T_i, T_i), \rho_{i-1,i}^{\text{Inf}}\}, \quad \kappa \in [\kappa_1, \kappa_2] \tag{8.49}$$

where κ_1 and κ_2 are functions of the other parameters.

For the zero-coupon volatility $\kappa = \psi_K(t, T_i, T_i)$ if $\rho_{i-1,i}^{\text{Inf}} \geq 1/2$, then:

$$\kappa \in]-\infty, \kappa_1] \cup [\kappa_2, +\infty[. \tag{8.50}$$

Finally, in another paper, Benhamou and Belgrade have used this framework to approximate the YoY swap's volatility and thus the pricing of inflation and real yield swaptions. As an analogy of the Vol Swap–Vol FRA relation for the interest rates, we can give the diffusion of a YoY swap maturity at time T_n seen at time $t \leq T_0$ year and stochastic volatility coefficient:

$$\frac{dS_n(t)}{S_n(t)} = \Sigma_n(t)dB(t), \tag{8.51}$$

where $B(t)$ is a Brownian motion under the terminal probability of the swap, the numeraire $\sum_{i=1}^{n} B(t, T_i)$ and $\Sigma_n(t)$ being function of

$\{\mathrm{CPI}(t,T_i), B(t,T_i), \sigma(t,T_i), \Gamma(t,T_i), 1 \le i \le n\}$. This volatility can be estimated by its value at time $t = 0$ and expressed in its normal form:

$$\Sigma_n(0) = \sqrt{\sum_{i,j=1}^{n} A_i A_i \xi_{i,j}^{A,B} + \sum_{i,j=1}^{n} B_i B_i \xi_{i,j}^{B,B} - 2\sum_{i,j=1}^{n} A_i B_i \xi_{i,j}^{A,B}}, \tag{8.52}$$

where the terms A_i and B_j are functions of $\{\mathrm{CPI}(0,T_i), B(0,T_i), \sigma(0,T_i), \Gamma(0,T_i)\}$, $\xi_{i,j}^{A,B}, \xi_{i,j}^{B,B}$, and $\xi^{A,B}$ are respectively the average correlations between $\{\mathrm{CPI}(t,T_i), \mathrm{CPI}(t,T_j)\}$, $\{\mathrm{CPI}(t,T_i), B(t,T_j)\}$ and $\{B(t,T_i), B(t,T_j)\}$. Many other forms of deterministic approximations of the YoY swap's volatility have been proposed in the paper but gave close results, as for Monte Carlo simulations. By this fact, we can give Black–Scholes formulas for the inflation swaptions pricing and weak numerical integration to compute real yield swaptions expressed as a spread option between the forward nominal swap rate and the YoY swap rate of a same maturity.

This model has many advantages. First, this approach is more intuitive for the traders and the market operators because, as for the BGM, it targets directly observed variables which allows linking them with correlation structure and no arbitrage relations. Second, it gives explicit formulas on the swap and the option values, from the forward CPIs and the volatility market data allowing link between zero-coupon and the YoY markets. Under weak but reasonable hypothesis on the volatility function, we could make coherence conditions between the market data and the model's parameters giving more consistency for the general framework. Finally, this model allows us giving first usual pricing of forward structures on inflation as swaptions.

However, due to its multidimensional nature, this approach, without specification of the volatility form, has too many degrees of freedom, through the intra-inflation and the inflation-interest rate correlation structures. This drawback is accented for the pricing of forward real yield structures for maturities making the procedure slow.

8.4.4. *Index equity model*

Pricing inflation derivatives with the Jarrow–Yildirim framework requires knowledge of the real yield. As explained above, this term structure is not accessible. A way to avoid this problem is to build a new model with a postulated relationship between real and nominal interest rates. An affine link gives

$$r_r(t) = ar_n(t) + b(t), \tag{8.53}$$

where the parameter a and the function b are deterministic. This relationship involves

$$i(t) = (r_n - r_r)(t) = (1 - a)r_n(t) - b(t). \tag{8.54}$$

The above equation implies

$$\begin{cases} \dfrac{d\mathrm{CPI}(t)}{\mathrm{CPI}(t)} = \alpha(r_n(t) - q(t))dt + \sigma_{\mathrm{CPI}}(t)dW_{\mathrm{CPI}}(t), \\ \dfrac{dB_n(t,T)}{B_n(t,T)} = r_n(t)dt + \Gamma(t,T)dW_n(t), \end{cases} \tag{8.55}$$

where the short rate without indices stand for nominal instantaneous rates, and

$$q(t) = \frac{b(t)}{1-a}, \tag{8.56}$$

$$\alpha = 1 - a, \tag{8.57}$$

Let us remark that $a = \mathrm{Cor}(r_n, r_r)(t)$ the instantaneous correlation between the two factors thanks to the linearity of the relation and by consequence a >0 because the real and the nominal rate involve in the same sense and surely thus $\alpha > 0$.

The dividend curve $q(t)$ is used to fit the forward CPI curve. Interest rates parameters are calibrated so as to fit market prices of interest rates products such as caps and swaptions; parameters of the inflation part of the model ($\sigma_{\mathrm{CPI}}(t)$ and α) are computed to match inflation product prices such as YoY swaps and swaptions.

In order to allow faster computation of inflation products prices, it is clever to choose a simple model for the interest rates part. A good candidate is the one factor Hull–White model. It is a matter of fact that with such a choice, the model is not smiled on YoY caps. This is due to the fact that $r(t)$ is a Gaussian process, and hence CPI(t) and CPI ratios are lognormal processes and produce un-smiled prices on caplets.

8.5. Summary

In this chapter, we first looked at vanilla products concerning inflation. There are two main categories: swaps (like YoY, zero-coupon swaps or bonds) and options (cap and floors, swaptions). Then, we have seen the pricing of these products. One important issue is to evaluate the YoY inflation swap rate. It can only be done by determining the forwards value of annual inflation cash flows. Inflation forwards exhibit seasonal patterns and is linked with economic phases, cyclic variations, energy price and food consumption. There are two types of estimation: the parametric and the non-parametric, both of them present advantages and drawbacks. The new tendency of markets to include more and more exotic products required the introduction of new hybrid models. So, we finally exposed the pricing of hybrid inflation vs. interest rates products. Jarrow–Yildirim's model was the first one; it is easy to understand and guarantees non-arbitrages but presents several drawbacks (historical estimations, no relation between zero-coupon and YoY swap rates). In this way, Mercurio's model introduces two different market models. Market model connects the inflation zero-coupon and YoY option markets. Thus it provides confidence intervals for zero-coupon volatility, YoY volatilities and implied correlation whereas index equity model is based on the relationship between real and nominal interest rates.

References

[1] Belgrade N and Benhamou E (2004). Impact of seasonality in inflation derivatives pricing. Ixis Capital Market Research Note.

[2] Belgrade N, Benhamou E and Koehler E (2004). A market model for inflation. CERMSEM Working Paper 2004.50 and SSRN Working Paper.

[3] Belgrade N and Benhamou E (2004). Reconciling year on year and zero coupon inflation swap: a market model approach. Ixis Capital Market Research Note.

[4] Belgrade N, Benhamou E and Schauly A (2005). Modeling inflation equity as an equity index. Ixis Capital Market Research Note.

[5] Belgrade N (2006). Modélisation de l'inflation dans la finance de marché. PhD Thesis, University of Paris.

[6] Benhamou E (2000). Pricing convexity adjustment with Wiener Chaos, London School of Economics Working Paper, FMG Dp351.

[7] Brondolo A (2005). Inflation Bonds Explained: Mastering Markets & Mechanics. Fixed Income Strategies Lehman Brothers.

[8] Brondolo A and Giani G (2005). Seasonality in price indices. Fixed Income Strategies Lehman Brothers.

[9] Bryan-Cecchetti (1995). The seasonality of consumer prices. NBER Working Paper No.W5173.

[10] Buys-Ballot C (1847). Les changements périodiques de températures, Utrecht, Kemink et Fils.

[11] Deacon M, Derry A and Mirefendereski D (2003). *Inflation-Indexed Securities*. Wiley Finance.

[12] Jarrow R and Yildirim Y (2003). Pricing treasury inflation protected securities and related derivatives using an HJM model. *Journal of Financial and Quantitative Analysis*, **2**, 337–358.

[13] Kerkho J (2005). Inflation Derivatives Explained. Lehman Brothers Fixed Income Quantitative Research.

[14] Mercurio F (2004). Pricing inflation-indexed derivatives. *Quantitative Finance*, **5**(3), 289–302.

Chapter 9

HYBRID MODELS

Over the years, complex structures have had a fast growth both in types and volumes. Nowadays single currency and single underlying products are considered as standard flow products with many competitors. As these vanilla products give small margins and investors are rather interested by high-rewarding products, hybrid products can be sold more easily to final customers. On the other hand the management of these complex and risky structures lead to very attractive margins for financial institution that summarize all the interest for them. But this management is winning if models used are able to tackle these new risks. In short, this consists in breaking down the complex risk project into more standard ones whose risk management is more accurately done.

In this chapter, some products that mix risks on different currencies and/or different markets are presented. All of them have an interest rate or forex rate component. These examples, obviously non-exhaustive, are above all a pretext to describe models to manage these new classes of risk.

9.1. Basis Hybrid

To speak about the basis of hybrid is quite difficult because generally in an hybrid product there are at least two economies. Cash flows are then discounted using basis curves to take into account the relative illiquidity between both currency flows. This effect is always in mind for every hybrid multi-currencies product valuation. However, in this section we will deal with exotic products whose underlying is based on the basis itself.

9.1.1. *Bermuda option*

In Chapter 5, vanilla basis swaps were described. Option on this type of swaps may appear in cancelable exotic structure.

The simplest product with such a feature is a bermuda swaption where the funding leg is expressed in one currency and the fixed one in another currency. The general convention for a cancelable structure is a nominal conversion decided at the beginning of deal. At each notification date, both legs and residual option are expressed in a single currency to be compared with each other.

Suppose that the funding is USD and the fixed rate pays JPY. Nominals are exchanged at each period. The present value is expressed in the domestic currency JPY. At the last exercise date T_N, for example, the option is worth:

$$(\text{Funding}^{\text{USD}}(T_N) \cdot S^{\text{JPY/USD}}(T_N) - \text{Fixed}^{\text{JPY}}(T_N))_+ \qquad (9.1)$$

The reference curve for basis margin is USD (i.e. an USD funding leg with notional exchange in a basis swap is always worth 0). Then an implicit FX option will appear if funding spread is not 0.

The PV at T_N could be rewritten as a function of the fixed JPY leg and the funding spread leg in USD converted at the payment date T_{N+1} using the forward FX rate:

$$F^{\text{JPY/USD}}(T_N, T_{N+1}) = S^{\text{JPY/USD}}(T_N)\frac{B^{\text{USD}}(T_N, T_{N+1})}{B^{\text{JPY}}(T_N, T_{N+1})}. \qquad (9.2)$$

In this expression the domestic JPY curve will be corrected to take into account basis spread against USD.

9.1.2. *Model*

9.1.2.1. *Forward forex rate*

As seen in the above section the FX risk exists and must be modeled. But it is related to the spread funding leg. The model uses the dynamics of forward FX rate and a yield curve rather than both yield curves and the spot FX rate. This approach leads to a 2-factors model quite simple to implement, where forward FX rate dymamics mixed spot FX, both interest rates and basis spread future evolutions.

The domestic yield curve, in our example the JPY one, will be modeled by a Hull–White (HW) model. Forward FX rates will follow a pure log-normal process. Under the risk neutral domestic probability, the diffusion equations is:

$$dX_r(t) = -\lambda_X X_r(t)dt + \sigma_r(t)dW_r, \tag{9.3}$$

$$r(t) = f(0,t) + \theta_r(t) + X_r(t), \tag{9.4}$$

$$\frac{dS(t,T)}{S(t,T)} = -\sigma_T(t)\Gamma(t,T)dt + \sigma_T(t)dW_S, \tag{9.5}$$

where

$$\Gamma(t,T) = -\sigma_X(t)\frac{1 - \mathrm{e}^{-\lambda_X(T-t)}}{\lambda_X} \tag{9.6}$$

is the HW zero-coupon bond volatility. The short rate deterministic drift correction $\theta_r(t)$ will be calibrated to fulfil the arbitrage free relation

$$B(0,t) = E_0^Q\left[\mathrm{e}^{-\int_0^t r(s)ds}\right]. \tag{9.7}$$

All forward FX rate must depend on only one stochastic process. So the dependency in time and maturity of the volatility is split:

$$\sigma_T(t) = a(t)b(T). \tag{9.8}$$

A classical decomposition is based on independence between the local volatility and the mean-reverting term:

$$\sigma_T(t) = \sigma_S(t)\mathrm{e}^{-\lambda_S(T-t)}. \tag{9.9}$$

Another form

$$\sigma_S(T)\mathrm{e}^{-\lambda_S(T-t)} \tag{9.10}$$

could also be used to simplify the forward FX volatility calibration process, but this will imply that local volatility is not any more time dependent.

The forward FX rate is then led by a mean reverting process $X_S(t)$ such that

$$dX_S = -\lambda_S X_S(t)dt + \sigma_S(t)dW_S \tag{9.11}$$

through an exponential mapping:

$$S(t,T) = S(0,T)\mathrm{e}^{-\int_0^t \sigma_T(s)\left(\Gamma(s,T)+\frac{\sigma_T(s)}{2}\right)ds + \mathrm{e}^{-\lambda_S(T-t)}X_S(t)} \tag{9.12}$$

The short-rate volatility $\sigma_X(t)$ is calibrated on a set of domestic European swaptions.

If it is chosen piecewise constant, an efficient bootstrap calibration will be used. The forward FX volatility $\sigma_S(t)$could be directly calibrated on market implicit FX option volatility $\bar{\sigma}_T$ because both are linked by

$$\bar{\sigma}_T = \frac{1}{T}\sqrt{\int_0^T \sigma_S^2(s)\mathrm{e}^{-2\lambda_S(T-s)}ds}. \tag{9.13}$$

So another bootstrap calibration process will extract $\sigma_S(t)$ expiry by expiry from $\bar{\sigma}_T$. For both interest rate and forward FX rate, mean reversions will be considered as an input.

A 2D lattice is built to sampled both correlated mean reverting process (X_r, X_S). This point will be described in details in section dealing with hybrid forex products.

Remark: Using a diffusion scheme under risk neutral probability, imposes a drift correction, the $\theta(t)$ calibration process, to ensure a sampled discounting process consistent with arbitrage free relation. In the presented model, zero-coupon bond volatility is deterministic because of HW model. An interesting alternative to the previous diffusion probability could be the forward neutral one. If its associated numeraire is the zero-coupon price $B(t,T^*)$ for a given time T^* (any time posterior to the last notification date is possible), the short rate does not appear in the discounting process and no more in the price evaluation. On the other hand both bond formula and forward FX rate are corrected by a convexity adjustment to take into account the probability change

$$\int_0^t \sigma_X(s)\Gamma(s,T^*)ds \text{ or } \int_0^t \sigma_S(s)\Gamma(s,T^*)ds \text{ on } (X_r, X_S), \quad (9.14)$$

respectively due to Girsanov theorem.

9.1.2.2. *Basis margin*

For markets where basis margin changes quite often as USD/JPY does, it could be interesting to access the specific risk on basis spread volatility. The model is then enhanced to diffuse two interest rate curves: a reference/fixing one to reset LIBOR JPY $B(t,T)$, and a spreaded/discount curve to pay JPY flows $\tilde{B}(t,T)$. An instantaneous basis margin $\alpha(t)$ is introduced to explain that the risk free return on shifted zero-coupon bond is no more $r(t)$ but

$$\tilde{r}(t) = r(t) - \alpha(t). \quad (9.15)$$

The shifted zero-coupon bond then writes:

$$\tilde{B}(t,T) = E_t^Q\left[e^{-\int_t^T r(s)ds} e^{\int_t^T \alpha(s)ds}\right] = B(t,T)E_t^{Q_T}\left[e^{\int_t^T \alpha(s)ds}\right] \quad (9.16)$$

introducing the forward neutral probability Q_T at time T.

The basis margin will follow under risk neutral probability a mean reverting process:

$$dX_\alpha = -\lambda_\alpha X_\alpha(t)dt + \sigma_\alpha(t)dW_\sigma, \quad (9.17)$$

$$\alpha(t) = \theta_\alpha(t) + X_\alpha(t). \quad (9.18)$$

The deterministic drift correction $\theta_\alpha(t)$ will be calibrated to fulfill the spot basis curve $\tilde{B}(0,t)$.

At this stage, one could use a 3D lattice to sample the hybrid model IR, FX and basis. But the impact of the basis margin volatility risk is supposed to be quite weak. So to avoid such a complicated implantation, some assumptions are made to simplify the framework and keep the initial 2D lattice:

(1) the volatility ratio between basis margin and reference yield curve is constant: $\sigma_\alpha(t) = k\sigma_r(t)$,
(2) mean reversions are identical: $\lambda_\alpha(t) = \lambda_r(t)$,

(3) Brownian motions are perfectly correlated: $\langle dW_r dW_\alpha \rangle = dt$.

The basis margin then rewrites

$$\alpha(t) = \theta_\alpha(t) + kX_r(t). \tag{9.19}$$

9.2. Forex Hybrids

9.2.1. *Market, products and models*

One of the most popular hybrid products on fixed income appears on JPY market in the last years. The so called power reverse dual currency (PRDC) note with callable or knock-out feature was estimated in June 2003 to more than US$ 50 bn. The typical underlying is a swap where a funding leg in USD is exchanged against an USD/JPY option strip.

Power reverse dual currency have been by far the most actively FX hybrid product traded because of its ability to offer very high coupons. This is achieved because of the very steep downward sloping FX forward curve (coming from the huge interest rates differential) and for a minor portion by the callability pick up. The first effect is stressed even more by the long dated structure, as power reverse dual standard maturity is around 20–30 years.

A typical structure will be the following: From year 1–30, the investor pays 3 month dollar LIBOR minus 50 basis points denominated in Yen quaterly on actual/360 day count convention.

In year 1, he receives a fixed coupon of 5%. Afterward, he receives 14.50%*(FX/104.5) – 10%, capped at 5% and floored at 0.20% annually, on the 30/360 day count convention, where FX is the FX spot 10 business day prior to each payment date, using JPNU ticker at 15:00 pm (International time). In addition, the bank has the option to cancel the swap from year 3 onwards, with 10 business days prior notice.

Without the call option from the bank, a PRDC is basically a strip of FX option. The cap and floor can be easily prices using call spread with a good modeling of the FX forward smile. Because of the strong FX smile, it is advisable to use a model to interpolate, extrapolate FX market points, given as 10 and 20% delta straddle and butterfly in addition to the FX ATM forward strike, using or not the option premium.

The callability introduces some additional risk that one needs to price accurately, with a model consistent with the pricing of the underlying. Because the underlying has to be priced consistently with the smile, so does the cancellation option. Another version is the power reverse dual knock out with a discrete USD/JPY knock clause such as the deal vanishes if the FX goes above a certain level.

Modeling of the PRDC has been a major challenge over the last few years because of the complexity of the product. Commonly, banks are using at least 3-factors models to model the two interest rates and the spot FX, with additional deterministic curve to handle the basis on the two interest rates. As a matter of fact, the most simple model to value PRDC is nowadays a 2-factor HW model for the two interest rates and a BS model for the FX. Tackling the smile can be done in a crude way by adjusting the cancellation option proportionally to the smile effect observed on the strip of FX options. Interpolating the FX smile can be done with a Heston model to account for the stochastic volatility nature of the FX market.

Another more model consuming approach is really to include FX smile in the hybrid model. Taking a non-lognormal diffusion such as CEV on the FX or shifted lognormal or with a mapping function can be done easily in a tree. The only constraint lies in fact in the fast calibration of the model. Some approximations can be derived (we are aware of various proprietary version of it) and are closed in the spirit to the ones found in Piterbarg (2005). It mainly targets to find a closed form solution for the volatility of the forward FX knowing the two interest rates and the spot FX diffusion.

Another important FX hybrid traded in Japanese market is FX-TARN. This product has similar coupon as PRDC. In the first year the coupon is 8.0%. For the year 2 up to year 30, it pays Max((SpotFX − 112.0) × 1.0%, 0.0%) coupon.

- *Early redemption condition*: On each respective coupon payment date, if the sum of coupons paid before reached the target coupon, the note is terminated before maturity.

- *Target coupon*: 16.0% per principal amount.
- *Principal redemption*: 12,500,000 $ * Redemption FX rate.
- *Redemption FX rate*: 5 business days prior to maturity.

9.2.2. *Short rates model*

In this section, we describe a 3-factor hybrid model based on short rates for yield curve modeling and a spot FX.

9.2.2.1. *Theoretical model*

The valuation of forex and interest rate hybrid product needs an accurate model for both foreign and domestic interest curves and the forex rate. To have a fully tractable model on each interest curve, the simplest and most efficient way is to use a HW model. Short rate evolution is led by a simple mean-reverting process (X_{d} or X_{f}) added to the forward instantaneous rate ($f_{\mathrm{d}}(0,t)$ or $f_{\mathrm{f}}(0,t)$). The spot forex dynamics are a pure lognormal process led by another normal process (X_S) and the forward FX rate ($S(0,t)$).

If r_{d} and r_{f} are the domestic and foreign short rate and S denotes the spot FX rate, under the domestic risk neutral probability, model equations are:

$$\begin{aligned}
dX_{\mathrm{d}}(t) &= -\lambda_{\mathrm{d}} X_{\mathrm{d}}(t)dt + \sigma_{\mathrm{d}}(t)dW_{\mathrm{d}}, \\
dX_{\mathrm{f}}(t) &= -(\sigma_{\mathrm{f}}(t)\sigma_S(t)\rho_{\mathrm{f},S}(t) + \lambda_{\mathrm{f}} X_{\mathrm{f}}(t))dt + \sigma_{\mathrm{f}}(t)dW_{\mathrm{f}}, \\
dX_S &= \left(r_{\mathrm{d}}(t) - r_{\mathrm{f}}(t) - \frac{1}{2}\sigma_S^2(t) \right) dt + \sigma_S(t)dW, \\
\langle dW_{\mathrm{d}}, dW_{\mathrm{f}} \rangle &= \rho_{\mathrm{d,f}}(t)dt, \quad \langle dW_{\mathrm{d}}, dW_S \rangle = \rho_{\mathrm{d},S}(t)dt, \\
& \langle dW_{\mathrm{f}}, dW_S \rangle = \rho_{\mathrm{f},S}(t)dt
\end{aligned} \tag{9.20}$$

linked to the real world variables :

$$\begin{aligned}
r_{\mathrm{d}} &= f_{\mathrm{d}}(0,t) + \theta_{\mathrm{d}}(t) + X_d(t), \\
r_{\mathrm{f}} &= f_{\mathrm{f}}(0,t) + \theta_{\mathrm{f}}(t) + X_f(t), \\
S(t) &= S(0,t)\mathrm{e}^{\theta_S(t) + X_S(t)},
\end{aligned} \tag{9.21}$$

where θ_{d}, θ_{f}, θ_S are deterministic drifts to fulfil arbitrage free condition on respectively domestic, foreign and FX curves. They will be calibrated as described in implementation details.

The diffusion takes place in a 3D space. The most popular and efficient numerical method to handle such a space is tree/lattice one. Each state of a slice in the future is defined by the state variables $(X_{\mathrm{d}}, X_{\mathrm{f}}, X_S)$.

9.2.2.2. *Underlying future price*

At each notification date, underlying flows that is to say funding leg including spread and a strip of European FX options, must be valuated. The future price at time t of domestic and foreign zero-coupon bonds is computed using the well-known HW linear exponential formula (we drop d or f subscripts for clarity):

$$B(t,T) = \frac{B(0,T)}{B(0,t)} \mathrm{e}^{A(t,T)} \mathrm{e}^{-\beta(t,T)X(t)} \tag{9.22}$$

with the decay factor

$$\beta(t,T) = \int_t^T \mathrm{e}^{-\lambda(s-t)} ds = \frac{1 - \mathrm{e}^{-\lambda(T-t)}}{\lambda} \tag{9.23}$$

and the deterministic drift correction:

$$A_{\mathrm{d,f}}(t,T) = -\frac{1}{2} \int_0^t \sigma^2(s)(\beta^2(s,T) - \beta^2(s,t))ds. \tag{9.24}$$

So any linear combination of zero-coupon bond as standard LIBOR or swap flows are easily valuated.

In order to get an FX option price by using the standard Black formula, one needs to get the forward forex rate and its volatility from reset date to the expiry. The arbitrage free relation gives the forward FX:

$$S(t,T) = \frac{B^{\mathrm{f}}(t,T)}{B^{\mathrm{d}}(t,T)} S(t). \tag{9.25}$$

Using Itô lemma and the log-normality of zero-coupon bonds and spot FX yields to the expression of the quadratic average of the forward FX volatility:

$$\bar{\sigma}_T^2 = \frac{1}{T}\int_0^t \sigma_T^2(s)ds = \frac{1}{T}\int_0^t \|\Gamma^{\mathrm{f}}(s,T) - \Gamma^{\mathrm{d}}(s,T) + \sigma_S(s)\|^2 ds, \quad (9.26)$$

that is,

$$\bar{\sigma}_T^2 = \frac{1}{T}\int_0^t \begin{pmatrix} \sigma_{\mathrm{d}}^2(s)\beta_{\mathrm{d}}^2(s,T) + \sigma_{\mathrm{f}}^2(s)\beta_{\mathrm{f}}^2(s,T) + \sigma_S^2(s) \\ -\,2\rho_{\mathrm{f,d}}\sigma_{\mathrm{f}}(s)\beta_{\mathrm{f}}(s,T)\sigma_{\mathrm{d}}(s)\beta_{\mathrm{d}}(s,T) \\ +\,2\rho_{\mathrm{f},S}\sigma_{\mathrm{f}}(s)\beta_f(s,T)\sigma_S(s) \\ -\,2\rho_{\mathrm{d},S}\sigma_s(s)\beta_s(s,T)\sigma_S(s) \end{pmatrix} ds. \quad (9.27)$$

Remarking that the forex FX is martingale under the domestic T forward probability (i.e. associated to numeraire $B^{\mathrm{d}}(t,T)$), the FX option price paying at time T is then given by the standard Black formula:

$$\mathrm{Call} = B_{\mathrm{d}}(0,T)(S(0,T)N(d_1) - KN(d_2)) \quad \text{with}$$
$$d_1 = \frac{\ln\left(\frac{S(0,T)}{K}\right) + \frac{\bar{\sigma}_T^2 T}{2}}{\bar{\sigma}_T\sqrt{T}}, \quad d_2 = d_1 - \bar{\sigma}_T\sqrt{T}. \quad (9.28)$$

9.2.2.3. *Implementation details*

9.2.2.3.1. Lattice calibration

The implementation must be arbitrage-free. This means that short rate and forward FX diffusion are sampled by the lattice so that:

$$B_{\mathrm{d}}(0,t) = E_0^{Q_{\mathrm{d}}}\left[\mathrm{e}^{-\int_0^t r_{\mathrm{d}}(s)ds}\right],$$
$$B_{\mathrm{f}}(0,t) = E_0^{Q_{\mathrm{f}}}\left[\mathrm{e}^{-\int_0^t r_{\mathrm{f}}(s)ds}\right], \quad (9.29)$$
$$B_{\mathrm{f}}(0,t)S(0) = E_0^{Q_{\mathrm{d}}}\left[\mathrm{e}^{-\int_0^t r_{\mathrm{d}}(s)ds}S(t)\right]. \quad (9.30)$$

A bootstrap calibration process is done. To be efficient in computation time, domestic and foreign curves are first calibrated in a mono-currency one factor version of the lattice.

The standard method uses Arrow–Debreu prices by a step by step calibration process. Assuming that Arrow–Debreu prices are already known at slice at time t_i, the arbitrage free requirement at next time step t_{i+1} rewrites to:

$$\begin{aligned} B(0, t_{i+1}) &= \sum_{\text{node}(t_i)} \text{AD Price}(t_i)\mathrm{e}^{-r(t_i)\Delta t_i} \\ &= \mathrm{e}^{-(f(0,t_i)+\theta(t_i))\Delta t_i} \sum_{\text{node}(t_i)} \text{AD Price}(t_i)\mathrm{e}^{-X(t_i)\Delta t_i}. \end{aligned} \tag{9.31}$$

The deterministic drift correction is then simply given by:

$$\theta(t_i) = \frac{\ln}{\Delta t_i}\left(\frac{\sum_{\text{node}(t_i)} \text{AD Price}(t_i)\mathrm{e}^{-X(t_i)\Delta t_i}}{B(0, t_{i+1})}\right) - f(0, t_i). \tag{9.32}$$

Back to the initial lattice, domestic 3D Arrow–Debreu prices are step by step computed and used to follow the same method to fulfil the arbitrage-free condition on the forex. Because this calibration is time consuming it will be done only at slices where spot FX is needed: at reset dates of the product.

9.2.2.3.2. Rotation of initial variables

This idea is to diffuse independent variables. So the transition probability computation is done independently in each direction and the global probability to connect a node is the product of elementary ones in each direction. At each time step of the lattice, the local variance–covariance matrix is computed.

$$\Omega^i = \begin{bmatrix} \begin{array}{l}\int_{t_i}^{t_{i+1}} \sigma_{\mathrm{d}}^2(s) \\ \times\, \mathrm{e}^{-2\lambda_{\mathrm{d}}(t_{i+1}-s)}ds\end{array} & \begin{array}{l}\omega_{\mathrm{d},f}^i = \int_{t_i}^{t_{i+1}} \sigma_{\mathrm{d}}(s)\sigma_{\mathrm{f}}(s)\rho_{\mathrm{d,f}}(s) \\ \times\, \mathrm{e}^{-(\lambda_{\mathrm{d}}+\lambda_{\mathrm{f}})(t_{i+1}-s)}ds\end{array} & \begin{array}{l}\omega_{\mathrm{d},S}^i = \int_{t_i}^{t_{i+1}} \sigma_{\mathrm{d}}(s)\sigma_S(s)\rho_{\mathrm{d},S}(s) \\ \times\, \mathrm{e}^{-\lambda_{\mathrm{d}}(t_{i+1}-s)}ds\end{array} \\ \omega_{\mathrm{d,f}}^i & \int_{t_i}^{t_{i+1}} \sigma_{\mathrm{f}}^2(s)\mathrm{e}^{-2\lambda_f(t_{i+1}-s)}ds & \begin{array}{l}\omega_{\mathrm{f},S}^i = \int_{t_i}^{t_{i+1}} \sigma_{\mathrm{f}}(s)\sigma_S(s)\rho_{\mathrm{f},S}(s) \\ \times\, \mathrm{e}^{-\lambda_{\mathrm{f}}(t_{i+1}-s)}ds\end{array} \\ \omega_{\mathrm{d},S}^i & \omega_{\mathrm{f},S}^i & \int_{t_i}^{t_{i+1}} \sigma_S^2(s)ds \end{bmatrix} \tag{9.33}$$

This matrix is diagonalized to rotate initial correlated $\mathbf{X}$ space to independent $\mathbf{Z}$ space:

$$\Omega^i = P(t_i)\begin{bmatrix} \alpha_1(t_i) & 0 & 0 \\ 0 & \alpha_2(t_i) & 0 \\ 0 & 0 & \alpha_3(t_i) \end{bmatrix} P^t(t_i), \tag{9.34}$$

where

$$P(t_i) = \begin{bmatrix} \vec{e}_1(t_i) & \vec{e}_2(t_i) & \vec{e}_3(t_i) \end{bmatrix}. \tag{9.35}$$

is the eigen vectors matrix and $(\alpha_k(t_i))_{k=1,3}$ the eigen values.

The diffusion takes place in **Z** space. For each direction, local variance $\alpha_k(t_i)$ and conditional expectation will be fitted by a trinomial local transition.

Conditional expectations are computed using rotations between **X** and **Z** spaces using the following steps:

(1) given state in **Z** space at current slice t_i, compute its associated **X** state

$$X(t_i) = P(t_i)Z(t_i), \tag{9.36}$$

(2) compute the conditional expectation

$$E^Q[X(t_{i+1})|X(t_i)] \tag{9.37}$$

using the theoretical diffusion of **X** processes,

(3) rotate back to **Z** space, that is, compute

$$P^t(t_i)E^Q[X(t_{i+1})/X(t_i)] \tag{9.38}$$

to get theoretical central connection at next slice t_{i+1}.

Once theoretical local variance and conditional expectation are known in **Z** space, transition probabilities are extracted direction by direction using standard method (see lattice method in Chapter 9).

9.2.2.4. *Basis curves*

The power reverse dual currency, product uses two currencies and a basis effect must be modeled. Previous model is enhanced to generate two more basis curves, one for the funding leg and another for the coupon leg. In case of USD, the reference for any basis margin, only one more curve is generated (funding or coupon). Basis curves $\tilde{B}_{\mathrm{d}}(t,T)$ and $\tilde{B}_{\mathrm{f}}(t,T)$ are generated from their reference ones $B_{\mathrm{d}}(t,T)$

and $B_{\mathrm{f}}(t,T)$ with the simple assumption that the instantaneous basis margin (Section 9.1) is deterministic:

$$\tilde{B}_{\mathrm{d}}(t,T) = B_{\mathrm{d}}(t,T)\mathrm{e}^{\int_t^T \alpha_{\mathrm{d}}(s)ds} = B_{\mathrm{d}}(t,T)\frac{\tilde{B}_{\mathrm{d}}(0,T)}{B_{\mathrm{d}}(0,T)}\frac{B_{\mathrm{d}}(0,t)}{\tilde{B}_{\mathrm{d}}(0,t)}. \tag{9.39}$$

We have the same relation in the foreign market.

9.2.2.5. *Smile issue*

One major drawbacks of the above model is the lack of smile effect on FX side. Unfortunately PRD structures are quite sensitive to it and the model must be enhanced to access this risk.

9.2.2.5.1. Q-model and other mappings

The idea of this framework is to map the underlying mean reverting process of the spot FX in a "Q mapping":

$$S(t) = S(0,t)\left(1 + \frac{\mathrm{e}^{\theta_S(t)+qX_S(t)} - 1}{q}\right). \tag{9.40}$$

Note that $q = 1$ restores the initial pure log-normal version. The Q mapping is equivalent to a displaced log-normal diffusion with an absolute shift parameter set to

$$m(t) = S(0,t)\left(\frac{1}{q} - 1\right). \tag{9.41}$$

For this reason, only skews between pure normal and pure log-normal distribution are possible with this mapping.

As for the non-smiled version, the drift correction $\theta_s(t)$ will be calibrated to fulfil the arbitrage free relation:

$$B_{\mathrm{f}}(0,t)S(0) = E_0^{Q_{\mathrm{d}}}\left[\mathrm{e}^{-\int_0^t r_{\mathrm{d}}(s)ds}S(t)\right]. \tag{9.42}$$

On volatility calibration side, the smiled market FX option volatility $\bar{\sigma}_T(K)$ is first converted to its Q model equivalent value $\bar{\sigma}_T^q$ using the standard and displaced Black formulae:

$$\begin{aligned}\text{Black}(S(0,T), K, \bar{\sigma}_T(K), T) \\ = \text{Black}(S(0,T) + m(T), K + m(T), \bar{\sigma}_T^q, T).\end{aligned} \quad (9.43)$$

This conversion is done ATM forward that is to say for $K = S(0,T)$.

In a second step, the relation linking instantaneous yield curve volatilities, spot FX volatility and forward one is still used to extract a stepwise constant spot FX volatility through a bootstrapping technique. For a q parameter quite near to 1, this pure log-normal relation works well and it is not needed to correct it adding a volatility term coming from the displaced log-normality of the spot FX rate.

Other mappings may be used to generate more complex smile shapes with more parameters to tune. The drift correction will always provide a non-arbitrageable model. But one must keep in mind that, to be efficient, the calibration must use closed form formulae. At least these ones will be approximations but they allow us to run a more complex calibration relying on better formulae and taking few zero search/optimization steps to converge.

9.2.2.5.2. Local volatility

The idea of local volatility approach is to extend or "fudge" as every Quant says, the quite simple diffusion model by an external FX option model. This one is inconsistent with the diffused hybrid model but it will allow an accurate spot price for any FX option of any expiry and strike. The PRD underlying is then ensured to be evaluated correctly even if the coupon is defined by a two strikes call spread FX option.

In fact, the initial hybrid model yields, for a given expiry and strike, have an implicit FX option volatility. This one is then corrected to fulfil the spot market one.

The classical approach is to use a standard Black model to evaluate FX option at a given notice date t_i. This imposes that all FX options maturing at T_j embedded in the coupon leg are priced analytically using this model (the best approach is to get the leg price by back-propagation of each option value at expiry). Because of the log-normality of both models (forward FX within the hybrid model and Black one), the volatility correction $\bar{\sigma}(t_i, T_j, K)$ is equal, in terms of integrated log-normal variance, to the difference between at strike implicit market volatility $\bar{\sigma}_{T_j}(K)$ and the forward FX volatility from spot date to the notice date:

$$\bar{\sigma}^2(t_i, T_j, K) = \frac{\bar{\sigma}^2_{T_j}(K)T_j - \int_0^{t_i} \sigma^2_{T_j}(s)ds}{T_j - t_i} \tag{9.44}$$

9.2.3. *LIBOR market model*

This section briefly describes an interesting alternative to diffuse both yield curves adding skew features. On both domestic and foreign curves, the shifted forward rate model is used (see Chapter 8 for more details on the model, a classical BGM framework with displaced log-normal forward rate distributions). This enables to manage without any additional cost skew on interest rate options.

9.2.3.1. *Diffusion overview*

A schedule synchronized with coupon leg $(T_k)_{k=1,N}$ defines the base of LIBOR set diffused by the model. In practice 1Y LIBOR are used.

The diffusion takes place under the spot probability that is using the rolling domestic numeraire $B_d(t, T_{n(t)})$ if $t \in]T_{n(t)-1}, T_{n(t)}]$. For more details, please see the BGM model presented in Chapter 8.

LIBOR rate equation diffusion is:

$$\frac{dL(t, T_k)}{L(t, T_k) + m_k} = \mu_k(t)dt + \sigma_L(t)\mathrm{e}^{-\lambda(T_k - t)}dW_L \tag{9.45}$$

using the classical volatility exponential shape. The stochastic drift $\mu_k(t)$ is a function of LIBOR volatilities and LIBOR rates themselves.

Spot FX values are known by using a one period or rolling forward FX rate as following:

(1) at a given schedule time T_k, suppose that spot FX is known $S(T_k)$, yield curves diffusion give both LIBOR $L_\mathrm{d}(T_k, T_{k+1})$ and $L_f(T_k, T_{k+1})$;

(2) compute the one period forward FX rate:

$$S(T_k, T_{k+1}) = S(T_k)\frac{B_f(T_k, T_{k+1})}{B_d(T_k, T_{k+1})} = S(T_k)\frac{1 + \delta_d L_d(T_k, T_{k+1})}{1 + \delta_f L_f(T_k, T_{k+1})}; \tag{9.46}$$

(3) because this forward FX rate has no drift up to next time T_{k+1} under the spot probability, its diffusion is quite simple. So we can easily compute in the interval

$$t \in]T_k, T_{k+1}] : S(t, T_{k+1}) = S(T_k, T_{k+1})e^{-\frac{1}{2}\int_{T_k}^{t}\sigma_S^2(s)ds + \int_{T_k}^{t}\sigma_S(s)dW_S}; \tag{9.47}$$

(4) the volatility $\sigma_S(t)$is the spot FX one. Here the yield curve volatility effect for short term forward FX diffusion is then reasonably neglected;

(5) take the forward FX at it expiry to get a new spot FX value

$$S(T_{k+1}) = S(T_{k+1}, T_{k+1}). \tag{9.48}$$

This diffusion is initialized by $S(0, T_1) = S(0)(B_f(0, T_1)/B_d(0, T_1))$. The start date of the PRDC is always closed to spot date and forward FX could diffuse with spot FX volatility from 0 to time T_1.

9.2.3.2. *Pricing method*

Due to the complexity of BGM stochastic drift, a lattice method, however possible, is not recommended here. In order to use BGM in a lattice, some approximations are indeed required and their accuracy decreases with the variance of the diffused process. On JPY market this constraint limits product maturity to around 7 years which is too short for PRDC structure.

The alternative is the Monte-Carlo method coupled with Andersen or Longstaff and Schwartz techniques to price early exercise feature (see Chapter 9 for more details). If Andersen method is used for instance, the exercise decision may be defined as a function of the exercise value, the PRDC residual swap. Exercise frontier values are maximized in a backward bootstrap process. The PRDC is then internally modified in the pricer to be triggered at these exercise levels using a second path set.

9.2.4. *Equity hybrids*

Another class of hybrids are the equity linked ones. While the first generation of equity linked product (such as barrier and digital options, corridor, forward start, cliquet, Hymalaya) requires very simple models, the second generation requires a real hybrid model as the risk are not only on the equity linked model but also potentially on interest rates and correlation risk. The hybrid nature of these products has many reasons:

(1) The basket includes different assets like equities commodities, foreign exchange index, interest rates and so on.
(2) The long dated nature of the product with capital guarantee protection embodies an interest rates risk.
(3) The various equities of the basket are in various currencies and a simple quanto correction may not be appropriate in all cases.

The typical products include the following:

1. *Accumulator or predator*: It pays at the maturity of the product the sum of the short-term capped performance of the equity index floored to a minimum coupon.

   At each date, it determines the short-term performance capped to the level p_i given by

$$\mathrm{Min}(S(T_{i+1})/S(T_i) - 1, p_i). \tag{9.49}$$

   At the maturity of the product, it pays the minimum between the sum of the capped short-term performance and a minimum coupon c given by

$$\mathrm{Max}\left(\sum_{i=1}^{N} \mathrm{Min}(S(T_{i+1})/S(T_i) - 1, p_i), c\right). \tag{9.50}$$

2. *Everest*: Call or put on the worst performing member of a large basket of stocks or indexes at maturity. Everest has a long maturity (10–15 year) and is written on many underlyings.
3. *Individual cap basket*: Call on a basket where each component of the basket is capped at a given level.
4. *Atlas*: This is a call on basket where at maturity some of the best and some of the worst performing assets are withdrawn.
5. *Himalaya*: Call option on the average of the maximum performers on a selection of assets.

   Altiplano: Option that pays a large coupon if no asset in a given selection hits a certain limit during a given period, otherwise pays a vanilla call on the basket of assets.

Standard equity hybrids models are based on a standard equity model with a stochastic drift embodied by a stochastic interest rate. A common simple model is based on a HW 1-factor model for the interest rates and a BS model for the equities. However, the same model with a non-BS model lies to a non-tractable model. The problem comes from the fact that the integrated volatility used in the pricing of call and put option is not given by a closed form solution.

Usually, dividends can be taken as deterministic. However, if one wants a more accurate modeling approach, one should as well model the dividends as discrete or continuous, proportional or absolute. This implies to calibrate the dividend to the forward curve of the equity asset. Stochastic interest rates with stochastic dividends lead to small convexity correction terms in the case of correlation between stochastic dividend and stochastic interest rates. Like any other hybrid models, the main challenge for risk managing hybrid products and consequently building the hybrid model.

9.2.5. *Credit hybrids*

In this section credit and interest rate hybrid product are presented pointing out embedded risks.

Market and products: It is commonly agreed that non-vanilla products include synthetic CDOs, nth-to-default baskets, synthetic callable credit default swaps, CDO squares and credit hybrids,... Giving an exhaustive view of all hybrid structures designed by unlimited imagination is merely impossible. However, one could give the following classification depending of risk origin:

(1) credit product sensible to yield curve risk;
(2) yield curve product sensible to credit risk;
(3) yield curve exotics contingent to credit event.

In any credit-linked product the main risk is that the reference entity can default. If there is no default of the reference entity, all expected cash flows will be received in full. If a default occurs, the investor will receive some recovery amount. Vanilla credit product as CDS depends on credit event and discounts their flows using the yield curve (see also the *Lehman Brothers Guide* to exotic credit derivatives for more details on CDS). The classical framework is to assume independency between default and interest rates. This leads to a defaultable bond value expressed as a product of the zero-coupon bond and a term depending only of credit risk.

The second set of products is typically vanilla interest rate products as swap, cap/floor or swaption where there is a counterparty risk. There is always a chance that one party get into financial problems and default in a swap contract. Interest cap provides insurance against the rise of interest rate above a certain level: the cap rate whereas floor provides insurance when the interest rate on the floating interest falls below a certain floor rate. This implies to compute the standard interest rates products on the risky zero-coupon curve.

Convertible bonds can be viewed as a special type of interest rates and credit hybrids. The holder has the option to exchange the bond for company's stocks at certain times in the future with a certain conversion ratio. They are almost callable: they give the right to buy them back at certain prices and at certain times (see Hull option futures and other derivatives). They are real hybrids as the risk comes both from the interest rates and the equity component as the bond may be converted to an equity asset in some scenarios while in the other it should be a simple bond. Therefore, we should take care of credit risk as it plays an important role in the valuation of convertibles.

The last category of credit equity hybrids have been the equity default swap (EDS). EDSs do not transfer credit risk but they transfer the risk of major diminution in the market value of shares. An EDS enable fund managers to change their risk exposure to an index without buying or selling stocks. In this contract, one party pays the return on an equity index on a notional while the other party pays a fixed or floated rate.

9.2.6. *Alternative structured products*

Alternative structured products cover a wide range of products coming from options on funds and funds of funds. This encompasses the option on CPPI, option on MEPI and so on. The problem lies mainly in the liquidity risk and the lack of daily data but rather only monthly data. Time horizon is therefore monthly and trading can have serious time delays.

Option on illiquid: Unlike liquid products, where pricing is based on hedging argument (dynamic or not), there exists no consensus on illiquid products. There exists consequently various ways of pricing illiquid products out of which the main ones are:

- quantile or VAR pricing,
- utility-based pricing.

9.3. Summary

In this chapter, we reviewed the models used for hybrid products. The main emphasis is on the PRDC product. We examined how this model should be implemented and the relationship between the various underlying models. We concluded on hybrid products such as equity and credit hybrids.

References

[1] Fries CP and Rott MG (2004). Cross-currency and hybrid Markov-functional models (May 4, 2004). Available at SSRN: http://ssrn.com/abstract=532122

[2] Piterbarg V (2005). A multi-currency model with FX volatility skew (February 7, 2005). Available at SSRN: http://ssrn.com/abstract=685084

Chapter 10

PRODUCT CATALOG AND USAGE

10.1. Typology

Derivatives have grown very rapidly over the last few years and encompass a very wide range of products. There is not one typology or classification of such products but rather many classifications according to various criteria.

10.1.1. *Investment vs. hedging*

The first criterion to classify derivatives lies in the distinction between investment or speculation and hedging products. Obviously, the same product may be for someone a hedging strategy and for someone else an investment. However, such a distinction is important for many reasons. First of all, from a client point of view, investment and hedging have very different motivations. The derivatives should be used with the full knowledge of what is aimed for. One should be able to know his/her risk and the motivation behind his/her trade. Hedging aims at reducing a particular risk by offsetting it with the particular derivatives while investing would rather increase a particular risk and reflect a view on the market or a speculation on a particular market scenario.

Second, the difference between investment and hedging is very meaningful for accounting reasons as it implies a different way of accounting and reporting for derivatives instruments. The two standard accounting standards are:

(1) The Financial Accounting Standard No 133, Accounting for Derivatives Instruments and Hedging Activities, generally referred to as FAS 133.
(2) The corresponding international accounting standard designated as IAS 39.

Third, from a seller's point of view, the marketing of investment and hedging instrument is very different. Selling hedging instrument should imply a real proof of the efficiency of the hedge (especially with the new accounting standards in order to qualify as an effective hedge). On the contrary, investment strategy should insist on the particular considered scenario where the bet should be very profitable and the worst cases in which it would lead to some losses.

For a derivative contract, to be qualified as a hedge has the great advantage of deferring the reporting of the gains and/or losses on the hedging instrument when the offsetting gain or loss on the hedge item is recognized. The ineffective component of the hedge results must be accounted in the current income while the effective portion of the hedge can be posted to the other comprehensive income and later on incorporated to the standard income. In addition, for any derivative, the new accounting standards impose to value the derivative contract at its fair value. Fair value reflects the market price of the derivative. The operation of reporting the value of the derivative at its market price is referred to as marked-to-market.

For instance, a typical hedge instrument would be:

(1) a standard swap contract where the received cash flows offset some liabilities;
(2) a vanilla cap or floor to hedge a standard one sided interest rate risk;
(3) a FX forward to hedge the FX risk and more generally any forward to offset a potential risk on the underlying;
(4) if the risk is only one sided, an option on the underlying.

Any more complex derivatives — such as for an interest rates hedge a trigger swap, a barrier cap — should be subject of a case study to see the effectiveness of the derivatives.

10.1.2. *Investment products: high-risk products*

A product should be called an investment product if it is used to get high returns. High returns means obviously to bear higher risk. Various strategies to invest in interest rates derivatives products are listed below.

10.1.2.1. *High-coupon strategy*

High-coupon strategy lies in products that pay potentially high coupons. We will see some of them in the product classification.

10.1.2.2. *Yield curve slope strategy*

Yield curve slope strategy is a bet on the slope of the interest rates curve. In a payer spread option, one bets that the curve should flatten and the opposite in a receiver one.

10.1.2.3. *Callability*

Adding some call option rights lowers the price of the instrument and can provide a more profitable strategy. Inversely, buying call options makes the product more expensive.

10.1.2.4. *Bet strategy: corridor and digital*

In a corridor or in a digital, one can bet that the interest rates will lie within a certain range and/or will not break some levels. This additional risk is rewarded by either higher coupons and/or better profitability of the trade.

10.1.2.5. *Quanto strategy*

Quanto products are derivatives whose payoff is paid in the domestic currency for asset denominated in foreign currencies. The foreign exchange (FX) risk of the underlying asset denominated in foreign currency is hence eliminated. An example of a quanto product is for instance a call on the Nasdaq but paid in Euro. The Nasdaq is denominated in USD. Hence, for a Euro investor, a call on Nasdaq would bear a risk from the exchange rate between the USD and the Euro. The same call quanto would basically means that we compute the payoff of the call on the Nasdaq and pay it in Euro. Quanto products imply a correlation risk as well as a risk to the FX volatility as the quanto correction can be computed in a BS model as follows:

$$\exp\left(\int_0^T \sigma_{\text{underlying}}(s)\sigma_{\text{FX}}(s)\rho_{\text{FX,underlying}}(s)ds\right), \tag{10.1}$$

where $\sigma_{\text{underlying}}(s)$ is the volatility of the underlying asset, $\sigma_{\text{FX}}(s)$ is the volatility of the FX and $\rho_{\text{FX,underlying}}(s)$ is the correlation between the FX and the underlying.

10.1.2.6. *Currency convergence trade*

Another investment strategy for interest rates is to play the convergence trade across various interest rates curves. Betting for convergence would mean that interest rates should converge on the long run to the same value. For the opposite view, one should bet that the two curves spread should widen.

10.1.2.7. *Hedging products: low premium*

The art of hedging correctly lies in two questions:

(1) Are we sure that the hedge is effective? One should consider a comprehensive set of scenarios to make sure that the product is hedging accurately. Simulations, mark-to-market envelope, mark-to-market forecasts are good tools to use.
(2) How much does the hedge cost? Obviously, one should take the cheaper hedging strategy.

10.1.2.8. *Zero-premium strategy*

A zero-premium strategy is a risk management strategy, that can take the investor to reduce the cost to "zero", for example, by selling an out-of-the money call option and using the premium received to purchase an-out of-the money put option. Besides this, zero-premium strategy often hides some risk. These risks may be offsetting some client risks, in which case, the client can consider that he/she is monetizing his/her risk. Or this may be additional risk taken in the strategy. In this case, the client should be aware that the attractive hedging strategy price lures additional risk and that he/she should be ready to bear it.

10.1.2.9. *Customized products*

Perfect hedging often means some customization of the products as standard product may not do the job. Customized products may mean some additional choice in the product characteristics such as the decision to exercise for a cap or the strike for an option ... and so on. But once again, greater choice means higher cost and one should be aware that a customized solution is often more expensive than a standard one.

10.1.2.10. *Barrier options*

Barrier options are a good strategy to get a product cheaper than the standard one. In the case of a knock out barrier product, the client may pay a cheaper option price in acceptance for the risk to have the structure cancelled and knocked out. The option will be cheaper if the expected value of the future cash flows that might be cancelled by the knock out feature is positive. In the case of negative cash flows, the knock out barrier should be even more expensive than the standard option.

Similarly, in a knock in product, the client is taking the additional risk that the structure will knock in. If this is not the case, the option ends worthless. Again, if the expected value of the stream of cash flows is positive, the knock in version of the option should be lower than the standard one. In short, barrier version of standard option is a good way of customizing a product to lower its price for additional known and controlled risk.

10.1.3. *Product typology*

Products could be classified into various categories as:

(1) Classification according to currency. This is the standard classification for trading desks structured according to currency and type of the underlying.
(2) Classification according to pricing techniques and risk. The differences of model, technology, risk are obvious and have lead to the distinction between vanilla, derivatives and exotic desks. Even more, within exotic, there are path dependent product (that depends on the trajectory), backward looking product (that needs numerical method that can evaluate backward like tree and PDE and so on).

10.2. Products Catalog

In this section, we will review a wide range of products. This consists in some of the blockbusters of the derivatives industry as well as some less famous but still interesting derivatives. This is the most interesting part of this chapter as we try to be fairly exhaustive.

10.2.1. *European options*

Although European options are quite simple in terms of their pricing, they already offer a variety of combinations. We list below the most common ones:

(1) Simple path dependent options: In addition to simple call and put option, the spectrum of European options encompasses the following options.
(2) Digital call and put option: Pays one if the underlying at maturity is above a certain level. As explained in the smile section, this product is not only sensitive to the level of the volatility but also the slope of the smile at the strike (see Chapter 6). It is then interesting to note that we can accurately value digital option in

BS model with accounting for the smile risk by taking the call spread limit. We obtain:

$$\text{Digital\_Call} = \text{Digital\_Call}_{\text{BS}} - \text{Vega}_{\text{BS}}{}^{*}\text{Slope}, \qquad (10.2)$$

where the Digital_Call is the standard price of the digital call, $\text{Digital\_Call}_{\text{BS}}$ is the digital call price of BS option, Vega_{BS} is the BS vega and Slope is the slope of the smile with respect to the strike. For more detailed see the part on smile risk.

(3) Contingent and reverse contingent options, pay-as-you-go/pay latter/cash-on delivery or Money back: This kind of option promises — at least on the paper — to pay only a premium in the case of a real use of the option. As there is no free lunch in finance, the fixed premium paid may be more than the payoff collected by the option to compensate for the potential large outcome of the option. This kind of option is fairly easy to value in BS as it is a call option with a digital option.

(4) Power options: These are options on the product of powers of several assets. This type of options has been over the years less marketable. Their pricing in BS is fairly easy because the power of a lognormal remains lognormal. For example, the payoff for an asset S_1, with power indice ω_1, at expiry date T and strike price K is:

$$\text{Max}(S_1(T)^{\omega_1} - K, 0). \qquad (10.3)$$

In the context of a smile, these options can be badly priced with a BS model as the power payoff implies to be sensitive to statistic moments of higher order than two. As a lognormal distribution is calibrated only on the first two moments, the misconception of the distribution is not taken into account and can be disastrous.

However, it is interesting to notice that like any European option, these options can be priced by replications using the following result: a payoff which is a function of an underlying on which options are available can be replicated with these options. If S_T is the value of the underlying at maturity T, and if $f(S_T)$ is the payoff, we have:

$$\begin{aligned} f(S_T) = f(\kappa) + f'(\kappa)((S_T - \kappa)^+ - (\kappa - S_T)^+) \\ + \int_0^\kappa f''(K)(\kappa - S_T)^+ dK \\ + \int_\kappa^\infty f''(K)(S_T - \kappa)^+ dK. \end{aligned} \tag{10.4}$$

Therefore, when we select a value of κ such that $f'(\kappa) = 0$. Taking the expectation of both members, we get:

$$\begin{aligned} E[f(F_T)] = f(\kappa) + \int_0^\kappa f''(K)\text{Put}(K,T)dK \\ + \int_\kappa^\infty f''(K)\text{Call}(K,T)dK. \end{aligned} \tag{10.5}$$

This formula tells us that knowing the price of vanilla options (call and put options at any strikes) should be enough to compute the arbitrage-free price of the payoff that pays $f(F_T)$.

10.2.2. *Asian options*

An option is referred to as an Asian option whenever its payoff is linked to some kind of averaging of one or more underlying. The averaging may be computed in various ways out of which the standard is arithmetic. Geometric averaging is not much used but is very useful to get a closed form pricing in a BS model. The arithmetic average should be given by:

$$A_t = \frac{S_1 + S_2 + \cdots + S_n}{n}, \tag{10.6}$$

while the geometric average is given by:

$$G_t = \sqrt[n]{S_1.S_2 \cdots S_n}. \tag{10.7}$$

Customized average may have specific unequal weights. In a BS model with an implied volatility σ, a short rate of r and a dividend of d, the solution for the geometric average Asian calls and puts are easy to compute and given by (Kemma and Vorst):

$$\begin{aligned} C_G &= S\mathrm{e}^{(b-r)(T-t)}N(d_1) - X\mathrm{e}^{-r(T-t)}N(d_2), \\ P_G &= X\mathrm{e}^{-r(T-t)}N(-d_2) - S\mathrm{e}^{(b-r)(T-t)}N(-d_1), \end{aligned} \tag{10.8}$$

when $N(x)$ is the cumulative normal distribution and

$$d_1 = \frac{\ln(S/X) + (b + 0.5\sigma_G^2)T}{\sigma_G\sqrt{T}}, \tag{10.9}$$

$$d_2 = d_1 - \sigma_G\sqrt{T}. \tag{10.10}$$

where the adjusted geometric average volatility and dividend are given by:

$$\sigma_G = \frac{\sigma}{\sqrt{3}}, \tag{10.11}$$

$$b = \frac{1}{2}\left(r - d - \frac{\sigma^2}{6}\right). \tag{10.12}$$

Therefore, a crude approximation for the volatility of an Asian option is to take the volatility of the standard European and divide it by $\sqrt{3}$.

For arithmetic average option, there exist various approximations, which are quite equivalent to our experience. A first approximation is the one of Turnbull and Wakeman. It basically computes the log-normal distribution that has the same first two moments as the sum of lognormals. The approximation is given by:

$$\begin{aligned} C_{\mathrm{TW}} &\approx S\mathrm{e}^{(b-r)(T-t)}N(d_1) - X\mathrm{e}^{-r(T-t)}N(d_2), \\ P_{\mathrm{TW}} &\approx X\mathrm{e}^{-r(T-t)}N(-d_2) - S\mathrm{e}^{(b-r)(T-t)}N(-d_1), \end{aligned} \tag{10.13}$$

where

$$d_1 = \frac{\ln(S/X) + (b + 0.5\sigma_A^2)T}{\sigma_A\sqrt{T}}, \tag{10.14}$$

$$d_2 = d_1 - \sigma_A\sqrt{T}, \tag{10.15}$$

where the adjusted arithmetic average volatility and dividend are given in terms of the first two moments:

$$\sigma_A = \sqrt{\frac{\ln M_2}{T} - 2b}, \tag{10.16}$$

$$b = \frac{\ln M_1}{T}. \tag{10.17}$$

The first two moments are given by

$$M_1 = \frac{e^{(r-d)T} - 1}{(r-d)T} \tag{10.18}$$

and

$$M_2 = \frac{2e^{(2(r-d)+\sigma^2)T}S^2}{(r-d+\sigma^2)(2r-2q+\sigma^2)T^2} + \frac{2S^2}{(r-d)T^2} \times \left(\frac{1}{2(r-d)+\sigma^2} - \frac{e^{(r-d)T}}{r-d+\sigma^2}\right). \tag{10.19}$$

Levy suggested another approximation based on Edgeworth expansion and given by:

$$\begin{aligned} C_{\mathrm{L}} &\approx S_{\mathrm{L}}N(d_1) - Xe^{-rT}N(d_2), \\ P_{\mathrm{L}} &\approx Xe^{-rT}N(-d_2) - S_{\mathrm{L}}N(-d_1), \end{aligned} \tag{10.20}$$

where

$$d_1 = \frac{1}{\sigma_{\mathrm{L}}\sqrt{T}}\left(\frac{\ln(L/X)}{2}\right), \tag{10.21}$$

$$d_2 = d_1 - \sigma_{\mathrm{L}}\sqrt{T}, \tag{10.22}$$

where

$$S_L = \frac{S}{(r-d)T}(e^{-dT} - e^{-rT}) \tag{10.23}$$

and

$$X_{\mathrm{L}} = X - S\frac{S}{(r-d)T}(e^{-DT} - e^{-rT}), \tag{10.24}$$

$$\sigma_{\mathrm{L}}^2 T = \ln(L) - 2(rT + \ln S_L), \tag{10.25}$$

$$M = \frac{M}{T^2}. \tag{10.26}$$

Asian options are indeed embedded in various other derivatives like TARN and Snowball and/or cumulative options. We will give more details about these products in the section specific to recent interest rates exotic derivatives products (later in this chapter).

10.2.3. *Hawaiian options*

Hawaiian options are a relatively new breed of exotic options that combine Asian and American features making them at the same time path dependent and callable. Typical illustration of Hawaiian options are callable snowball, callable TARN product in the interest rates derivatives area. Because these options are path dependent and callable, they require either sophisticated extension of tree and PDE method to diffuse the path dependent variable or American Monte Carlo method (see, for instance, Longstaff Schwartz and Andersen methods in Appendices A–C).

10.2.4. *Barrier options*

In the BS world, the density of the maximum and minimum can be computed closed forms, hence a closed forms pricing of barrier options. However, BS pricing of barrier option misses the real behavior of the volatility and can be misleading. Better management is to have some calibrated real smile models either like local volatility and jump models or combination of local and stochastic volatility models (see Chapter 6).

Good risk management of the barrier risk should precisely accounts for the fact that the volatility level may vary at the barrier level. European barrier option can be valued easily with call spread. But for real barrier option, the risk management is not as trivial as that.

There exists a wide variety of barrier options such as:

(1) *floating barrier*, where the barrier changes according to a deterministic rule, usually using a barrier yield;
(2) *Asian barrier options*, where the barrier option is written on the average of the underlying, making it an Asian barrier option;

(3) *forward start and other partial window barrier*, where the period of activity of the barrier is not the full life of the barrier but rather a given window of time;
(4) *outside barrier/equity triggered options* are barrier options where the underlying used to observe the barrier activation or cancellation is different from the one paid.

Typical example is, for instance, an option paying an equity index with a barrier on the interest rates level. Remark that it becomes increasingly difficult to price barrier option when the number of the assets increases. In order to solve this problem, we can use, for example, Monte Carlo methods, precisely shifting barriers for multi-assets.

10.2.5. *Lookback and extensions*

Some equity derivative variations around the lookback option like the Napoleon, and worst-off option have made these options popular. However, standard lookback options are in general not that popular as they can be expensive. Closed form pricing exists for lookback option in Black Scholes and various distributions like CEV as the density of the maximum is well known.

10.2.6. *PNL and passport options*

The PNL or passport option used to be quite popular as it offers the ability to lock in the profit of optimal trading on a given asset subject to position limit. Basically, a passport option is a zero-strike call option on the balance of a trading account. The holder of the option trades dynamically on the underlying. At exercise, if the trades have realized a profit, he receives the proceeds of the trading account, otherwise, it is the seller of the option that pays the losses. In return for the option, the investor makes an upfront payment, equal to the option premium. For a concrete valuation, see, for instance, Henderson and Hobson.

10.2.7. *Simple correlation/multi-asset options*

We will denote by ω a flag to say whether the payoff is a call or put with the convention that $\omega = -1$ for a put, respectively $\omega = 1$ for a call. Simple correlation trades encompass the following options:

Quotient option: These options are call or put options on the ratio of two assets. It basically pays

$$\text{Max}\left(\omega\frac{S_2}{S_1} - \omega K, 0\right). \tag{10.27}$$

Exchange option: This option gives the holder the right to exchange one asset for another one, given by:

$$\text{Max}(\omega S_2 - \omega S_1, 0). \tag{10.28}$$

Correlation digital: It a digital option that pays a call on a given asset with a digital condition on another asset:

$$\text{Max}(\omega S_1 - \omega K, 0) \cdot 1_{\{\omega_2 S_2 > \omega_2 K\}}, \tag{10.29}$$

where ω_2 is equal to 1 or -1 depending if the digital is an up or down digital.

Best-of/Worst-of for two assets, it pays a call or a put option on the best or the worst of the two assets. In the case of the best performing asset, this is given by:

$$\text{Max}(\omega\text{Max}(S_2, S_1) - \omega K, 0), \tag{10.30}$$

while this is given by:

$$\text{Max}(\omega\text{Min}(S_2, S_1) - \omega K, 0) \tag{10.31}$$

in the case of the worst performing asset.:

Spread option: It is a call or *put* option on the spread of two assets and this is given by

$$\text{Max}(\omega(S_2 - S_1) - \omega K, 0). \tag{10.32}$$

10.2.8. *Option on options*

Options on options are a typical product that depends on volatility of volatility as they imply some big convexity in the volatility itself. These options encompass the following types:

Compound options: They are constructed in four ways:

- call of call,
- call of put,
- put of call,
- put of put.

If we consider a BS model, the payoff for a compound option is given as:

$$\max\{\varphi PV_t[\max(\theta S^* - \theta K_u|T,0)]^- \varphi K, 0\}, \tag{10.33}$$

where S^* is the value of the stock underlying the underlying option, K_u is the underlying strike price and K is the compound strike, t is the expiry date of the compound and T is the expiry date of the underlying option.

The variables φ and θ are binary variables in that they take either values of 0 or 1. θ is given as 1 when the underlying option is a call, and -1 when the underlying option is a put option. φ is given as 1 when the compound is a call and -1 when the compound is a put.

Options on options are very sensitive to volatility of volatility. Their pricing should therefore used models that account for the curvature of the volatility smile, like stochastic volatility models.

10.2.9. *Chooser options*

Chooser options are options which allow the holder to choose whether their option is a call or a put at a particular date. Indeed, there are two kinds of chooser options: the simple and the complex one, but we detail here only the payoff of the simple chooser which is like:

$$\max\{C_t(S_t, K), P_t(S_t, K)\}, \tag{10.34}$$

where K is the strike, S is the underlying, $C(P)$ are the price respectively of the call (put).

Floortion on the interest rates side: A Floortion is simply the right but not the obligation to buy or sell an interest rate floor at some defined point in the future for a defined premium.

10.3. Equity Derivatives

Equity derivatives have developed extensively in the arena of innovative correlation products. Stronger correlation products, based on more than two underlyings, have been very popular among investors as they offer correlation pick up, coming from more diversifications. Their range includes the mountain range type products as well as the various equity exotic basket options.

10.3.1. *Complex correlation/multi-asset options*

Stronger correlation products, based on more than two underlyings have been very popular among investors as they offer correlation pick up, coming from more diversification. Their range includes the mountain range type products as well as the various equity exotic basket options. This has been the subject of intensive innovation from equity derivatives dealers. In particular, the complex correlation multi-asset options includes the following common options.

Outperformance options: These options basically pay the out performance of one asset over a basket of assets.

$$\text{Max}\left(\omega\frac{S_2(T)}{S_2(0)} - \omega\frac{1}{N}\sum_{i=1}^{N} S_i(T), 0\right). \tag{10.35}$$

Rainbow and variants: A standard rainbows pays the best or the worst performing asset in a given basket. Variant of this includes the difference between the best performing and the worst performing asset, known as spread rainbow, or the difference between the first best performing asset and the third best performing asset known as ranked rainbow and so on. In the equity derivatives world, a rainbow paying the worst performing asset on a large basket and on long-term maturities (10–20 year) is sometimes referred to as an Everest option.

$$\mathrm{Max}(\omega \mathrm{Max}(S_1, \ldots, S_n) - \omega K, 0) \tag{10.36}$$

for a rainbow on the max, and

$$\mathrm{Max}(\omega \mathrm{Min}(S_1, \ldots, S_n) - \omega K, 0) \tag{10.37}$$

for a rainbow on the min, and

$$\mathrm{Max}(\omega \mathrm{Max}(S_1, \ldots, S_n) - \omega \mathrm{Min}(S_1, \ldots, S_n), 0) \tag{10.38}$$

for a rainbow spread.

Corridor: It is a sum of digital options and pays a fixed coupon times the number of days that a certain underlying is within a range. Its payoff is given by, for a low strike K_1 and an upper strike K_2, for n observations of the underlying at the dates $T_1, \ldots, T_n$:

$$C\frac{1}{n}\sum_{i=1}^{n} 1(K_1 < S(T_i) < K_2), \tag{10.39}$$

where $1(K_1 < x < K_2)$ equals 1 when $K_1 < x < K_2$, and 0 otherwise.

Boost: It is a variation around the corridor payoff, where in fact the lower strike of the range applies to one underlying while the upper range applies to another underlying asset:

$$C\frac{1}{n}\sum_{i=1}^{n} 1(K_1 < S_1(T_i))1(S_2(T_i) < K_2). \tag{10.40}$$

Scoop: It is a basically a boost payoff with only one period:

$$C.1(S_2(T_i) < K_2).1(K_1 < S_1(T_i)), \tag{10.41}$$

Cliquet call: It is a (usually capped) call on the performance of an asset.

$$\mathrm{Min}\left(\mathrm{Max}\left(\omega\frac{S_1(T)}{S_1(0)} - \omega K, 0\right), P_{\mathrm{Min}}\right), \tag{10.42}$$

where P_{Min} is the minimum performance.

Accumulator or predator. It is in a sense a variant of the Cliquet option as the principle remains the same (Cliquet call on the performance at each payment date, but with a floor on the sum of the short-term capped performance as opposed to the Cliquet where the floor applies only to each performance). It is therefore given for each payment date by a coupon C_i:

$$C_i = \text{Max}\left(\omega \frac{S_1(T_{i+1})}{S_1(T_i)} - \omega K, 0\right), \tag{10.43}$$

and at maturity it pays the following floor:

$$\text{Max}\left(C_{\min}^{\text{Tot}} - \sum_{i=1}^{N} C_i, 0\right). \tag{10.44}$$

Atlas: Call on a basket where at maturity some of the best and worst performing assets are withdrawn. In the case of withdrawing the best and worst performing asset, this is given by:

$$\begin{aligned}\text{Max}\Bigg(\omega \frac{1}{N-2} \sum_{i=1}^{N} S_i(T) &- \text{Max}(S_1(T), \ldots, S_n(T)) \\ &- \text{Min}(S_1(T), \ldots, S_n(T)) - \omega K, 0\Bigg).\end{aligned} \tag{10.45}$$

Himalaya: At each payment date, it is a call on best performing asset of a given basket. After the asset is paid, it is withdrawn from the basket. So for the first payoff, we pay

$$\text{Max}(\omega \text{Max}(S_1, \ldots, S_n) - \omega K, 0), \tag{10.46}$$

and knowing that the first asset was S_{i_0}, the second payoff is given by:

$$\text{Max}(\omega \text{Max}(S_i)_{i=1..n, i \neq i_0} - \omega K, 0), \tag{10.47}$$

Altiplano: It pays a large coupon if no stock on a given basket hits a certain limit, otherwise it pays a vanilla call on a given basket:

$$C_1 1_{\left\{\omega\frac{1}{N}\sum_{i=1}^{N} S_i<\omega K\right\}} + 1_{\left\{\omega\frac{1}{N}\sum_{i=1}^{N} S_i>\omega K\right\}} \text{Max}\left(\omega_2\frac{1}{N}\sum_{i=1}^{N} S_i - \omega_2 K, 0\right). \tag{10.48}$$

Everest: Call on a basket where at maturity some of the best and worst performing assets are withdrawn. In the case of withdrawing the best and worst performing asset, this is given by:

$$\text{Max}\left(\omega\frac{1}{N-2}\sum_{i=1}^{N} S_i(T) - \text{Max}(S_1(T),\ldots,S_n(T)) - \text{Min}(S_1(T),\ldots,S_n(T)) - \omega K, 0\right). \tag{10.49}$$

Annapurna. Call on the best performing asset of a basket with a participation rate growing with the time at which one of the asset crosses a predetermined barrier level. Its payoff is basically given by:

$$\text{Max}\left(\omega\text{Max}\left(\frac{S_1(\tau)}{S_1(0)},\ldots,\frac{S_n(\tau)}{S_n(0)}\right) - \omega K, 0\right) * P(\tau), \tag{10.50}$$

where $P(\tau)$ is the participation rate and given by ladder level:

Lifetime condition	Participation rate $P(\tau)$
$0 < \tau < 1$y	$x_1\%$
1y $< \tau <$ 2y	$x_2\%$
2y $< \tau <$ 3y	$x_3\%$
3y $< \tau <$ 4y	$x_4\%$
4y $< \tau <$ 5y	$x_5\%$

With $x_1\% < x_2\% < x_3\% < x_4\% < x_5\%$ and τ is the time at which a given asset (it can be the best performing asset) reaches the level L.

Galaxy: This pays at maturity the lowest performance of a basket of assets over the life of the product. This option is closed in spirit to

Napoleon option. Its payoff is basically given by:

$$\text{Min}\left(\frac{S_1(T)}{S_0(0)}, \ldots, \frac{S_n(T)}{S_n(0)}\right). \tag{10.51}$$

Podium. This option pays out a call on a return based on a series of coupons. The value of each coupon is determined by the number of assets that meet certain performance criteria. The coupons are rolled up and paid out at maturity. Its payoff is basically given by for the first coupon:

$$GC_1 = c_1 \sum_{i=1}^{n} 1\left(\frac{S_i(T_1)}{S_i(T_0)} > K_1\right), \tag{10.52}$$

and after for the period, j between T_{j-1} and T_j by:

$$GC_j = c_j \sum_{i=1}^{n} 1\left(\frac{S_i(T_j)}{S_i(T_{j-1})} > K_j\right). \tag{10.53}$$

These coupons are paid at maturity, hence the payoff at maturity is given by

$$\sum_{j=1}^{m} GC_j. \tag{10.54}$$

Napoleon. This option pays out a coupon in each period augmented by the worst performance of the underlying in different sub-periods. The underlying is typically a single index. Coupons can be paid at each period-end or aggregated and paid at maturity In the case of a Napoleon where the coupon is paid at each period, j between $T_{j=1}$ and T_j, the payoff is given by:

$$c_j\left[x_j\% + \text{Min}\left(\frac{S_1(T_j)}{S_0(0)}, \ldots, \frac{S_n(T_j)}{S_n(0)}\right)\right]. \tag{10.55}$$

Over the last years, various equity derivatives on volatility, variance and correlation have been created. This includes the following two options.

Volatility, variance swaps and other volatility products where one agrees to swap a stream of cash flows proportional to a given volatility in exchange for fixed coupon. The payoff of the variance swap is the realized variance of the daily log returns of the underlying stock between start date and maturity, minus a squared strike.

Crash puts refer to very out of the money put option. This option bears the risk of financial market meltdown.

10.4. Exotic Interest Rates Products

Exotic interest products are variation of the standard products but adapted to the interest rates. Techniques of averaging, implicit barrier and Bermudan lead to the most standard products. Here is a list of standard interest rates products.

10.4.1. *Averaging amortizing compounding/accreting swap*

Compared to a standard swap, the notional of the swap is accreting according to the level of the LIBOR. Typical structure is given by the following two legs:

(1) The pay leg pays the Euribor 6M Semi annually on a daycount basis of Act/360.
(2) The receive leg receives a fix lef.

The notional is 100 at the start of the swap and afterwards it is equal to previous notional times (100-previous LIBOR). This kind of structure is easy to price as it only includes payment lag and convexity correction. These are the only risks.

10.4.2. *Autocap, chooser cap, flexi cap*

Compared to a standard cap, an autocap is an interest rates product that allows its holder to exercise a given number of caplets out of the total range of caplets.

Let us say, for instance, that the total number of caplets is 20 caplets and that the chooser cap allows the exercise of only 10 caplets. These 10 caplets can be the first 10, in which case, the product is referred to as an automatic cap or auto cap, or it allows to choose the best 10, in which case, it is referred to as a chooser cap.

This product is very sensitive to correlation risk and should be priced with a 2-factors model to capture the correlation risk. Because correlation is captured in short rates models like the HW and its extension by the mean reversion, a 1-factor version could match the two factors but with very strong mean reversion. However, although a 1-factor model may price similarly the product, its Greeks will be very different.

10.4.3. *Callable reverse floater*

In a reverse floater swap, one pays a reverse floater given by a strike minus a floating rate, usually a money market rate given by K-LIBOR. Usually, the reverse floater coupon is capped and floored with a max coupon and a min coupon. A typical structure is given by the following deal.

In a 10-year structure,

(1) we pay:
  (a) on year 1: a fixed coupon of 5% on a daycount basis of 30/360;
  (b) then for the next year, the coupon will be reverse floater coupon with a floor:

Years	Reverse floater	Additional conditon
2	07.00% – 1Y LIBOR	Floored at 2.00%
3	07.50% – 1Y LIBOR	Floored at 2.50%
4	08.00% – 1Y LIBOR	Floored at 3.00%
5	08.50% – 1Y LIBOR	Floored at 3.50%
6	09.00% – 1Y LIBOR	Floored at 4.00%
7	09.50% – 1Y LIBOR	Floored at 4.00%
8	10.00% – 1Y LIBOR	Floored at 4.00%
9	10.50% – 1Y LIBOR	Floored at 4.50%
9	11.00% – 1Y LIBOR	Floored at 4.50%

(2) we receive the 6 month LIBOR — 5 bps on a daycount Act/360.

Typical modeling should account for accurate pricing of the reverse floater capped and floored leg and a good way to capture the callability on equivalent strikes.

Best is to force our model to correctly reprice the underlying either by computing equivalent strikes in the reverse floater leg or by using closed form pricing of the underlying with external volatilities to value accurately the reverse floater leg.

This strategy is referred to as a local model where the model is only responsible for the pricing of the Bermudan option and the closed form pricing guarantees accurate modeling of the underlying.

10.4.4. *Target redemption note (TARN)*

Target accrual redemption note refers to any financial product whose life is conditional to the fact that the sum of the paid coupons is below a target level. When the target level is crossed, the deals get cancelled.

Potentially, there exists a target floor level. This means that if at maturity, the target level is not reached, we pay a last coupon to reach the target floor level. Mathematically, this last coupon is equal to

$$\mathrm{Max}(\mathrm{TFL} - \mathrm{SPC}, 0), \tag{10.56}$$

where TFL stands for target floor level, SPC for the sum of previously paid coupons.

TARN structures can be with very different coupons:

(1) coupon on the difference between swap rates referred to as the CMS rates;
(2) coupon on reverse floater;
(3) any variation on the coupons.

In a steep increasing curve environment, reverse floater coupons may be worth very little for long maturities. Investors may therefore want to exit the structure whenever these are worthless. Simultaneously, they may have in mind a certain target level for the sum of all the received coupons. Reverse floater TARN answers this problem elegantly. This features basically means that the structure is knocked out whenever the sum of all the paid coupons has reached a certain target level.

For instance, in a standard TARN structure

(1) we receive a floater + spread funding leg.
(2) we pay a fixed coupon over a certain period,

afterwards, a reverse floater coupon potentially capped and floored.

The structure is cancelled whenever the sum of all the paid coupons is above a certain level called the lifetime cap. In addition to the lifetime cap, the structure may have as well a lifetime floor. A lifetime floor will guarantee the investor that the sum of the paid coupons should reach a certain level. If not the case, the structure pays at maturity the difference between the lifetime floor and the sum of all paid coupons. The latter is the standard.

Last but not least, whenever the lifetime cap target is reached, the structure may pay either the full coupon or just the difference between the cap and the sum of previously paid coupon.

There exist many variations around the reverse floater TARN. TARN on CMS spread are TARN product but with a coupon based on the difference of swap rates referred to as CMS rates. TARN CMS Spread have developed more recently than reverse floater TARN as they are more difficult to price.

Product example: Typical deal for a TARN on CMS spread. The paid coupon is based on CMS spread. It is usually capped and floored. The data necessary for computing the price are given in the following table:

Start date	7 Sep 2006
End date	7 Sep 2012
Coupon	5*(Cms10y-Cms2y) + 7bps, capped and floored to 1% and 6% respectively
Frequency	Annual
Daycount	30/360
Calendar	TARGET

TARN condition	If SumPaidCoupon<= 13%, pays the current coupon Else Max(13% – SumPaidCoupon(t–1),0) At maturity, pays Max(11% – SumPaid-Coupon,0)

The TARN condition means the following condition:

(1) If the sum of the paid coupons up to the current period is above the target of 13%, the structure stops to exist.
(2) If the sum of the paid coupons up to the current period has just crossed the target of 13%, we pay the difference between 13% and the sum of the paid coupons up to the previous period. This difference is such that the sum of the paid coupons reaches 13%.
(3) In the other cases, if the sum of the paid coupons up to the current period is below the target of 13%, we pay the current coupon.

If at maturity the sum of the paid coupons is below the target floor level of 11%, we pay the complement to reach 11% for the sum of paid coupon.

The product interest in the case of the TARN on CMS spread is obvious. The client expects first coupon to be big ones and does not want to an exotic structure as soon as his/her target has been reached. CMS spread coupons indicate that he/she anticipates that the yield curve will remain steep during the first years. More precisely, the structure is very adequate if the first years, the yield curve remains steep and flattens after a few years when the structure has been knocked out. TARN on CMS spread is therefore a bet on the steepness of the curve. The TARN feature can make the deal cheaper.

This kind of product can be easily priced with the software of Pricing Partners, Price-it®. Using the generic pricer api, we can described the TARN as follows:

(1) provide a good description of the dates, given in Table 10.1;
(2) describe the swap rates used in the product, the coupon and the sum of the coupon as follows: coupon = Min(Max(CMS_Spread, 1%),6%)*Interest_Terms of the period, with the CMS_Spread given by 5*(CMS10[i]-CMS2[i])+0.0007.

Table 10.1: Details of the dates in the deal.

Reset_Dates	Start_Dates	End_Dates	Interest_Terms
5-Sep-06	7-Sep-06	07-Sep-07	1
5-Sep-07	7-Sep-07	08-Sep-08	1.002777778
4-Sep-08	8-Sep-08	07-Sep-09	0.997222222
3-Sep-09	7-Sep-09	07-Sep-10	1
3-Sep-10	7-Sep-10	07-Sep-11	1
5-Sep-11	7-Sep-11	07-Sep-12	1

Within Price-it®, we should say that the CMS10 is given by a swap rate on the Euro model with a start date given by the value of the start date in the previous table (Table 10.1) and with a tenor of 10Y. To refer to a specific cell, the user of Price-it®, can give the name of the column (in this case Start_Dates) followed by [and the line number]. For instance, to refer to the first line of the column Start_Dates (which is 7-Sep-06), one should say Start_Dates[1]. We should write for the first line: SwapRate(EUR,Start_Dates[1],2Y).

Similarly, for the CMS 2 years, that starts at each start dates of Table 10.1, we should write for the first line: SwapRate(EUR,Start_Dates[1],2Y) (Table 10.2).

Describe the TARN condition, saying that the deal cancels as soon as the sum of the coupon is above 13%, with a target floor of 11%. This is given in Table 10.3.

Table 10.2: Details of the component of the deals.

CMS10	CMS2	CMSS	Coupon
SwapRate(EUR, Start_Dates[1],10Y)	SwapRate(EUR, Start_Dates[1],2Y)	5*(CMS10[1]-CMS2[1])+0.0007	Min(Max(CMSS[1], 1%),6%)*IT[1]
SwapRate(EUR, Start_Dates[2],10Y)	SwapRate(EUR, Start_Dates[2],2Y)	5*(CMS10[2]-CMS2[2])+0.0007	Min(Max(CMSS[2], 1%),6%)*IT[2]
SwapRate(EUR, Start_Dates[3],10Y)	SwapRate(EUR, Start_Dates[3],2Y)	5*(CMS10[3]-CMS2[3])+0.0007	Min(Max(CMSS[3], 1%),6%)*IT[3]
SwapRate(EUR, Start_Dates[4],10Y)	SwapRate(EUR, Start_Dates[4],2Y)	5*(CMS10[4]-CMS2[4])+0.0007	Min(Max(CMSS[4], 1%),6%)*IT[4]
SwapRate(EUR, Start_Dates[5],10Y)	SwapRate(EUR, Start_Dates[5],2Y)	5*(CMS10[5]-CMS2[5])+0.0007	Min(Max(CMSS[5], 1%),6%)*IT[5]

Table 10.3: Details of the TARN condition.

SumCoupon	Alive	TARNCoupon
Coupon[1]	SumCoupon[1] < 0.13	Min(0.13)
SumCoupon[1] + Coupon[2]	SumCoupon[2] < 0.13	(Alive[2]*Coupon[2] + (1−Alive[2])* Max(13%−SumCoupon[1],0))*DF[2]
SumCoupon[2] + Coupon[3]	SumCoupon[3] < 0.13	(Alive[3]*Coupon[3] + (1−Alive[3])* Max(13%−SumCoupon[2],0))*DF[3]
SumCoupon[3] + Coupon[4]	SumCoupon[4] < 0.13	(Alive[4]*Coupon[4] + (1−Alive[4])* Max(13%−SumCoupon[3],0))*DF[4]
SumCoupon[4] + Coupon[5]	SumCoupon[5] < 0.13	Max(11%−SumCoupon[4],0)*DF[5]

B	C
Calendar	TARGET
Start_Dates	7-Sep-06
EndDate	7-Sep-12
Frequency	A
DayCount	30/360
Min	1%
Max	6%
Leverage	5
Spread	0.07%
CMS1	10Y
CMS2	2Y
TargetLevel	13.00%
TargetFloor	11.00%
DateTable	DateTable

E	F
Gensecuri	GenSecurity
Model	HW1FMod
Pricer	GenPricer

Average Coupon	
Regular	TARN
3.61%	

	RegularCoupon	TARNCoupon	DF	NotTARNED?
Price	18.411%	11.874%	5.11	
StdDev	0.003%	0.002%	0.00	
Intermedia	4.998%	4.998%	94.413%	
	3.954%	3.954%	90.868%	
	3.075%	2.921%	87.120%	
	2.488%	0.000%	83.265%	
	2.123%	0.000%	79.400%	
	1.772%	0.000%	75.602%	

				SwapRate(EUR,Start_Dates[i],	SwapRate(EUR,Start_Dates[i],2Y)	5*(CMS10[i]-CMS2[i])+0.0007	Min(Max(CMSS[i],0.
Reset_Dates	Start_Dates	End_Dates	IT	CMS10	CMS2	CMSS	Coupon
5-Sep-06	7-Sep-06	07-Sep-07	1	SwapRate(EUR,Start_Dates[i],	SwapRate(EUR,Start_Dates[i],2Y)	5*(CMS10[i]-CMS2[i])+0.0007	Min(Max(CMSS[i],0.0
5-Sep-07	7-Sep-07	08-Sep-08	1.002778	SwapRate(EUR,Start_Dates[i],	SwapRate(EUR,Start_Dates[i],2Y)	5*(CMS10[i]-CMS2[i])+0.0007	Min(Max(CMSS[i],0.0
4-Sep-08	8-Sep-08	07-Sep-09	0.997222	SwapRate(EUR,Start_Dates[i],	SwapRate(EUR,Start_Dates[i],2Y)	5*(CMS10[i]-CMS2[i])+0.0007	Min(Max(CMSS[i],0.0
3-Sep-09	7-Sep-09	07-Sep-10	1	SwapRate(EUR,Start_Dates[i],	SwapRate(EUR,Start_Dates[i],2Y)	5*(CMS10[i]-CMS2[i])+0.0007	Min(Max(CMSS[i],0.0
3-Sep-10	7-Sep-10	07-Sep-11	1	SwapRate(EUR,Start_Dates[i],	SwapRate(EUR,Start_Dates[i],2Y)	5*(CMS10[i]-CMS2[i])+0.0007	Min(Max(CMSS[i],0.0
5-Sep-11	7-Sep-11	07-Sep-12	1	SwapRate(EUR,Start_Dates[i],	SwapRate(EUR,Start_Dates[i],2Y)	5*(CMS10[i]-CMS2[i])+0.0007	Min(Max(CMSS[i],0.0

PRICING PARTNERS

Figure 10.1: Spreadsheet for the TARN using Price-it®.

All the details provided in Table 10.3 can be easily implemented within Price-it®, the pricing tool developed by Pricing Partners. An example of a spreadsheet that computes the TARN will look like Fig. 10.1.

Let us look in more details about the result provided by Price-it®. In Price-it®, we can get the details of the product. The TARN leg price is worth 11.874%, while the leg without TARN condition is 18.411% (Regular coupon's price). The average coupon is in this case 4.33% for the TARN structure, while it is 3.61% for the average coupon. It proves one of the advantage of the product. With Price-it®, we can also compute the average life of the deal. We have found 3.02 years (Fig. 10.2.)

We can also be interested by the sensitivities (Greeks) of the deal (Fig. 10.3) to the interest rates market data point. We find the following result:

- the negativity of delta for the short maturity (from 3Y to 7Y) which is due to the CMS 2Y;

	E	F	G	H	I	J	K
1							
2	Gensecurity	GenSecurity		Average Coupon			
3	Model	HW1FMod					
4				Regular	TARN	PRICING PARTNERS	
5	Pricer	GenPricer		3,61%	4,33%		
6							
7		RegularCoupon	TARNCoupon	DF	NotTARNED?		AverageLife
8	Price	18,411%	11,874%	5,11	2,74		3,02
9	StdDev	0,003%	0,002%	0,00	0.00		
10						Proba	
11	IntermediatePrice	4,998%	4,998%	94,413%	94,413%	100,000%	
12		3,954%	3,954%	90,868%	90,868%	100,000%	
13		3,075%	2,921%	87,120%	87,120%	100,000%	2,9445
14		2,488%	0,000%	83,265%	1,542%	1,851%	0,0741
15		2,123%	0,000%	79,400%	0,000%	0,000%	-
16		1,772%	0,000%	75,602%	0,000%	0,000%	-

Figure 10.2: Result for the TARN using Price-it®.

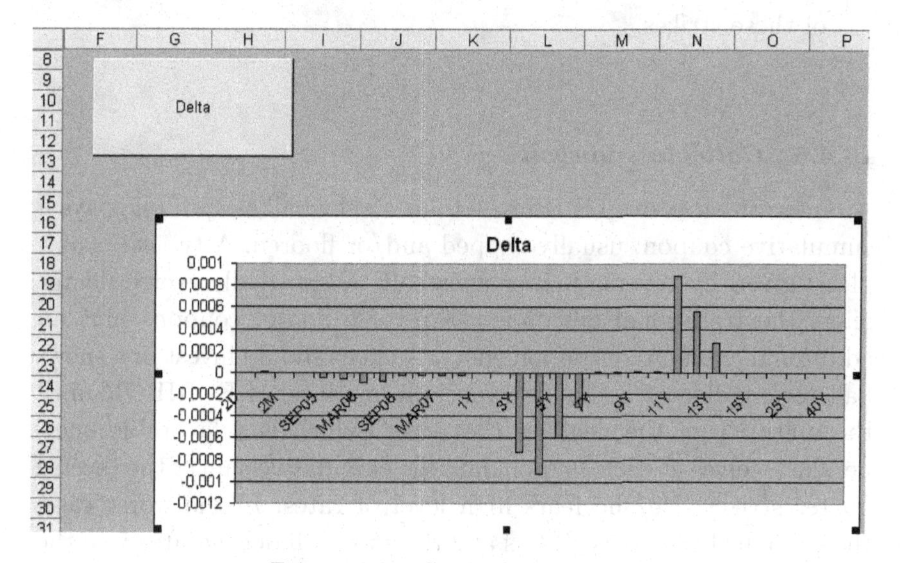

Figure 10.3: Greeks for the deal.

- the delta changes of sign because of the influence of the CMS 10Y;
- from 14 Y to 16Y, the delta is truncated to null due to the TARN effect.

In terms of pricing, the product's risk lies in the following:

(1) The Good valuation of the underlying. This means to find the accurate pricing of the CMS spread, which leads to convexity correction on CMS. Market practice is to use SABR model with replication (see, for instance, Chapter 6).
(2) The accurate modeling of the CMS spread distribution at the various strikes.
(3) The accurate modeling of the joint distribution of the CMS, indicating that this is a real correlation trade.
(4) The TARN condition is an implicit put on the sum of the paid coupons. This Asian feature implies that an accurate modeling should correctly repriced the corresponding Asian option. With respect to the TARN condition, one can compute the estimated strikes by taking the forward value of the coupon and see at each dates what is the corresponding average strike. The drawback of this method is that it does not account for the stochastic nature of these strikes.

10.4.5. *Callable snowball*

An interest rates swap is referred to as a snowball swap if one pays a cumulative coupon, usually capped and/or floored. A typical example is given by reverse floater snowball. A snowball reverse floater offers the traditional advantage of reverse floater coupons plus an additional cumulative coupon effect. A client should invest in a snowball reverse floater if he/she expects a low level of EURIBOR until the maturity of the deal. In that case (which is a favorable one), he/she receives at each coupon date the cumulative sum of the reverse floater strikes. He/she fears high level of rates. In the worst case, the coupons become worthless and the snowball degenerates into the corresponding classical reverse floater, usually quite in the money and consequently worthless. Compared to a standard reverse floater swap, the snowball feature magnifies the coupon because of the cumulative effect on the reverse floater strikes which are increased by previous paid coupons. In addition, the right to call given to the issuer leads to higher coupon for the investor. This explains why this structure has been very popular in the last few years.

A typical example of a callable snowball is the following (example taken from Pricing Partners):

- Start date: 7 sep 2006
- End date: 7 sep 2011
- Schedule of coupon.

Semester	Issuer pays the following coupon
1	6%
2	Coupon[i−1] + [3.80%] − 6m LIBOR
3	Coupon[i−1] + [3.80%] − 6 m LIBOR
4	Coupon[i−1] + [3.80%] − 6 m LIBOR
5	Coupon[i−1] + [4.00%] − 6 m LIBOR
6	Coupon[i−1] + [4.00%] − 6 m LIBOR
7	Coupon[i−1] + [4.20%] − 6 m LIBOR
8	Coupon[i−1] + [4.20%] − 6 m LIBOR
9	Coupon[i−1] + [4.40%] − 6 m LIBOR
10	Coupon[i−1] + [4.40%] − 6 m LIBOR

- Funding leg:
  - the issuer receives: LIBOR 6 m + 10 bps;
  - the frequency of the funding led is semi annual:
  - the daycount is 30/360.

The calendar for the two legs is TARGET. The issuer has the right to call after 2 years.

In terms of pricing analysis, the product's risk lies in:

(1) Good valuation of the underlying. The underlying is a reverse swap. Its good valuation lies in the good valuation of the various reverse floater coupons depending on integrated volatility at the implied strikes of the cap and potentially the floor of the reverse floater coupons. These effects are easy to value as they are just payment lags.
(2) Good valuation of the exercise boundary. The callable snowball is priced with a BGM model with an American Monte Carlo using Longstaff Schwartz algorithm. It is, therefore, important to calibrate the model on swaption with meaningful strikes to capture the callability risk.

Like the TARN product, the callable snowball product can be easily priced within Price-it®. Figure 10.4 provides the delta profile done with Price-it®.

We can notice in Fig. 10.4 that the snowball effect tends to reduce the delta except for the last delta (7 -years) that accumulate (in a sense) all the trajectories that are not called.

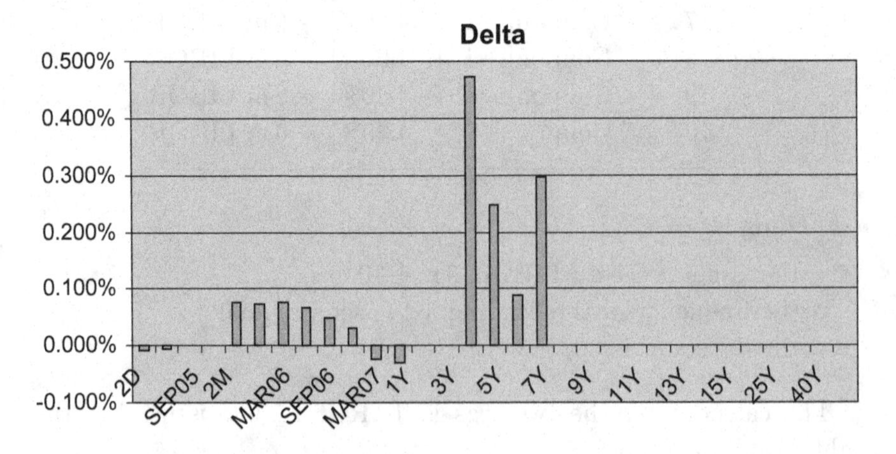

Figure 10.4: Delta of a callable snowball.

10.4.6. *Callable spread options*

In the interest rates derivatives market, callable spread options are products that pay a coupon based on the spread of CMS versus a funding. A typical structure is the following: the client receives 6.75% the first year and the year after a fixed coupon times the number of business days the spread CMS30–CMS10 is above 0% over the total number of business days over the period. In return, he pays LIBOR minus 10 basis points.

This kind of product can either be priced with a rich model that generates a coherent correlation structure such as a forward market LIBOR or uses the concept of local model where the pricing of the spread options is computed analytically such as to match the pricing of the underlying as seen from today.

10.4.7. *New types of underlying and options*

Over the last few years, a wide range of derivative products on energy, inflaton, tax, funds, weather have been created. These markets share in common an important liquidity risk as well as very high bid offer making it a juicy market for banks ready to bear the liquidity risk and ready to warehouse the risk.

Constant proportion portfolio insurance (CPPI): It refers to a trading strategy that is designed to ensure that a fixed minimum return is achieved either at all times or, more typically, at a set date in the future. It involves continuously rebalancing the portfolio of investments during the term of the product between so called risky assets and non-risky assets (usually the bonds or the cash).

10.4.8. *Management of a trading book*

With the increasing complexity and increased professionalism of the derivatives industry, managing a trading book can be a major challenge if not done properly with the proper method and proper tools.

10.4.9. *Risk class*

First of all, any trading strategy should aim at analyzing correctly the various risks included in the product. Standard risk classification should aim at looking at the various risk classes and decomposing the book risk according to these risk classes.

10.4.9.1. *Interest rate risk*

Interest rate risk lies almost in every book due to cash balance variation as well as funding cost of the desk. Usually, this risk is monitored by the funding group in charge of financing the various desks of the bank. More important interest rate risks comes from interest rates derivatives trading desk, where the risk is a delta or gamma risk since it comes from the underlying itself.

10.4.9.2. *Volatility risk*

Volatility risk is quite different from delta risk as this is in common cases more a model risk than a market risk as volatility is just the translation of the reality into the BS model. Assuming the world were BS, this risk would not exist while the delta and gamma risk would still remain. Volatility risk is therefore of a different nature from delta risk although these two risks can have some correlation that can be captured in the vanna.

10.4.9.3. *Basis risk*

Basis risk comes from the fact that floating rates in different currencies bear some additional risk traded as basis risk through the basis swap market. Because of a discrepancy of demand and offer in receiving floating funding, a floating leg in one currency is not equal to a floating leg in another currency. Basis risk is mainly strong on currencies like JPY vs. USD, or JPY vs. EUR.

10.4.9.4. *Smile*

Smile risk is strong on digital products as these products depend not only on the level of the volatility but also the slope of the volatility at the digital strike. This can be simply seen from the fact that a digital option is the limit of a call spread at a strike K and $K + \varepsilon$. This is given by

$$\text{Digital\_Call} \approx \frac{\text{Call}(S_0, K) - \text{Call}(S_0, K + \varepsilon)}{\varepsilon},$$

where Digital_Call is an option paying one for $S_T > K$ and 0 otherwise.

If we value the call option in BS, and we have a smile, hence an implied volatility depending on the level of the strike $\sigma(K)$, we have

$$\text{Digital\_Call} \approx \lim_{\varepsilon -> 0} -\frac{\partial C\,(S_0, K)}{\partial \varepsilon} - \text{Vega}.\frac{\partial \sigma}{\partial K},$$

which says that a digital call is correctly valued with a BS digital option corrected by the vega times the slope of the smile at the digital strike.

10.4.9.5. *Correlation*

Correlation risk is quite particular in the sense that it is very hard to hedge the correlation risk as there is in most cases (most products) no underlying for the correlation.

10.4.9.6. *Foreign exchange*

In addition to FX risks from FX books, FX risk comes naturally in multi-currency books. This is the case of hybrid books where the various underlyings are denominated in different currencies.

10.4.10. *Risk management*

Proper risk management of derivatives books is to get the various Greeks, understand the potential colinearity between these factors and choose to hedge the components where you see some risk and take the risk on the rest.

10.4.10.1. *Delta, Gamma, Vega hedging*

Standard hedging of option lies in delta hedging. Gamma hedging should be done to hedge against large move while vega hedging should be undertaken for anticipated implied volatility move. For example, in BS world, gamma hedging is more robust than delta hedging because of minimizing the tracking error, and that we do not need a lot of trading dates to have a good estimation of the price, so it minimize also the cost of the transactions.

10.4.11. *PNL explanations (Greeks)*

Last but not the least, Greeks should be used to explain PNL as hedging lies in first or second order hedging. Hence, with stable hedge and with small market moves, the PNL explanation coming from Greek prediction should be closed to the realized PNL. This PNL explanation with Greeks should be the opportunity to back test not only the stability of the hedge but also the small importance of second order hedging if ignored.

10.5. Summary

In this chapter, we have seen many exotic derivatives; we detailed some recent products, like TARNs, callable snowball, moutain range and so on. For these derivatives, we must take in consideration of the structure of the payoff, the different Greeks, the correlation between the assets (the more riskiest asset, ...), the eventual correction (for the pricing, like convexity adjustment for the TARN), the different risks (interest, volatility, FX, ...). Finally, we observed that the market of derivatives becomes more and more complex, because of the apparition of new domains like inflation, weather, and the expansion of the exotics derivatives.

References

[1] Ahn H, Penaud A and Wilmott P (1999). Various passport options and their valuation. Preprint, Mathematical Finance Group, Mathematical Institute, Oxford University.

[2] Cont R, Copinot R, Jaeck C and Bruyère R (2004). *Les produits dérivés de credit.* Paris: Economica.
[3] Das S (2005). *Credit Derivatives: CDOs and Structured Credit Products.* John Wiley.
[4] Martellini L and Priaulet P (2004). *Produits de Taux d'Intérêt: Méthodes dynamiques d'évaluation et de couverture.* Paris: Economica.
[5] Piterbarg V (2005). A multi-currency model with FX volatility skew. Available at http://ssrn.com/abstract=685084, February 7, 2005.
[6] Rebonato R (2002). *Modern Pricing of Interest-Rate Derivatives: The LIBOR Market Model and Beyond.* Princeton University Press.
[7] Taleb NN (1996). *Dynamic Hedging: Managing Vanilla and Exotic Options.* John Wiley.

Chapter 11

THIRD GENERATION TRADING SYSTEM AND ITS UNDERMINING COPERNICAN REVOLUTION

11.1. The New Generation of Trading Systems

Like many other technology-based systems, trading systems have undergone major evolution over the last few years. Although it takes relatively some time to build up a complete trading system, at least a couple of years, banks as well as software companies have pushed up the limits of existing trading systems to new frontiers.

(1) First of all, pricing libraries have been rewritten with the aim of fast upgrade and genericity of their key components. Third generation pricers referred to as generic pricers are becoming more and more common. We will draw more on generic pricing in the rest of this chapter.
(2) Second, development of plug and play numerical methods that are not linked to particular pricers and models but rather adaptable to various models and pricers are becoming common architecture philosophy of pricing libraries. This means that Monte Carlo engines are not related to any particular model but rather can deal with any model and any payoff if designed from day one to be both model and payoff free. The same applies obviously for American version of Monte Carlo engine, PDE solver and tree methods.
(3) Third, performances have come under scrutiny as hybrids structures require modeling many factors. Because of high performance requirements, C++ keeps its dominating situation as a very efficient object-oriented language. C lacks genericity and has

been out fashioned. Java still needs serious performance enhancements to come to the competition with compiled languages and is used still mainly as a graphic user interface (GUI) language, competing with proprietary windows based solutions such as C# or Visual Basic. Also Java cannot be easily integrated within Excel which remains the primary tool for end users like traders.

On the technological side, in addition, the key new features have been:

(1) Support of web based interface and distributed solutions with the use of application service providing technology (e.g. PHP or ASP.Net).
(2) With the emergence of hybrid books that requires heavy pricing, distributed computing that consists in distributing the various deals across various computers of a cluster of PCs enables us to price very efficiently these books. Use of distributed/grid computing with clusters of PCs either under Linux or Windows provide fast pricing of trading books by splitting the computing load over various machines, with the use or not of MPI support. And eventually, on a minor scale, the use of recent technologies like XML format for data exchange between various applications.

Innovation of third generation trading systems comes from genericity. With the genericity and the ability to describe financial product payoffs, end users (traders, structurers and sellers) have the possibility to use the same system/library as it is tailor made for structuring and pricing any new financial products. Using a generic system/library is also interesting for risk model control, as it is easy to interchange models and numerical methods for pricing (Fig. 11.1).

In this chapter, we will first give a better overview of generic pricing as this constitutes in our opinion the major evolution of the last few years. We will in particular review the motivation for generic pricing and give the critical tasks for implementing a generic pricing library. We will then conclude with the risk integration of a pricing library and its connection with distributed computing.

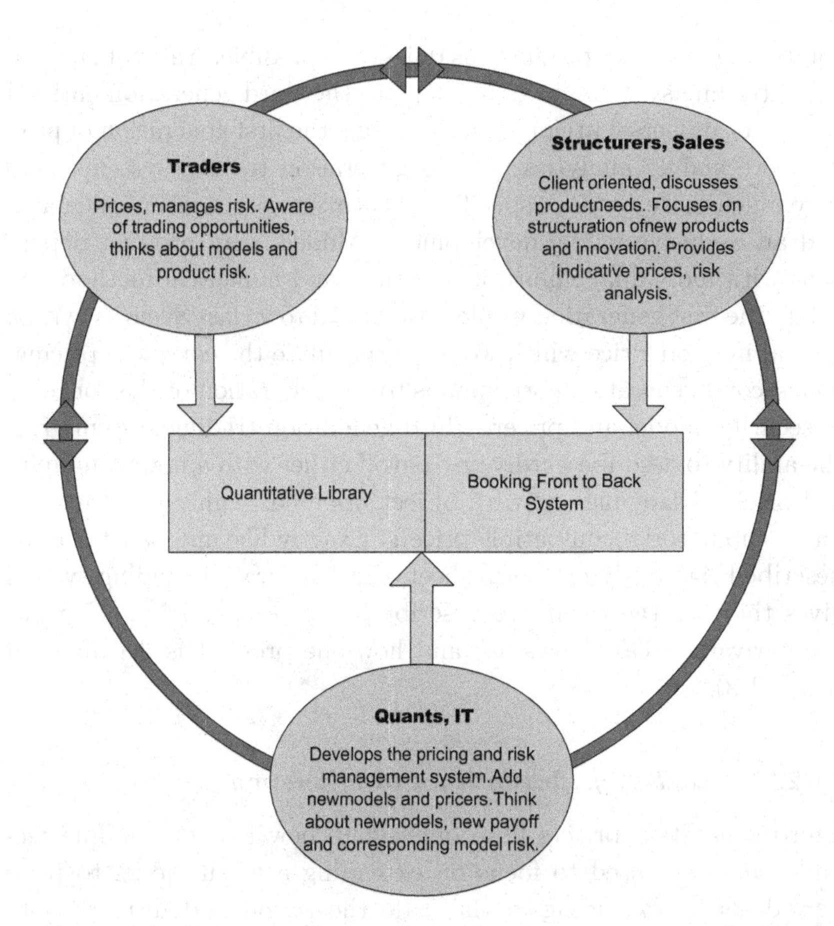

Figure 11.1: Interaction between quants, traders, structurers and sales.

11.2. Motivation for a Generic Pricer

11.2.1. *Product innovation*

In the exotic derivatives business, innovation is a key to success. And while financial engineers have kept up pushing the limits of products to more and more tailor made and computing demanding pricing products, information technology systems have come under pressure to innovate simultaneously. The game has been to produce trading systems that can incorporate these highly customizable and computing intensive products as quickly as possible. And not surprisingly, trading systems have migrated to the third generation entitled generic or universal pricing system. While the first generation of pricing were mainly analytics, second generation trading systems were very much more product specific for each security a given calculator and an exclusive model development. Models were product dependent with individual calibration method and numerical methods. In C++, the first generation would correspond to a class Security with a virtual method Price which would encapsulate the complete pricing. The second generation corresponds to the separation of the concepts of security, model and pricer. Third generation trading system gives the ability to describe hardly any payoff either with a powerful financial oriented language or with object oriented, highly customizable and combinatorial compatible pricers. Exactly like any payoff can be described rigorously in a term sheet, third generation trading system gives the user the ability to describe in almost plain English what the derivatives cash flows are and how the product is decomposed (Fig. 11.2).

11.2.2. *Reactivity, flexibility and efficiency*

Third generation pricers have opened up new horizons. While second generation used to focus on extending a given model to price new deals by extending in the code the payoff and from time to time the numerical method or even worse a twist of the model, third generation pricers require no development for new products. Changing the library for new products could be very risky operationally and requires expertise in the pricing library. But, the ever growing imagination of financial engineers with an increasing pressure to release new products very rapidly have urged to review the process

Rate	DF	CF
LIBOR(EUR,StartDate,6M)	DF(EUR,PayDate)	Rate*Accrual*DF

Figure 11.2: Example of a payoff.

of creating new products in the pricing library and to standardize this in a very effective procedure. Following other industries, like in the airplane industry, where suddenly pilots have received assistance from automatic pilot system, traders have asked for the help of new companions such as a cash flow analyzer and risk analyzer to price any instrument without the need for any heavy quantitative development.

Generic pricers enable reactivity in terms of payoffs, calibration and models. Flexibility is key to the system as all the components have to be designed and implemented to be plug and play. Efficiency in terms of the integration of new products or new pricing of existing products is dramatically improved as the various building blocks are already tested and the integration process consists only in using these tested components.

11.2.3. *Maintenance and evolution*

Not only has this dramatically shortened the process of new product development but it also has given the opportunity to innovate very rapidly. In addition, generic pricers have changed dramatically the relationships between traders, quants and structurers as their role are becoming more and more intricate. But more is to come with potentially standardization for the description of products, like certain try on standardization like FpML. In addition, maintenance and evolution is made simple because generic infrastructure centralizes the behavior of the system. Any change of core components impacts the whole system and propagates to the tiny elements of the systems.

11.2.4. *Front to back, enterprise wide*

Generic system enables us to have standardized behavior for not only the front office but also the middle and back office systems. As the various events of the life of the product can easily be incorporated into the product description itself, the deal management in terms of back office events can be streamlined into standardized ageing processes. For instance, the fixing of cash flows (reset events) or the management of events like exercising the option or not, paying cash flows, etc. can be directly connected to the payoff description. From the payoff description, one can create standardized spreadsheet taking exactly the payoff description as saved in the system. This lowers down dramatically any operational risk when entering in the system the product characteristic as the spreadsheet uses the same framework as the pricing system.

11.3. Example of an Architecture

11.3.1. *Price-it®, the Pricing Partners' generic pricer solution*

In order to demonstrate concretely what should be a generic pricer, following examples and spreadsheets are taken from the Price-it® solution of Pricing Partners.

Pricing Partners develops cutting-edge quantitative solutions for pricing and structuring financial products. Founded in 2005, Pricing Partners teams up experienced senior professionals that worked previously in quantitative research and front IT in leading investment banks such as Goldman Sachs, Société Generale, Commerzbank, Ixis CIB and IMI San Paolo. It has more than 20 years of experience in pricing library development.

Pricing Partners offers a unique solution for independent valuation of financial products thanks to Price-it®, its state-of-the-art software dedicated to valorize derived financial products:

(1) Price-it® technology is based on an innovative concept of generic description of financial cash flows using an intuitive language adapted to the financial products.
(2) Price-it® new and original architecture allows us an easy interchange between the mathematical models, the numerical tools and the calibration parameters used, enabling the user to model

and develop complex financial products according to various assumptions.

11.3.2. *Cash flows vs. events*

A generic pricer is a tool that enables to describe hardly any payoff without any new programming development. Products are completely described in a simple, intuitive but almost exhaustive meta-language whose description structuring abilities and capacities are infinite and encompass the whole range of existing traded products and much more. Like any programming language, its potential is infinite, leaving freedom for imagination. For any financial product, the description part is completely separated from the pricing part. This enables to plug various models to assess model risk as well as model performance simply. In order to have plug and play models as well as numerical methods, library developers need to specify a standardized interface to communicate between pricing models, numerical methods and the payoff or security part.

There are various ways to develop such a tool, out of which, to our experience, the most natural, intuitive and fruitful relies on the concept of cash flows and language. Indeed, a generic pricer can also be built upon the concept of events or combinatorial objects with defined interaction, pricing primitives and/or a script language that supports call back to other languages. FpML has also contributed to spread the overall ideas that products could be completely specified by a general structured language. Taking back the official definition, FpML aims to streamline the process supporting trading activities in the financial derivatives domain through the creation, maintenance and promotion of an e-business language for describing these products and associated business interactions based on industry standards.

A financial product is simply various events happening to some cash flows. So, at one side of the spectrum, the generic pricer can focus on cash flows while on the other side of the spectrum, it can focus on event. We will not go any further on the various directions

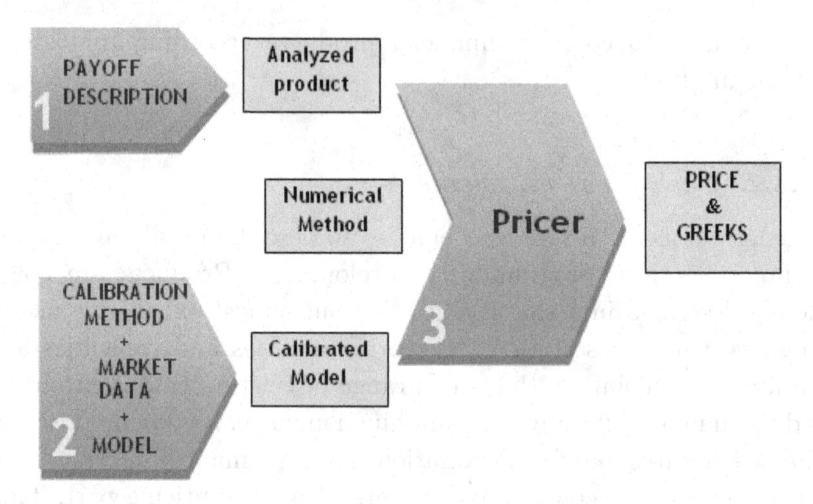

Figure 11.3: Interaction between the core components of a generic pricer.

and will concentrate on the language aspect of a generic pricer (Fig. 11.3).

11.3.3. *Decomposition: security, model and numerical methods*

A pricer is composed of different components such as:

(1) A model that can be a simple model such as a BS model in the equity world or a HW model, or a more complex one like a BGM model with stochastic volatility or any fancy model.
(2) A numerical method that will handle the time discretization and the numerical representation of the model. In order to be model free, the generic numerical method has to interact with the model on characteristic data such as drift and volatility terms, locally and globally.
(3) A payoff analyzer that will combine the payoff description, the model and the numerical method and will be in charge of properly initializing the model, the numerical method and precomputing various elements to have a fast infrastructure.

11.3.4. *Critical tasks for a generic pricer*

In greater details, in order to build a generic pricing engine, we must have:

(1) a simple and intuitive language description, referring to plain English words;
(2) a language independent of any model;
(3) the ability to define rapidly any new keyword;
(4) a great flexibility in terms of combining the various keywords;
(5) the ability to calibrate generically a model;
(6) the ability to split model and numerical methods.

In order to use the generic pricer, the steps are as follows (keeping in mind that these various steps can be fastened with the use of template spreadsheets):

(1) Choose the various market data and the model to use. This creates a default model like an empty shell. The art of pricing and risk management is precisely to choose appropriate model, choosing the appropriate diffusion among the category of models.
(2) Choose the calibration instruments and the model parameters to calibrate and those to freeze.
(3) Describe the product's cash flows to price in the meta language of the generic pricer.
(4) Choose the pricing methodology (if applicable) among Monte Carlo, PDE solver, generic tree and any other numerical methods.

Risk assessment is done by changing some of the inputs and measuring the impact on the price. It can be automated by providing the definition of the security and the input used as market data. A smart system should reassess the dependency links to eliminate any market data non-relevant for the pricing and hence bump only meaningful parameters.

The calibration part can be itself built generically to allow easily new model to be plug and play components of the infrastructure.

In order to get a price, we will need to define the following key variables:

(1) choice of the currency in which to discount (for the multi-asset version);
(2) description of the cash flows;
(3) description of the parameters of the model with their correlation for multi-asset models;
(4) mapping table between keywords and model;
(5) description of the numerical method to use.

11.3.5. *Parsing system and financial language*

11.3.5.1. *Interest of a meta language*

Meta language refers to any language whose syntax is not completely determined once for all. Rather, the language allows the addition of new keywords. Standard language such as C, C++, Java, C#, VB have a number of fixed keywords that have been determined once for all. To include a new concept, the programmer needs to use the combinatorial power of the primitive keywords to build up the new concept.

In contrast, a meta language allows us to incorporate a new keyword for the new concept in the language itself. This allows much more flexibility in the evolution of the language itself, exactly in the same way as English receives new words each year and widens its dictionary. Meta language is very useful in an innovative and fast moving environment as it provides a simple way to cope with the creation of new concept without the need to develop lengthy periphrases for new concept.

11.3.5.2. *Description of the cash flows with a meta language*

The central structure of the generic pricer is a table of cash flows described by the generic pricer language.

A table of cash flows should contain at least two necessary elements:

(1) The event dates, which are the dates at which a cash flow will be examined. The cash flow may not be paid, it may simply fix

and be observed for future referencing. Most of the time those dates will be "reset dates".

(2) Linked to these event dates, a set of cash flows that represents the event of the deal.

As any derivative is really only a collection of cash flows, the description of the deal will be simply the series of event date and the corresponding cash flows.

In order to illustrate this, we will examine the following three real live examples.

First example: TARN CMS spread

TARN CMS spread corridor (see Chapter 12 for more information on this product).

Start	spot+1M
Maturity	10 years
Currency	Euro
Notional	100 million
Issuer swap	
Pays	Euribor16M
Receives	1st year 6% semi-annually 30/360 Then 5.5%* n/N semi-annually where N is the number of days in the semester and n is the number of days when: Euribor6M $\leq$ 3.85 for the 2nd year Euribor6M $\leq$ 4.25 for the 3rd year Euribor6M $\leq$ 4.50 for the 4th year Euribor6M $\leq$ 5.00 for the 5th year Euribor6M $\leq$ 5.25 for the 6th year Euribor6M $\leq$ 5.50 for the remaining time
Condition	Whenever the sum of the received coupon reaches 12%, the swap is cancelled and the last coupon paid is such that the sum of paid coupons equal to 12%
Business days	Target, modified following

Fees upfront 0.50%

Second example: Callable CMS spread snowball
Callable snowball on CMS spread

Start	spot+3M
Maturity	10 years
Currency	Euro
Notional	100 million
Issuer swap	
Pays	4% semi-annually 30/360
Receives	Annually 30/360: 1st year: 3% + Max[1.20-(CMS10Y-CMS2Y),0] 2nd year: Previous coupon + Max[1.55-(CMS10Y-CMS2Y),0] 3rd year: Previous coupon + Max[1.50-(CMS10Y-CMS2Y),0] 4th year: Previous coupon + Max[1.45-(CMS10Y-CMS2Y),0] 5th year: Previous coupon + Max[1.40-(CMS10Y-CMS2Y),0] Afterwards: Previous coupon + Max[1.35-(CMS10Y-CMS2Y),0]
Condition	The swap is callable after year 2 by the issuer Fixing of the CMS in arrears.
Business days	Target, modified following

Third example: Callable CMS spread range note
10 year non-call 1 CMS spread accrual
Callable Snowball on CMS spread

Start	spot+3M
Maturity	7 years
Currency	USD
Notional	100 million
Issuer swap	
Pays	4% semi-annually 30/360
Receives	Annually 30/360:

1st year: 7% annually 30/360
2–7 years: 6.50%* n/N, where n is the number of days where CMS30Y–CMS10Y is above 0%, paid quarterly 30/360, unadjusted

Condition The swap is callable after year 2 by the issuer.
Business days London, New York, Sydney
Modified following, business day convention

From the above three examples, we can see that we should have at least:

(1) a good description of the dates both fixing and payment dates;
(2) a good description of the underlyings: Euribor, swap rates, etc.;
(3) a way of describing the cash flow specification as well as the ability to call or to cancel the structure according to a certain condition. This compels for the use of variables and conditions in the meta language of the generic pricer.

Let us consider a generic pricer based on the table of cash flows where the left most upper column will be the event date at which the cash flow are priced, while the right upper most column will describe the cash flow. Because this cash flow may be quite complex, we may need some intermediate variables for intermediate computation. We could have explicit variables declaration but this would lead to a payoff description language that looks like a programming language.

Rather, our target is to provide a language that is as close as possible to plain English and looks very similar to Excel functionalities. Plain English language is much easier to learn and use for non-programmer users. Having a solution with functionalities close to Excel would give to the language an amazing compositional power, which is exactly the aim of a generic tool.

Based on these considerations, we come up with a simple but still very flexible way of describing cash flows:

(1) A table of cash flows is defined by a matrix with the upper left column being the various event dates of the deal and the upper right column being the cash flows.

(2) Intermediate cells are referred to by a text convention simple to read. Variables are not typed to provide an easy interface.

11.3.5.3. *Referencing*

A column is entitled by a column name. To denote the current row of a given column, we refer the cell by its column name followed by $[i]$. This convention is largely inspired by the one of standard programming language. Hence, to refer a cell of a column entitled ColumnA, we should use ColumnA$[i]$ for the current row, ColumnA$[i+1]$ for the next row, and ColumnA$[i-1]$ for the previous row.

We have given an example of referencing (Fig. 11.4) the current row, the next row, the previous row.

The obvious result are given in Fig. 11.5.

In addition, the rules about the formatting of the table of cash flows impose the following rules. The first row of the cash flow table contains column names and enables us to refer to these columns. The upper left column contains dates in increasing order. The upper right column refers to cash flows. Intermediate cells are used to define implicitly variables. This leads to the following rules:

Rule number 1 about the formatting of the deal description:

(1) The first row should contain name of column which are different. These names will be used to refer cells according to their column title.

Dates	A	CF	CF2	CF3
4-Jan-05	20	0	0	0
4-Jan-06	50	A[i]	A[i+1]	A[i-1]
4-Jan-07	100	0	0	0

Figure 11.4: Example of referencing variables.

Dates	CF	CF2	CF3
4-Jan-05	0	0	0
4-Jan-06	50	100	20
4-Jan-07	0	0	0

Figure 11.5: Result of the referencing variables.

(2) The upper left column should contain dates in increasing order to be used as event dates for the corresponding cash flows.
(3) The upper right column refers to cash flows description. By convention, these cash flows are paid at the event dates of the upper left column.

This leads to the type of cash flows as described in Fig. 11.6.

We will now detail the keywords of the meta language to describe a cash flow and will explain how to describe simple cash flows.

11.3.5.4. *Components of the meta language: Functions and operators*

We can decompose operators used in the cash flow matrix in three categories with respective subcategories:

(I) Standard programming operators:

  (1) Arithmetic
  (2) Comparison
  (3) Logic
  (4) Assignment operator

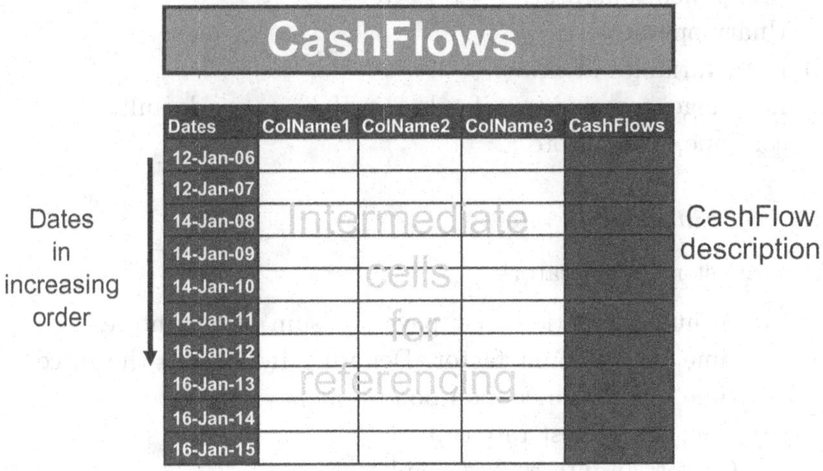

CashFlows

Dates	ColName1	ColName2	ColName3	CashFlows
12-Jan-06				
12-Jan-07				
14-Jan-08				
14-Jan-09				
14-Jan-10				
14-Jan-11				
16-Jan-12				
16-Jan-13				
16-Jan-14				
16-Jan-15				

Dates in increasing order

Intermediate cells for referencing

CashFlow description

Figure 11.6: Example of generic cash flows.

   (5) Unary operator
   (6) Data variables identifier
   (7) Language core syntax

(II) Financial operators:

   (1) Interest rates operators
   (2) Equity, FX and commodities operators
   (3) Inflation operators
   (4) Credit operators
   (5) Miscellaneous functions and variables
   (6) Dates operators

(III) Mathematical:

   (1) Base mathematical
   (2) Advanced mathematical tools

11.3.5.5. *Standard programming operators*

This includes standard operators such as:

(1) Arithmetic: +,−,/,*
(2) Comparison: ==, !=, <, <=, >, >=
(3) Logic or, and, xor
(4) Assignment operator: + =, − =, *=, /=
(5) Unary operator -
(6) Data variables identifier
(7) Language core syntax if, else, switch, case, default, do, for, continue, break, goto

11.3.5.6. *Financial operators*

(1) Interest rates operators:

   (a) Annuity: function that pays the sum of the interest terms time the discount factor. Denoting by $B(t, T)$ the price at time t of a zero-coupon bond that pays 1 at time T.
   (b) Cap: an interest rate cap.
   (c) Caplet: an interest rate caplet.
   (d) DF: discount factor an interest rate risk free discount factor.

   (e) Digital: digital option.
   (f) Libor: the libor rate.
   (g) SwapRate: the swap rate.
   (h) Swap: a simple swap.
   (i) Swaption: a swaption.
   (j) PriceToYield: the yield of a given bond given a price.
   (k) YieldToPrice: the price of a bond given its yield.

(2) Equity, FX and commodities operators:
   (a) Call
   (b) Equity digital
   (c) Forward
   (d) Spot

(3) Inflation operators:
   (a) CPI: the CPI rate
   (b) InfSwap: a simple inflation swap
   (c) InfSwapRate the inflation swap rate

(4) Credit operators:
   (a) Default date
   (b) Loss: cumulated loss
   (c) MarginalLoss: marginal loss
   (d) Recovery

(5) Miscellaneous functions and variable:
   (a) Exercise: to exercise an option. Used mainly for American Monte Carlo type payoffs.
   (b) PV: to take the present value of a cash flow at a certain date.
   (c) Pays: tells to pay a cash flow at a certain date.
   (d) Various closed forms addins to do local modelization.

(6) Dates operators:
   (a) Period for the description of a period of time.
   (b) Date for a given date.
   (c) AddMonths, AddYears, AddDays and so on for date manipulation.
   (d) Calendar for the management of holidays.

(e) DateStrip for the creation of a strip of dates with some accessors methods on the dates.
(f) DayCountFraction.

(7) Mathematical

(a) Base mathematical:

a. Max for the maximum of two values.
b. Min for the minimum of two values.
c. RunningMax for the running maximum of values.
d. RunningAvg for the running average of values.

(b) Advanced mathematical tools (like trigonometric functions and standard gaussian functions):

(a) Exp for exponential.
(b) Log for logarithm.
(c) Sqrt for square root.
(d) Pow for power.
(e) Abs for absolute value.

This leads to keywords shown in Fig. 11.7.

11.3.5.7. *Parsing the grammar and creating a syntax*

The theory of lexing and parsing has been greatly simplified by the usage of street tools such as the GNU lex (flex for the new version), yacc (bison for the new GNU version) or ANTLR. More simple tool such as regex or regular expression enables to control that a certain text is done according to a grammar specification.

Creating a language requires to use lex to "tokenize" in a sense the text while yacc gives its structure. In short, lex helps write programs whose control flow is directed by instances of regular expressions in the input stream. Lex receives as an input a table of regular expressions and the corresponding program action. Lex generates the C code of the deterministic finite automaton to partition the input text into tokens which match the given expressions. Yacc generates the C code for the parser part of the program. The parser receives a collection of tokens and gives semantic meaning to them by executing some codes.

Keywords		
Name	Type	Description
DF	1.Financial	Discount factor
Exercise	1.Financial	Exercise
FwdVal	1.Financial	Remove discounting
ModelFactor	1.Financial	Model Factor
PV	1.Financial	Present value
Trigger	1.Financial	Trigger
Annuity	2.Interest Rates	Annuity (PVBP)
Cap	2.Interest Rates	Cap
Caplet	2.Interest Rates	Cap/Floor-let Option
Digital	2.Interest Rates	Digital Cap/Floor-let
Libor	2.Interest Rates	Libor Rate
PriceToYield	2.Interest Rates	Price to yield.
Swap	2.Interest Rates	Swap
SwapRate	2.Interest Rates	Swap Rate
Swaption	2.Interest Rates	Swaption
YieldToPrice	2.Interest Rates	Yield To Price.
DefaultDate	3.Credit	DefaultDate
DefaultNb	3.Credit	Default Nb
Loss	3.Credit	Cumulated Loss
MarginalLoss	3.Credit	Marginal Loss
Recovery	3.Credit	Recovery
Call	4.Equity	Call or put
EqDigital	4.Equity	Equity digital
Fwd	4.Equity	Fwd
Spot	4.Equity	Spot
CPI	5.Inflation	CPI
InfSwap	5.Inflation	Inflation Swap
InfSwapRate	5.Inflation	Inflation Swap Rate
CFCall	6.Closed Forms	Closed Form Call
CFGreek	6.Closed Forms	Closed Form Greek
If	7.Conditional	If/Then/Else
DCF	8.Dates	Day Count Fraction
Abs	9.Mathematical	Absolute value
Bound	9.Mathematical	Bound Elem
CumNorm	9.Mathematical	Cumulative Normal!
Exp	9.Mathematical	Exponential
FirstElem	9.Mathematical	First Elem
Log	9.Mathematical	Natural Logarithm
Max	9.Mathematical	Maximum of two values

Figure 11.7: Example of keywords.

Other tools such as open source languages like interpreted scripting language like Python or Mathematic or compiled Caml can help to create a generic pricer by using directly their parser and just extending the grammar of the language. Python is easy to connect and has been, for instance, used with success for the Google search engine and is believed to have been used in various banks in a private usage. In this chapter, we will look at a very simple generic pricer.

11.3.5.8. *Example of table of cash flows in a meta language*

Let us start by simple examples such as a LIBOR leg, a cap. We will gradually increase the complexity of the product as we get a feeling of the philosophy behind this generic pricer. This generic pricer is inspired from various conversations as well as experience of our quantitative group across various banks as well as various teams.

LIBOR leg
A LIBOR leg would simply be described by the table of cash flows shown in Fig. 11.8.

Where we indeed says that we pay the LIBOR rate times the interest terms at the end date, hence, at the event date, this has to be discounted from the end date to the event date (in this case the reset date).

Cap
The table of cash flow is shown in Fig. 11.9.

Bermuda swaption.
The Bermuda swaption table of cash flow is shown in Fig. 11.10.

ResetDate	Start	Pay	Acc	Rate
1-Sep-05	5-Sep-05	6-Sep-05	0,0028	Libor(EUR,Start[i],6M)
2-Sep-05	6-Sep-05	6-Sep-06	1,0139	Libor(EUR,Start[i],6M)
4-Sep-06	6-Sep-06	6-Sep-07	1,0139	Libor(EUR,Start[i],6M)
4-Sep-07	6-Sep-07	8-Sep-08	1,0222	Libor(EUR,Start[i],6M)
4-Sep-08	8-Sep-08	7-Sep-09	1,0111	Libor(EUR,Start[i],6M)
3-Sep-09	7-Sep-09	6-Sep-10	1,0111	Libor(EUR,Start[i],6M)

Figure 11.8: Example of a LIBOR leg.

ResetDate	Start	Pay	Acc	Rate	CF
1-Sep-05	**5-Sep-05**	**6-Sep-05**	**0,0028**	**Libor(EUR,Start[i],6M)**	**Max(Rate[i]-3%,0)*DF[i]*Acc[i]**
2-Sep-05	6-Sep-05	6-Mar-06	0,5028	Libor(EUR,Start[i],6M)	Max(Rate[i]-3%,0)*DF[i]*Acc[i]
2-Mar-06	6-Mar-06	6-Sep-06	0,5111	Libor(EUR,Start[i],6M)	Max(Rate[i]-3%,0)*DF[i]*Acc[i]
4-Sep-06	6-Sep-06	6-Mar-07	0,5028	Libor(EUR,Start[i],6M)	Max(Rate[i]-3%,0)*DF[i]*Acc[i]
2-Mar-07	6-Mar-07	6-Sep-07	0,5111	Libor(EUR,Start[i],6M)	Max(Rate[i]-3%,0)*DF[i]*Acc[i]
4-Sep-07	6-Sep-07	6-Mar-08	0,5056	Libor(EUR,Start[i],6M)	Max(Rate[i]-3%,0)*DF[i]*Acc[i]
4-Mar-08	6-Mar-08	8-Sep-08	0,5167	Libor(EUR,Start[i],6M)	Max(Rate[i]-3%,0)*DF[i]*Acc[i]
4-Sep-08	8-Sep-08	6-Mar-09	0,4972	Libor(EUR,Start[i],6M)	Max(Rate[i]-3%,0)*DF[i]*Acc[i]
4-Mar-09	6-Mar-09	7-Sep-09	0,5139	Libor(EUR,Start[i],6M)	Max(Rate[i]-3%,0)*DF[i]*Acc[i]
3-Sep-09	7-Sep-09	8-Mar-10	0,5056	Libor(EUR,Start[i],6M)	Max(Rate[i]-3%,0)*DF[i]*Acc[i]
4-Mar-10	8-Mar-10	6-Sep-10	0,5056	Libor(EUR,Start[i],6M)	Max(Rate[i]-3%,0)*DF[i]*Acc[i]

Figure 11.9: Example of a cap.

Pay	Swap	Fixed	Annuity
8-Jan-05	SwapRate(EUR,Pay[i],7y,12m)	0.04	Annuity(EUR,Pay[i],7y,12m)
8-Jan-06	SwapRate(EUR,Pay[i],6y,12m)	0.04	Annuity(EUR,Pay[i],6y,12m)
8-Jan-07	SwapRate(EUR,Pay[i],5y,12m)	0.04	Annuity(EUR,Pay[i],5y,12m)
8-Jan-08	SwapRate(EUR,Pay[i],4y,12m)	0.04	Annuity(EUR,Pay[i],4y,12m)
8-Jan-09	SwapRate(EUR,Pay[i],3y,12m)	0.04	Annuity(EUR,Pay[i],3y,12m)
8-Jan-10	SwapRate(EUR,Pay[i],2y,12m)	0.04	Annuity(EUR,Pay[i],2y,12m)
8-Jan-11	SwapRate(EUR,Pay[i],1y,12m)	0.04	Annuity(EUR,Pay[i],1y,12m)

Swaption	Option	CF
Max(Swap[i]-Fixed[i],0)*Annuity[i]	0	Max(Swaption[i],PV(Option[i+1]))
Max(Swap[i]-Fixed[i],0)*Annuity[i]	Max(Swaption[i],PV(Option[i+1]))	0
Max(Swap[i]-Fixed[i],0)*Annuity[i]	Max(Swaption[i],PV(Option[i+1]))	0
Max(Swap[i]-Fixed[i],0)*Annuity[i]	Max(Swaption[i],PV(Option[i+1]))	0
Max(Swap[i]-Fixed[i],0)*Annuity[i]	Max(Swaption[i],PV(Option[i+1]))	0
Max(Swap[i]-Fixed[i],0)*Annuity[i]	Max(Swaption[i],PV(Option[i+1]))	0
Max(Swap[i]-Fixed[i],0)*Annuity[i]	Swaption[i]	0

Figure 11.10: Example of a Bermuda swaption.

11.3.5.9. *Split between models and numerical methods*

In order to have flexible numerical methods, one should provide a certain number of default methods that allow us to have plug and play numerical methods. This means in particular that a numerical method should know how to induct (either backward or forward given the status of the world, in a Monte Carlo, the current paths, or in a tree the current slice). In addition, it may be useful to have objects representing the numeraire as an object. Numeraire and model should dialog in order to change the representation of the model under different models.

In addition, to have the ability to describe generically the calibration method on an object, one should use as building blocks the model parameters. Hence a calibration method should say that it calibrates (optimizes or solves) a given target function modifying some model parameters.

11.3.5.10. *Generic code and design*

Re-usable code is a boon. Its easy to manage, because it only needs to be written once; then, every time you want to do the same thing, you just call it up again. However, it does require a bit of thought and planning up front. In order to have a single code that can handle all the numerical method, one need to create a good representation of the world (the famous pricing states) that would represent the current status of our numerical method. In a Monte Carlo, the key information would be the values of the current factors while in a tree, it would be the current slice. Model and numerical method should use these pricing states to induct from the current state to the next step. The generic pricer will communicate to the model and the numerical method to which time it should induct. Obviously, depending on the numerical method, the generic pricer should either go backward or forward. A numerical method should therefore communicate to the pricer the induction direction. The parser will organize the various event time and sort them according to the induction direction. It will then communicate to the model and the numerical method to initialize themselves and then will induct from one time to the other. At each induction time, it will evaluate the generic payoff according to the payoff description. Once the total induction times are finished, it should finalize the pricing. In Monte Carlo, this should be the step at which to compute the average according to all the paths.

11.3.5.11. *Split between models and calibration*

The art of pricing complex derivatives lies in the calibration of the model. Simple model can do a lot if appropriately calibrated while complex models can perform badly if not appropriately calibrated. Hence, it becomes clear that the calibration framework in a derivatives pricing library should be as flexible as possible.

Calibration can be represented by an object summarizing all the parameters in the calibration method. It should contain at least:

(1) The description of the target function (either in price or volatility space or any other appropriate functional). This target function may have some weights to give relative importance to certain market prices.
(2) A portfolio of market instrument on which the model will be calibrated.
(3) The model parameters to change in the calibration procedure.
(4) The type of calibration: best fit, perfect fit, leading to a bootstrap method or a minimization procedure or any other solving or optimization procedure. This could be a modified version of a Newton–Rhapson for a perfect fit where fitting equations are solved sequentially, or a best fit with a conjugate gradient method, or a BFGS algorithm or any other similar method.
(5) The calibration method may imply itself other calibration method. Hence the calibration method object should have a recursive structure to be able to link with embedded or successive calibration method.

In order to have a very powerful pricing library, interaction between the various elements should assume very little. This means in particular that

(1) The numerical method should not assume any particular model but rather communicate with a model to know its local and global drift and variance and then knowing these key elements discretize the model accordingly. This would mean that the Monte Carlo, the tree and the PDE engines should be able to diffuse many different types of models with the same piece of code. Common functions should be summarized in the base object of the infrastructure.
(2) As seen before, the calibration part should be as model free as possible. It should always use the base class for any function to price. Hence any generic model should be able to price the calibration instruments.

Appendix A

TECHNICAL TOOLBOX

A.1. Stochastic Calculus

The stochastic calculus applied in finance is used to compute actual and future prices of derivatives payoffs. This discipline of mathematics has been developed using the modern probability theory and it was applied in physics for a long time. Using the same random context in finance and physics the quantitative financial analysts use rigorous tools which require some knowledge in probability. We describe in this appendix the most basic of these tools employed to evaluate asset prices.

A.1.1. *Itô formula*

Most of all payoffs of derivatives are functions of time and their underlying. In order to calculate the expected value of a payoff we must know at least the law of the underlying that may be determined from its dynamics. The aim of Itô formula is to obtain the dynamics of the payoff of a derivative from that of its underlying.

The Taylor formula provides a development of an n-differentiable deterministic function into a polynomial of n degree in the vicinity of 0. This principle made for composted functions, that is, functions of functions allows us to have quick results by embrocating developments. The Itô formula has the analogous role for functions of stochastic processes.

Let $(X_t)_{t\geq 0}$ be an Itô process following the diffusion:

$$dX_t = \mu(t, X_t)dt + \sigma(t, X_t)dW_t, \tag{A.1}$$

where W is a standard Brownian motion. We know that if $\mu(.,.)$ and $\sigma(.,.)$ are real Lipschitz functions, then there exists a unique solution. Let us consider f a real 2-differentiable (C^2) function, than the process $(f(t, X_t))_{t\geq 0}$ is an Itô process and its diffusion is:

$$df(t, X_t) = \left\{\frac{\partial f}{\partial t} + \frac{\partial f}{\partial x}\mu + \frac{1}{2}\frac{\partial^2 f}{\partial x^2}\sigma^2\right\}(t, X_t)dt + \left\{\frac{\partial f}{\partial x}\sigma\right\}(t, X_t)dW_t, \tag{A.2}$$

$(f(t, X_t))_{t\geq 0}$ has as drift partial derivatives of f, the drift and the square of the volatility term of $(X_t)_{t\geq 0}$ and its volatility coefficient contains the drift and the square of the volatility of $(X_t)_{t\geq 0}$. We obtain then the dynamics of $(f(t, X_t))_{t\geq 0}$.

As a corollary of Itô formula, we define infinitesimal generator of the diffusion of $(X_t)_{t\geq 0}$ associated to f, the operator:

$$(A_t f)(t, x) = \mu(t, x)\frac{\partial}{\partial x}f(t, x) + \frac{1}{2}\sigma^2(t, x)\frac{\partial^2}{\partial x^2}f(t, x). \tag{A.3}$$

Example A.1:

(1) Find the diffusion of $g(W_t)$. Where g is a (C^2) function?
(2) Let be the Itô process $dX_t = \alpha dt + \sigma dW_t$; what is its law? and that is of $\exp(X_t)$?

A.1.2. *Girsanov theorem*

In the cases where we have to compute the expectation of an asset under the risk probability is too complicated, we may change the probability measure of valuation using Girsanov theorem under some conditions. This operation modifies the underlying and brings us to compute simpler expectation with a modification of the drift of the underlying process.

The idea of this theorem is that if we dispose of a process $(M_t)_{t\geq 0}$ of the form:

$$M_t = e^{-\frac{1}{2}\int_0^t H_s^2 ds + \int_0^t H_s dW_s}, \tag{A.4}$$

where $(H_t)_{t\geq 0}$ is a process of finite square integral $\int_0^T H_t^2 dt < \infty$ and $(W_t)_{t\geq 0}$ is standard Brownian motion under a probability $\mathbf{P}$. Thus M_t is a martingale under $\mathbf{P}$. We could define an associated probability $\mathbf{P}_T$ by:

$$\frac{\partial \mathbf{P}_T}{\partial \mathbf{P}} = M_T, \tag{A.5}$$

which satisfies for every random variable X:

$$\mathbf{E}^{\mathbf{P}_T}[X] = \int_\Omega X d\mathbf{P}_T = \int_\Omega X M_T d\mathbf{P} = \mathbf{E}^P[M_T X]. \tag{A.6}$$

Furthermore, we can define the process $(Z_t)_{t\geq 0}$ by $Z_t = W_t + \int_0^t H_s\, ds$ which is a Brownian motion under this P_T probability. The process $(M_t)_{t\geq 0}$ is called the Radom–Nikodym derivative of P_T respect to P.

The results of theorem are powerful for many reasons. At first, because its hypothesis are weak and often checked in reality. We can define too different term-probabilities for different values of T. And finally, in the case where X is defined through $(W_t)_{t\geq 0}$, we could keep the same stochastic structure by modifying only the drift, as we would recover a Brownian motion $(Z_t)_{t\geq 0}$ under the new probability. Changing of probability in Eq. (A.5) implies a modification of the probability distribution and thus the moments of the variable we want to evaluate.

Example A.2: We will give a direct application of Girsanov theorem to define the risk neutral probability in Section A.1.6.

A.1.3. *Feynman–Kac*

The actual price of an asset is its expected value under the risk neutral probability. In the two previous sections, we see how to solve this problem with probabilistic tools: first by determining the diffusion and if possible the law of the underlying and then by changing the measure into the adequate probability.

The Feynman–Kac formula transposes the probabilistic problem into a numerical equivalent one. It results from a property of the infinitesimal generator of diffusion.

Let us explain how the Feynman–Kac formula works by a simple example. Let $f(T, X_T)$ be the payoff at time T of an underlying $(X_t)_{t\geq 0}$ and let r be the short interest rate; so its expected value at time $0 \leq t \leq T$ is equal to the following conditional expectation

$$e^{-r(T-t)}\mathbf{E}^Q[f(T, X_T)|\mathsf{F}_t], \tag{A.7}$$

which is a function of time and the initial condition X_t. The infinitesimal generator of the diffusion of $(X_t)_{t\geq 0}$ associated to f defined in Eq. (A.3) satisfies that the following process associated is a martingale:

$$M_t^{(f)} = e^{-tr}\left(f(t, X_t) - \int_0^t \left(\frac{\partial f}{\partial s} + A_s f - rf\right)(s, X_s)ds\right). \tag{A.8}$$

Now let us suppose that there exists a function u satisfying $u(T, x) = f(T, x)$ for every x and which is the solution of the partial differential equation:

$$\left(\frac{\partial u}{\partial t} + A_t u - ru\right)(t, x) = \left(\frac{\partial u}{\partial t} + \frac{\partial u}{\partial x}\mu + \frac{\partial^2 u}{\partial x^2}\sigma^2 - ru\right)(t, x) = 0, \tag{A.9}$$

for each $0 \leq t \leq T$ and for every x. Thus, remark that the process $(M_t^{(\cdot)})_{t\geq 0}$ associated to u for each $0 \leq t \leq T$ is reduced to:

$$M_t^{(u)} = e^{-(T-t)r}u(t, X_t) \tag{A.10}$$

and M_t^u being a martingale we will have:

$$\mathbf{E}^Q\left[M_T^{(u)}|\mathsf{F}_t\right]^{=}M_t^{(u)} = u(t, X_t). \tag{A.11}$$

Finally we have the direct result by Eqs. (A.7), (A.10) and (A.11):

$$e^{-r(T-t)}\mathbf{E}^Q[f(X_T)|\mathsf{F}_t] = u(t, x). \tag{A.12}$$

Then, to find the value $u(T, x)$ we should solve the PDE (A.9). This numerical problem is the equivalent of the initial probabilistic by Eq. (A.7). There exists an extension of the Feynman–Kac formula when the short interest rate is stochastic, characterized by the underlying $(r(T, X_t))_{t \geq T}$. Many other extensions are provided to solve local and barriers payoffs.

The resolution of this PDE requires numerical methods among which belongs the Theta (Θ) schema described in Chapter 9 (Section 9.3).

A.1.4. *Tanaka and local time*

The Tanaka formula on Itô processes may be seen as an extension of the Itô formula for larger category of function. It uses what we called the local time. Let us give an example; what happens if we want to apply Itô formula for $f(x) = |x|$ for a standard Brownian motion $(W_t)_{t \geq 0}$? Remember in Section A.1.1, that f should be differentiable which is not the case here at $x = 0$. So we do have a strong solution for the SDE $df(W_t) = \cdots$

Let us modify f in to the extension function g_ε in the vicinity of $x = 0$, defined as:

$$g_e = f(x)\,\mathbf{1}_{\{|x| \geq \varepsilon\}} + \frac{1}{2}\left(\varepsilon + \frac{x^2}{\varepsilon}\right)\mathbf{1}_{\{|x| < \varepsilon\}}, \tag{A.13}$$

where $\varepsilon > 0$. Remark that g_ε is convex inside the interval$]-\varepsilon, \varepsilon[$ which is a method of placing ourselves in a differentiable context (see Fig. A.1):

Now applying Itô to the process $(g_\varepsilon(W_t))_{t \geq 0}$ gives the solution:

$$g_e(W_t) = g_e(W_0) + \int_0^t g_e'(W_t) dW_s + \frac{1}{2\varepsilon}\int_0^t \mathbf{1}_{[-\varepsilon,\varepsilon]}(W_s) ds. \tag{A.14}$$

Remark that comparatively with the result of the exercise of Section A.1.1, we recover the additional term $(1/2\varepsilon)\int_0^t \mathbf{1}_{[-\varepsilon,\varepsilon]}(W_s) ds$. From this derivation, we generalize gradually the Itô formula for any function f using and defining the Brownian local time process at a specified point. We call the Brownian local time process at the point

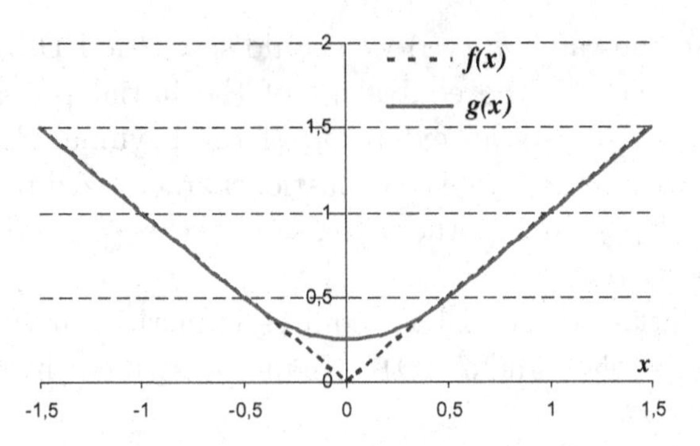

Figure A.1: The function f and its extension g_ε for $\varepsilon = 0.5$.

$x \in \mathbf{R}$ the integral:

$$\mathsf{l}(t,x) = \lim_{\varepsilon \downarrow x} \frac{1}{2\varepsilon} \int_0^t \mathbf{1}_{(x-\varepsilon, x+\varepsilon)}(W_s)ds, \qquad \text{(A.15)}$$

which is adapted to $(\mathsf{F}_t)_{t\geq 0}$ the natural filtration of $(W_t)_{t\geq 0}$. It measures the time during which the Brownian motion $(W_t)_{t\geq 0}$ remains near the point x. First, we define the Tanaka formula for the Brownian motion:

$$|W_t - x| - |W_0 - x| = \int_0^t \operatorname{sgn}(W_s - x)dW_s + \mathsf{l}(t,x), \qquad \text{(A.16)}$$

where sgn is the sign function.[1] Then, we extend this to linear combination of convex functions. Considering function like F, there exists a function f of finite variation such that:

$$F(x) = F(0) + \int_0^x f(y)\, dy. \qquad \text{(A.17)}$$

Let $f(dy)$ be the signed function measure associated to f via its representation as the difference of two increasing functions, we have:

$$F(W_s) - F(W_0) = \int_0^x f(W_s)dW_s + \frac{1}{2}\int_{\infty}^{+\infty} \mathsf{l}(t,y)f(dy). \qquad \text{(A.18)}$$

[1] $\operatorname{sgn}(x) = \mathbf{1}_{\{x>0\}} - \mathbf{1}_{\{x<0\}}$.

When this is only locally bounded and Borelian we would have:

$$F(W_s) - F(W_0) = \int_0^x f(W_s)dW_s + \frac{1}{2}\int_{\infty}^{+\infty} f(y)d\mathrm{l}(t,y), \qquad \text{(A.19)}$$

and finally, when f is locally square integrable we can write:

$$F(W_s) - F(W_0) = \int_0^x f(W_s)dW_s + \frac{1}{2}[f(W), W]_t, \qquad \text{(A.20)}$$

where

$$[f(W), W]_t := \lim_{n\to\infty} \sum_{i=1}^{\infty} \left(f\left(W_{t_{i+1}^{(n)}}\right) - f\left(W_{t_i^{(n)}}\right)\right)\left(W_{t_{i+1}^{(n)}} - W_{t_{i+1}^{(n)}}\right). \qquad \text{(A.21)}$$

A.1.5. *Markov property*

The Markov property of a stochastic process means it is "amnesic" for short past. The amnesia of the process signifies that at time t, it depends only on some of its last observation and not all of them. The memory of a process $(X_t)_{t\geq 0}$ is characterized by its all past realizations thus by its natural filtration $(\mathsf{F}_t)_{t\geq 0} = \sigma(X_s, s \leq t)$. This property is important in finance because it allows economizing the time and information in future pricing.

For discrete time processes, this property is written mathematically as follows:

$$\mathbf{E}[X_t|\mathsf{F}_{t-1}] = \mathbf{E}\Big[X_t|\underbrace{X_{t-1}, \ldots, X_{t-n}}_{n}\Big]. \qquad \text{(A.22)}$$

This means that the projection of $(X_t)_{t\in\mathbf{N}}$ on all its past realizations is equal to its projection on only a subset of them: the n lasts $(X_t)_{t\in\mathbf{N}}$ is called a Markov process of order n. We say wrongly that $(X_t)_{t\in\mathbf{N}}$ is Markovian if $n = 1$. In this case, we can express the process $(X_t)_{t\in\mathbf{N}}$

at each time t by an initial observation x_0 and a sequence of independent random variables of the same law $(\varepsilon_t)_{t\in\mathbf{N}}$:

$$X_t = g(x_0, (\varepsilon_s)_{0\leq s\leq t}). \tag{A.23}$$

In continuous time, for the solutions of SDE we can provide an equivalent expression. If we consider the following SDE, with the initial condition:

$$\begin{cases} dX_t = \mu(t, X_t)\,dt + \sigma(t, X_t)\,dW_t, \\ X_s = x, \end{cases} \quad \text{for } s > t, \tag{A.24}$$

then its solution denoted by $(X_t^{s,x})_{t\geq s}$ verifies $\mathbf{P}$-a.e. for each $0 \leq s \leq t$ that:

$$\mathbf{E}[f(X_t)|\mathsf{F}_s] = \phi(X_s), \tag{A.25}$$

This last being equal to $\phi(x) = \mathbf{E}[f(X_t^{s,x})]$. This is a consequence of the flow property of a stochastic diffusion solution of Eq. (A.25) than we have the relation $\mathbf{P}$-a.e. for each $0 \leq s \leq t$ that:

$$X_t^{s,x} := X_t^{0,x} = X_t^{s,X_t^{s,x}}. \tag{A.26}$$

The proof of the result (A.25) uses the independence of the Brownian increases. The main idea is that $X_t^{s,x}$ is a function of the initial condition and independent Gaussian variables:

$$X_t^{s,x} = \phi(X_s^{s,x} = x, (W_{s+u} - W_{s+t})_{u\geq 0}). \tag{A.27}$$

Note that $((W_{s+u} - W_{s+t})_{u\geq 0})$ are independent of the tribe F_s. Remark that the expression (A.27) may be seen as the analogous in continuous time of the result (A.23).

A.1.6. *Risk neutral and numeraire*

In finance convention, we suppose that the historical evolution of an asset is compound of a "sum" of a trend and volatility terms. In a BS model the asset price process $(S_t)_{t\geq 0}$ follows under the historical probability **P** a geometric Brownian motion:

$$dS_t = S_t \mu \, dt + S_t \sigma \, W_t, \tag{A.28}$$

where $S_t\mu$ is the instantaneous drift representing the expected term target and $S_t\sigma$ is the instantaneous volatility representing the local random risk.

In order to compare at short and long term the expected yield of different assets, these lasts should have the same trend to have a same scale risk comparison; we have to write then with the same drift (even if null!) (see Fig. A.2).

The expected value of the asset S at time t is equal to $\tilde{S}_t = \mathrm{e}^{-rt} S_t$, with r is constant short interest rate. Applying Itô formula we can find its diffusion:

$$d\tilde{S}_t = \mathrm{e}^{-rt} dS_t + S_t d\mathrm{e}^{-rt} = \tilde{S}_t \{(\mu - r)dt + \sigma dW_t\}. \tag{A.29}$$

One way to remove the new drift is to integrate it among the stochastic term: in the Brownian motion. By the Girsanov theorem to define the new Brownian motion by

$$B_t = W_t + \frac{\mu - r}{\sigma} t, \quad t \geq 0, \tag{A.30}$$

and the correspondent the so-called Risk Neutral probability denoted by **Q** under which $(\tilde{S}_t)_{t \geq 0}$ is an exponential martingale and then we

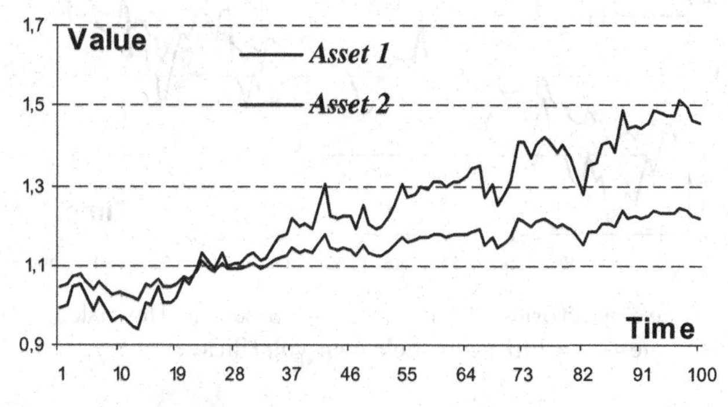

Figure A.2: Historical trajectories of two assets with different trends and volatilities.

can know its expected value at time 0:

$$\mathbf{E}^{Q}[\tilde{S}_t] = \tilde{S}_0 = S_0. \tag{A.31}$$

We call the parameter $\lambda = (\mu - r)/\sigma$, as the market risk premium which allows us to pass into the correspondent[2] risk neutral probability $\mathbf{Q}$ because it verifies the boundary Girsanov theorem conditions. The dynamics of under the risk neutral probability becomes (see Fig. A.3):

$$dS_t = S_t r dt + S_t \sigma B_t, \tag{A.32}$$

where we used the un-risky value of the asset dividing by the no risked asset $(e^{rt})_{t \geq 0}$. We could define any other numeraire from which we can define a forward probability under which our asset would be a martingale. When we would have an expectation $\mathbf{E}^{\mathbf{A}}[S_t X_t]$ to compute under a probability $\mathbf{A}$, we will try to write the product X_t as a product of $Y_t Z_t$ such that:

(1) Z_t is a martingale of the form of Eq. (A.4),
(2) $S_t Y_t$ written with the new Brownian motion is a martingale,

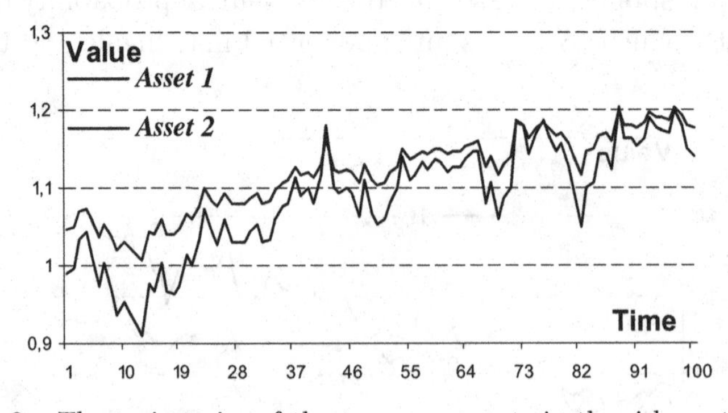

Figure A.3: The trajectories of the two same assets in the risk neutral world: they have the same trend but keep their own volatilities.

[2]The risk neural probability is defined via a unique market risk premium.

to change the probability measure into **B** to have the solution:

$$\mathbf{E}^{\mathbf{A}}[S_t X_t] = \mathbf{E}^{\mathbf{A}}[S_t Y_t Z_t] = \mathbf{E}^{\mathbf{B}}[S_t Y_t] = S_0 Y_0, \tag{A.33}$$

where the numeraire is $1/Y_t$ and the probability **B** is called the t-terminal neutral probability.

The Asset 1 is less risky than Asset 2 at long term and inversely at short term.

A.1.7. *Copula*

The recent copula theory was developed to model multivariate dependence structure between variables with dissociating the correlation structure from the marginal distributions. Mathematically, it is a function that joins univariate distribution functions to form multivariate distribution functions. The dependence notion is important in finance especially for assets portfolios pricing in particular in credit.

The linear Pearson correlation is not equivalent to dependence except for some cases. Furthermore, it exhibits several limits as its exclusion of nonlinear correlation, its invariance only by linear transformation and the fact that given margins, all the linear correlation between -1 and 1 cannot in general be obtained by a suitable choice of the joint distribution (see the Fig. A.4).

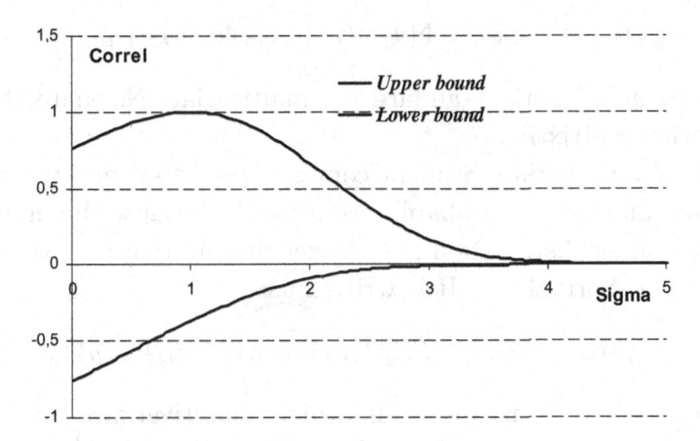

Figure A.4: An example of upper and lower bounds of lognormal correlation between LN(0, 1) and LN(0, Sigma).

In order to give functional comprehension of copula, we give immediately the existence Sklar's theorem. Let F be N-dimensional distribution with margins $F_1, \ldots, F_N$. Then, there exists a unique corresponding copula representation C given for each $(x_1, \ldots, x_N) \in \mathbf{R}^N$ by:.

$$F(x_1, \ldots, x_N) = C(F_1(x_1), \ldots, F_N(x_N)), \tag{A.34}$$

C can be then seen as the function of reparation of uniform variables:

$$C(x_1, \ldots, x_N) = F\big(F_1^{-1}(x_1), \ldots, F_N^{-1}(x_N)\big). \tag{A.35}$$

The copula of independent variables is written as follows:

$$C^{\perp}(X_1, \ldots, X_N) = X_1 \times \cdots \times X_N.$$

The most important of them is that a copula is invariant by increasing transformation, that is, for each increasing function g_i, $1 \leq i \leq N$:

$$C(X_1, \ldots, X_N) = C(g_1(X_1), \ldots, g_2(X_N)). \tag{A.36}$$

The following two special copulas are the most used in finance:

(1) The N-multivariate Normal copula is the most used in finance because it is also the copula of log-normal random vector (the log being an increasing function!). It is written as:

$$C_\rho(u_1, \ldots, u_N) = \Phi(\Phi^{-1}(u_1), \ldots, \Phi^{-1}(u_N); \rho), \tag{A.37}$$

where $\Phi(.;\rho)$ is the standardized multivariate Normal with correlation matrix ρ.

(2) The N-multivariate Student copula is used to compute extreme values and default probabilities in Credit because the multivariate Student distribution has thicker distribution tails then multivariate Normal one. It is written as:

$$C_{\nu,\rho}(u_1, \ldots, u_N) = T(t_\nu^{-1}(u_1), \ldots, t_\nu^{-1}(u_N); v, \rho), \tag{A.38}$$

where $T(.;\nu,\rho)$ is the standardized multivariate Student of degree of freedom ν and correlation matrix ρ and t_ν^{-1} the inverse of the unvaried Student distribution of degree of freedom ν.

In order to compute the expected value of a portfolio of N-ρ-correlated assets $(X_1, \ldots, X_N)$ following a multivariate lognormal process, have to calculate:

$$\mathbf{E}[f(X_1, \ldots, X_N)] = \int_{\mathbf{R}^N} f(x_1, \ldots, x_N) dC_\rho(F(x_1), \ldots, F(x_N)), \tag{A.39}$$

A.2. Econometrics

Econometrics is a result from inference statistic and it was developed to analyze economic data. Its aim is to set a model which reproduces as best as possible observed data to interpret them and for temporal series moreover, to provide a forecast expected series. Econometrics of time series is used in finance to mainly simulate and estimate stochastic volatility. We give in the following section some financial applications of Econometrics and its consistence with stochastic calculus.

A.2.1. *Time series*

Let $\{X_t, 0 \leq t \leq T\}$ be a set of temporal data. The first principle of time series is to decompose the series into hidden components through a separable schema f:

$$X_t = f(T_t, S_t, \varepsilon_t), \quad 0 \leq t \leq T, \tag{A.40}$$

where $\{T_t, 0 \leq t \leq T\}$ is the trend component, $\{S_t, 0 \leq t \leq T\}$ is the seasonality or secular component and $\{\varepsilon_t, 0 \leq t \leq T\}$ is a sequence of independent identical random variables, considered mainly as Gaussian.

The two first components could themselves be function of the past observations $\{X_s,\, 0 \leq s \leq t-1\}$. So define a model for the series $(X_t)_{0 \leq t \leq T}$ is equivalent to specify it as a function of the random series $\{\varepsilon_s,\, 0 \leq s \leq t-1\}$:

$$X_t = g(X_0, \{\varepsilon_s,\, 0 \leq s \leq t\}), \quad 0 \leq t \leq T. \tag{A.41}$$

As mentioned earlier, the models must replicate the data characteristics and then we should adapt each model for a "category" of series. We distinguish two specific processes: the stationary ones with unique mean and finite homogeneous temporal covariance matrix and the non-stationary ones.

For the stationary processes $(X_t)_{0\leq t\leq T}$, the simplest framework is to suppose a linear relation with $(\varepsilon_t)_{0\leq t\leq T}$. The underlying idea is that a moving average (of some past observations) of $(X_t)_{0\leq t\leq T}$ is equal to another moving average of $(\varepsilon_t)_{0\leq t\leq T}$. This is written as follows:

$$X_t + \sum_{i=1}^{p} \phi_i X_{t-i} = \varepsilon_t + \sum_{j=1}^{q} \theta_j \varepsilon_{t-j}, \quad 0 \leq t \leq T, \tag{A.42}$$

for some specified coefficients $\{\phi_i, 1 \leq i \leq p\}$, $\{\theta_j, 1 \leq j \leq q\}$, p and q.

Equation (A.42) models called the auto regressive moving average (ARMA) (p, q) family, it implies that at each time t, the conditional moments of X_t, $E[X_t/X_{t-p}, X_{t-(p-1)}, \ldots, X_{t-1}]$, $0 \leq t \leq T$, are constant so independent of the filtration $\sigma(\varepsilon_s, s \leq t)$ and that X_t is Gaussian.

Empirically, the log of asset prices is non-Gaussian and its historical volatility is equal to its conditional standard deviation (see Fig. A.5). Actually, in markets, it is stochastic and strongly agitated. Consequently, the asset prices could not be modeled by the ARMA models.

As we can see in Fig. A.5, the ARMA model does not fit with market observations.

A.2.2. *GARCH and Nelson result*

The generalized auto regressive conditional heterocedastic (GARCH) models developed for asset prices series are nonlinear and model the asset price and its volatility simultaneously. With these models, we simulate the volatility. These models assume that at each time the asset price is function of its volatility whose squared follows

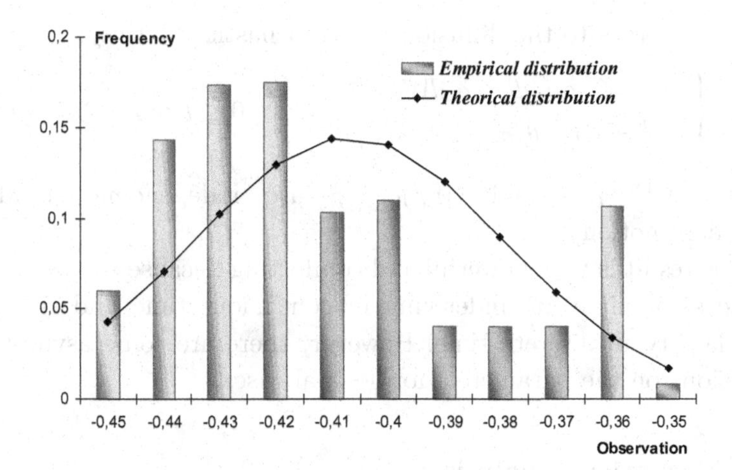

Figure A.5: Gaussian normal distribution and Empirical distribution of logarithm of 10-year zero-coupon bond over 600 observations.

an ARMA model. A GARCH (p, q) is so written as:

$$\begin{cases} X_t = \varepsilon_t \sigma_t = \varepsilon_t \sqrt{V[X_t|\varepsilon_{t-1}, \ldots, \varepsilon_0]}, \\ \sigma_t^2 = \sum_{j=1}^{q} \theta_j \sigma_{t-j}^2 + \phi_0 + \sum_{i=1}^{p} \phi_i X_{t-i}^2, \end{cases} \quad 0 \le t \le T. \tag{A.43}$$

The volatility is restricted only to the past elements of the asset price.

Once all the parameters $\{\phi_i, 1 \le i \le p\}$, $\{\theta_j, 1 \le j \le q\}$ are estimated, we can provide a forecast series of a horizon h, with a trust interval:

$$\begin{cases} \hat{X}_t = \varepsilon_{t-T} \hat{\sigma}_t, \quad T \le t \le T + h, \\ \hat{\sigma}_t^2 = \hat{\phi}_0 + \sum_{i=1}^{p} \hat{\phi}_i X_{t-i}^2 + \hat{\theta}_0 + \sum_{j=1}^{q} \hat{\theta}_j \hat{\sigma}_{t-j}^2, \quad 1 \le h \le T. \end{cases} \tag{A.44}$$

In order to be consistent with the continuous time models in stochastic calculus for stochastic volatility, Nelson showed in 1990 that some GARCH models are approximations of some stochastic diffusions.

For example, an exponential GARCH (1,1), we get:

$$\begin{cases} X_t = X_{t-1} - \alpha \sigma_t^2 + \sigma_t \varepsilon_t, \\ \hat{\sigma}_t^2 = \beta + \theta_1 \sigma_{t-1}^2 + \phi_1 \, \sigma_{t-1}^2 \varepsilon_{t-1}^2, \end{cases} \quad 0 \le t \le T, \tag{A.45}$$

which converges to the diffusion with stochastic volatility:

$$\begin{cases} dX_t = -\alpha\sigma_t^2 dt + \sigma_t dW_t^1, \\ d\sigma_t^2 = (\beta + \theta_1\sigma_t^2)dt + \phi_1\,\sigma_t^2 dW_t^2, \end{cases} \quad 0 \le t \le T, \tag{A.46}$$

where $(W_t^1)_{0\le t\le T}$ and $(W_t^2)_{0\le t\le T}$ are independent standard Brownian motions.

This result is very powerful and comforting because we can model at the same time an underlying in continuous time and simulate its volatility in discrete time. However, there are some asymptotic conditions on the parameters for general cases.

A.3. Numerical Analysis

The numerical analysis is used in finance mainly in the so called calibration of models. It consists in determining the parameters of the model such that market prices would be replicated from the closed formulas implied. Also numerical analysis is used to reduce the number of parameters to have more robustness and speed for large portfolios. The tools used are several algorithms of linear and nonlinear optimization.

A.3.1. *PCA, Cholesky*

Suppose that we want to diffuse a n-dimensional asset $(S_t)_{0\le t\le T}$ following the stochastic diffusion:

$$dS_t = \mu(t, S_t)dt + \sigma(t, S_t)dW_t, \quad 0 \le t \le T, \tag{A.47}$$

where $(W_t)_{0\le t\le T}$ is a n-dimensional standard Brownian of correlation matrix:

$$\mathbf{E}[dW_t dW_t^{\perp}] = \Xi\, dt. \tag{A.48}$$

The usage of Ξ matrix for n large, for pricing and hedging purpose, becomes non-stable and heavy, as we have to control $(1/2)n(n+1)$ elements. In addition, the information contained in Ξ is redundant. Could we write Eq. (A.47) under a basis of non-redundant or independent elements of a dimension $k \le n$?

First, could we extract non-redundant factors functions of the original variables? This is the aim of the principal components analysis (PLA): to analyze the correlation points in a space of lower dimension with synthetic orthogonal components which correspond in linear algebra to a basis changing.

The matrix Ξ being symmetric positive definite can be written as a diagonal matrix Λ of dimension $p \leq n$ of eigen values and its associated orthogonal matrix U of p eigen n-vectors:

$$\underset{(n\times n)}{\Xi} = \underset{(n\times p)}{U} \cdot \underset{(p\times p)}{\Lambda} \cdot \underset{(p\times n)}{U^{\perp}} . \tag{A.49}$$

Each eigen value is positive and corresponds to the relative importance of a factor. By sorting them, we choose the $k \leq p$ firsts, the greatest of them.

Now, instead to diffuse the n processes with the original n factors, we use the k new factors and replace the n Brownian motions $(W_t^i)_{0\leq t\leq T}$ in Eq. (A.47) by the real standard Brownian motion $(Z_t)_{0\leq t\leq T}$:

$$dZ_t = \sum_{i=1}^{k} \frac{u_i}{\sqrt{\lambda_i}} dZ_t^i, \tag{A.50}$$

where $(Z_t^1)_{0\leq t\leq T}, \ldots, (Z_t^k)_{0\leq t\leq T}$ are independent standard Brownian motions. The algorithm used to transform the matrix Ξ is Gauss–Jacobi transformation.

Second, could we write our asset diffusion under a basis of independent elements of the same dimension n? It would be a way to de-correlate the n Brownian motions $(W_t^i)_{0\leq t\leq T}$. How could we keep the correlation by changing the writing?

The linear algebra tool to use is the Cholesky factorization which writes a definite symmetric positive matrix like Ξ in to a lower matrix L and its transpose:

$$\underset{(n\times n)}{\Xi} = \underset{(n\times n)}{L} \cdot \underset{(n\times n)}{L^{\perp}} . \tag{A.51}$$

The diagonal elements of L are positive and the whole elements of L are found recursively by the Cholesky algorithm.

We define now the new basis, $(Z_t)_{0\leq t\leq T}$ a n-dimensional Brownian motion of independent elements $(Z_t^1)_{0\leq t\leq T}, \ldots, (Z_t^n)_{0\leq t\leq T}$ verifying:

$$W_t = L.Z_t \iff W_t^i = \sum_{j=i}^{i} L_{ij}.Z_t^j, \quad 1 \leq i \leq n. \tag{A.52}$$

By writing the diffusion with this new Brownian motion, we could integrate the L elements among the volatility coefficient:

$$dS_t = \mu(t, S_t)dt + (\sigma(t, S_t).L).dZ_t, \quad 0 \leq t \leq T. \tag{A.53}$$

We can simulate $\frac{1}{2}n(n+1)$ independent random variables by keeping all correlation structure of our asset since we have always:

$$\mathbf{E}[L.dZ_t dZ_t^{\perp} L^{\perp}] = L.\mathbf{E}[dZ_t dZ_t^{\perp}].L^{\perp} = L.\text{I}.L^{\perp}dt = \Xi\, dt. \tag{A.54}$$

A.3.2. *Minimizing the errors: Newton–Raphson search*

Modeling consists in assuming that a liquid asset price follows a stochastic function of time and parameters. Therefore, it is important to determine the model parameters to price products depending on this asset price. How do we determine them? In reality, we dispose of a set of historical prices (potentially volatilities and more generally function values) depending on this asset:

$$y_i^{\text{market}}. \tag{A.55}$$

These function values can be computed within our modeling assumptions as follows:

$$y_i^{\text{model}} = f(i, \underline{x}), \quad 1 \leq i \leq n. \tag{A.56}$$

We search the m-vector of parameters $(m \leq n)$ such that the model reproduces the historical prices (and more generally function values) as best as possible. For this, we look at the difference, often taken as the squared spread series and given by:

$$e_i(\underline{x}) = \left(y_i^{\text{market}} - f(i, \underline{x})\right)^2, \quad 1 \le i \le n. \tag{A.57}$$

We would like the sum of these spreads (positive or negative) be minimum, that is, we want to minimize the sum of squares function:

$$g(\underline{x}) = \sum_{i=1}^{n} e_i(\underline{x}). \tag{A.58}$$

In the case of one parameter x, under the hypothesis of smoothness of the g (i.e. it is twice differentiable by parameters x), we could use the Newton–Raphson method. This method is an iterative algorithm that uses the two first terms of the Taylor series of a function in the vicinity of a point x_0:

$$g(x) \approx g(x_0) + g'(x_0)(x - x_0) + \frac{1}{2} g''(x_0)(x - \underline{x}_0)^2, \tag{A.59}$$

This development is the best quadratic approximation of g in the vicinity of x_0. The minimum of g denoted by x^* verifies:

$$g'(x^*) = 0, \quad g''(x^*) > 0. \tag{A.60}$$

By differencing the quadratic expression once, if x^* is near than x_0, it should verify the first condition (it is a stationary point) as follows:

$$g'(x_0) + g''(x_0)(x^* - x_0) = 0. \tag{A.61}$$

We search then by a successive series of points verifying Eq. (A.60), until we find the one that verifies the second condition of Eq. (A.59). The algorithm is then written as:

$$x_{k+1} = x_k - \frac{g'(x_k)}{g''(x_k)}. \tag{A.62}$$

(see Fig. A.6). This method is simple to implement and is convergent but its speed depends on the choice of the initial point x_0. For the case of maximizing a function h, it is enough to use the same method by minimizing the function $-h$.

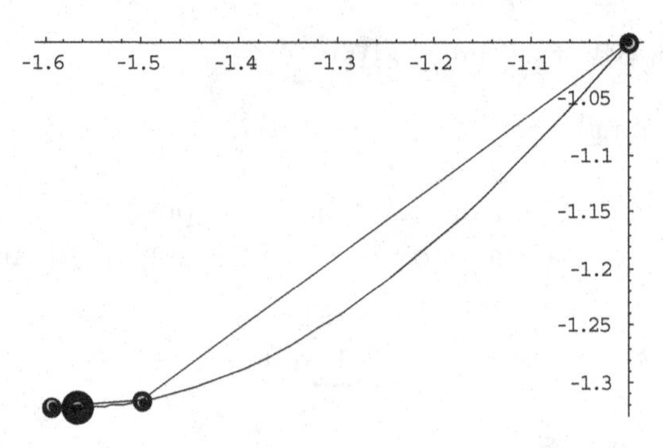

Figure A.6: The scenario of steps taken by the Newton–Raphson method to find a local minimum of the function $g(x) = x\cos(x+1)$ starting at the point $x = -1$.

A.3.3. *Optimizer: BFGS, conjugate and steepest gradient*

In the multi-dimensional case, we should use a minimizing algorithm among the descent methods family. The common idea is to determine the descent direction that minimizes the first derivative of target function as fast as possible. We suppose that g is twice differentiable that and denote by $\nabla g(\underline{x})$ and $\mathrm{Hg}(\underline{x})$ respectively the gradient and the Hessian matrix of g by $\underline{x}$.

A m-vector $\underline{d}$ is called descent vector if there exists a positive real α such that:

$$g(\underline{x} + \alpha\underline{d}) < g(\underline{x}). \tag{A.63}$$

The principle of a descent method consists in making the following iterations:

$$\underline{x}_{k+1} = \underline{x}_k + \alpha_k \underline{d}_k, \tag{A.64}$$

and guaranteeing the decreasing property:

$$g(\underline{x}_{k+1}) < g(\underline{x}_k). \tag{A.65}$$

This is the general algorithm of a steepest descent method. The steepest descent methods differ by their choice of descent direction $\{\underline{d}_k,\ k \geq 0\}$ and steps $\{\alpha_k,\ k \geq 0\}$.

First, we can choose the best descent vector that decreases the function g as quickly as possible is at the kth iteration. It can be easily proved that it is:

$$\underline{d}_k = -\text{Hg}(\underline{x}_k)^{-1}\nabla g(\underline{x}_k). \tag{A.66}$$

The steepest ascent method takes this descend vector and a constant value for all α_k. It requires a great many iterations for functions having long, narrow valley structures (see Fig. A.7).

For quadratic form functions which are written as:

$$g(\underline{x}) = \underline{c} + \nabla g(\underline{x})^{\perp}.\underline{x} + \underline{x}^{\perp}.\text{Hg}(\underline{x}).\underline{x}, \tag{A.67}$$

the conjugated gradient algorithm is more appropriated. The successive descent directions are conjugated mutually ($\underline{d}_i^{\perp}.\underline{d}_j = 0$, $\forall i \neq j \geq 0$) and the steps are specified as follows:

$$\begin{cases} \alpha_k = -\dfrac{\nabla g(\underline{x}_k)^{\perp}.\underline{d}_k}{\underline{d}_k^{\perp}.\text{Hg}(\underline{x}_k).\underline{d}_k}, \\ \underline{d}_{k+1} = -\nabla g(\underline{x}_{k+1}) + \beta_k \underline{d}_k, \quad \text{with } \underline{d}_0 = -\nabla g(\underline{x}_0), \\ \beta_k = \dfrac{\|\nabla g(\underline{x}_{k+1})\|^2}{\|\nabla g(\underline{x}_k)\|^2}. \end{cases} \tag{A.68}$$

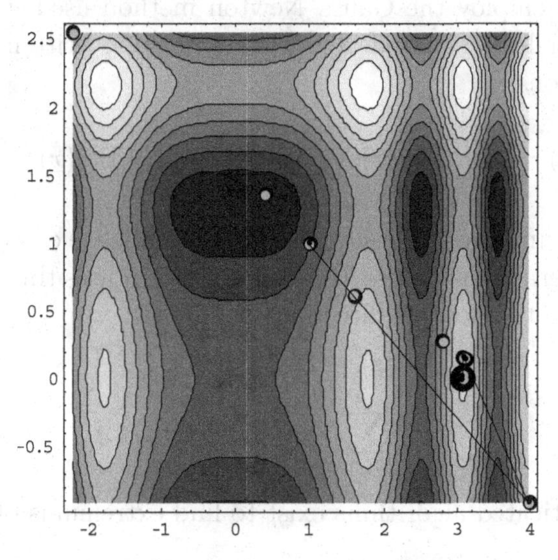

Figure A.7: The scenario of steps taken by the steepest ascent method to find a local minimum of the function $f(x, y) = \cos(x^2) + \sin(y^2)$ starting at the point $\{x, y\} = \{1, 1\}$.

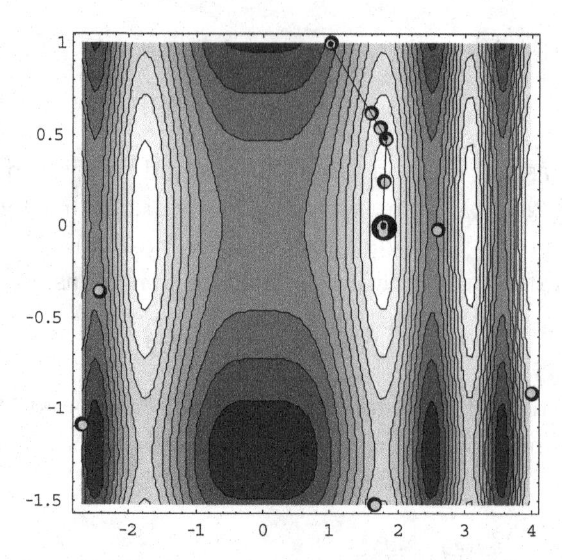

Figure A.8: The scenario of steps taken by the conjugated gradient method is faster to find a local minimum of the f function starting at the same point $\{x, y\} = \{1, 1\}$.

This method can be generalized for any functions and its convergence is much faster than the gradient method (see Fig. A.8).

We could employ the Gauss–Newton method used especially for the nonlinear least squared. The Hessian of g is written from those of the functions e_i:

$$\mathrm{Hg}(\underline{x}) = 2\sum_{i=1}^{m} \nabla e_i(\underline{x}).\nabla e_i(\underline{x})^{\perp} + \nabla e_i(\underline{x}).\mathrm{He}_i(\underline{x}). \tag{A.69}$$

At the optimum, the e_i are supposed to be small, we can then neglect the second term. The Gauss–Newton algorithm uses this approximation to define the descent directions as follows:

$$\begin{aligned} \underline{x}_{k+1} &= \underline{x}_k - H_k^{-1}\nabla g(\underline{x}_k), \\ H_k &= \sum_{i=1}^{m} \nabla e_i(\underline{x}_k).\nabla e_i(\underline{x}_k)^{\perp}. \end{aligned} \tag{A.70}$$

More sophisticated algorithms exist to find extremums of g function by solving:

$$\nabla g(\underline{x}) = 0. \tag{A.71}$$

The Newton family offers many derivative algorithms adapted for specific case but take all the descent directions as:

$$\underline{d}_k = -\text{Hg}(\underline{x}_k)^{-1}\nabla g(\underline{x}_k). \tag{A.72}$$

They differ by the way in which they replace the Hessian $\text{Hg}(\underline{x}_k)$ by an approximation H_k (if possible definite positive), built during iterations. For example, the BFGS algorithm proposes a successive actualization of the matrixes $\{H_k\}$ such that they would be definite positive, function of the successive gradients $\{\nabla g(\underline{x}_k)\}$ and the points $\{\underline{x}_k\}$ (see Fig. A.9).

Finally, note that all the algorithms presented are made to find local extremums and their convergences depend than from the proximity of the initial point $\underline{x}_0$ than the extremum $\underline{x}^*$. The stability and the efficiency of this method depend on their implementation too.

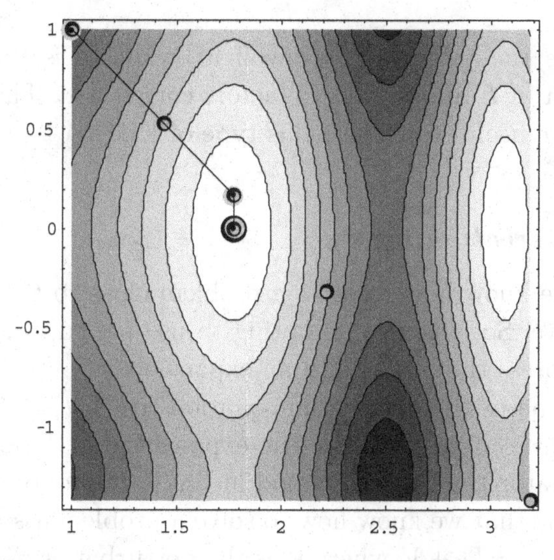

Figure A.9: The scenario of steps taken by the Gauss–Newton method is more direct to find a local minimum of f function starting at the same point $\{x, y\} = \{1, 1\}$.

A.4. Expansion Technique

In Section A.1.3, we have seen that we could find the actual value option by solving a PDE. In the case of stochastic volatility, the PDEs are hard to solve even numerically. For example, in Black–Scholes with dividend model:

$$\begin{aligned} dS_t &= (r_t - q_t)S_t dt + \sigma_t S_t dW_t, \\ d\sigma_t^2 &= \varphi(\sigma_t^2)dt + \eta(\sigma_t^2)dZ_t, \end{aligned} \tag{A.73}$$

the European option PDE to resolve is:

$$\begin{aligned} -\frac{\partial F}{\partial t} &= -rF + \tilde{A}F, \\ \tilde{A}F &= (r-q)D\frac{\partial F}{\partial S} + \frac{1}{2}VS^2\frac{\partial^2 F}{\partial S^2} + \varphi(\sigma^2)\frac{\partial F}{\partial \sigma^2} \\ &\quad + \frac{1}{2}\eta^2(\sigma^2)\frac{\partial^2 F}{\partial(\sigma^2)^2} + \rho\eta(\sigma^2)\sigma S\varphi(\sigma^2)\frac{\partial^2 F}{\partial S \partial \sigma^2} \end{aligned} \tag{A.74}$$

$F(S_t, \sigma_t^2, t)$ is the value of the payoff at t and A is the infinitesimal generator of F and ρ the two factors correlation. The expansion techniques are useful to resolve this type of PDE.

A.4.1. *Perturbation theory*

Very often we know how to solve a problem close to the one we are trying to solve. So it is very tempting to use perturbation operator theory to express the result with an expansion.

The solution of the differential equation $(\partial K/\partial t) = LK$ can formally be expressed as $e^{tL}K(0)$. The exponential of L here is a very peculiar operator that is understood in the semi-group way.

Let assume that we know how to solve a problem associated with L_0 such that $L = L_0 + V$ where V is the perturbation. The problem here is that L_0 and V do not commute in general and $L(s)$ and $L(t)$ do not commute either, so we cannot use ordinary Taylor series.

It is possible to show the following formula:

$$K(s) = Q(s)e^{L_0}K(0), \tag{A.75}$$

where

$$Q(s) = 1 + \sum_{1\leq n\leq\infty} \int_{0\leq s_1\leq\cdots\leq s_n\leq s} e^{s_1 \mathrm{ad}(L_0)(V)} \ldots$$
$$e^{s_{n1} \mathrm{ad}(L_0)(V)} ds_1 ds_2 \cdots ds_n, \tag{A.76}$$

where $\mathrm{ad}(L_0)(V)$ the adjoint of L_0 is defined by

$$\mathrm{ad}(L_0)(V) = L_0 V - V L_0. \tag{A.77}$$

At order 2, this development looks:

$$Q(s) = 1 + sV + \frac{1}{2}s^2(\mathrm{ad}(L_0)(V) + V^2) + O(s^2) \tag{A.78}$$

This is used (e.g. Hagan *et al.*, 2004) to adjust the difference between the Laplacian equation we want to solve and the Beltrami equation we know to solve using differential geometry for the SABR model. It can also be used to adjust for the drift interest rate models. The perturbation formula is detailed in Dunford and Schwartz (1958).

A.4.2. *Taylor series with Malliavin weights*

It is possible to stay inside the stochastic world and perform the expansion, using a trick coming from Malliavin calculus. The basic idea is to compute

$$E[\Phi(X_{\varepsilon,t})] = E[\Phi(X_0)] + \left.\frac{dE[\Phi(X_{\varepsilon,t})]}{d\varepsilon}\right|_{\varepsilon=0} \varepsilon + \cdots$$
$$+ \left.\frac{d^2 E[\Phi(X_{\varepsilon,t})]}{d\varepsilon^n}\right|_{\varepsilon=0} \frac{\varepsilon^n}{n!}, \tag{A.79}$$

where $X_{\varepsilon,t}$ follows the SDE ($X_{\varepsilon,t}$ can be a vector):

$$dX_{\varepsilon,t} = \mu(t, X_{\varepsilon,t}, \varepsilon)dt + \sigma(t, X_{\varepsilon,t}, \varepsilon)dW_t \tag{A.80}$$

and Φ is a pay-off function which takes as argument the state vector $X_{\varepsilon,t}$. ε is a parameter that can be a vector and which parameterizes our family of models. It is assumed that we know how to compute $E[\Phi(X_0)]$. Then the problem is to compute terms like

$$g_n = \left. \frac{d^n E[\Phi(X_{\varepsilon,t})]}{d\varepsilon^n} \right|_{\varepsilon=0}. \tag{A.81}$$

Here Malliavin calculus allows us to express g_n as $g_n = E[\Phi(X_{0,t})Y_{n,t}]$, where $Y_{n,t}$ is the Malliavin weight associated with the integrand and the derivation order. This weight is usually given by a Skohorod integral of the ratio of a Malliavin derivative and the Malliavin covariance times the product of derivatives of $X_{\varepsilon,t}$ (for exact calculations, see e.g. Hubalek *et al.*).

The interest of using Malliavin weights here is to handle the discontinuous shape of the usual pay-off functions (e.g. vanilla option or digitals) and the problems they usually induce in the calculation of the derivatives. An interesting indication for using this expansion is when the exactly known model (here $X_{0,t}$) relies on Gaussian processes and polynomials of them. It appears that the Skohorod integral and the Malliavin derivative respect the structure of polynomial of Gaussian processes. Therefore, the expectation to compute for the derivatives involves just expectation of such polynomials of Gaussian processes. This can be implemented extremely efficiently, using Hermite polynomials based integration techniques.

It does not come as a surprise that we have this simplification, because it is in essence what the decomposition of processes on the Wiener Chaos does regarding the Skohorod integral. Malliavin derivative can be seen as shifting operators on the Chaoses (see e.g. Nualart, 1995) Very Good introduction to Malliavin calculus techniques can be found in Bally (2003), Kohatsu-Higa and Montero (2004) and Friz (2003). A comprehensive reference is Nualart (1995).

A.4.3. *Singular perturbation and the WKB expansion*

When we built multifactor models, it happens that we know how to solve models with less factors. In that case the perturbation we introduce in the parabolic equation by adding one or more factors is singular because even for small t, the pricing kernel does not have the same dimensionality. So the idea is to try a solution of the form:

$$K(x) = A(x)\mathrm{e}^{1/\varepsilon\left(\sum_{i\geq 0} \varepsilon^i B_i(x)\right)}. \tag{A.82}$$

We introduce this expansion into the equation to solve and match the terms, generating a sequence of differential equation of smaller order. Usually the first equation that drive the first order is of the form

$$(B_0')^2 = Q(x). \tag{A.83}$$

This type of equation is called an Eikonal equation. In the context of electromagnetic problems, it generates geometric optics. It is why Hagan *et al.* (2004) and Berestycki *et al.* (2004) use this terminology, while deriving their approximate option formula.

The second equation that drives B_1 is called a transport equation. It is this type of equation that describes the neutronic flow in a nuclear reactor. It generates the first-order term with respect to maturity in Hagan *et al.* (2004).

The method is well explained in Bender and Orszag (1999), it is applied, for example, in Berestycki *et al.* (2004) for deriving equations of first-order analytic formula of options in a stochastic volatility model. These formulas can be easily solved because of the smaller dimensionality. For example, for one-dimensional problem, spectral expansion methods are very powerful.

A.4.4. *Spectral expansion*

The pricing of European options in the context of local volatility models and/or stochastic volatility models is a domain where the pricing formula can be represented by the solution of a differential equation. Spectral methods are very powerful methods to solve these problems and are especially efficient for the stochastic one-dimensional case or as soon as the problem has been reduced or approximated to a one(space)-dimensional case.

Very often, the problem can be formalized as (K is the pricing kernel):

$$\frac{\partial K}{\partial t} = LK \equiv a_2(x)\frac{\partial^2 K}{\partial x^2} + a_1(x)\frac{\partial K}{\partial x} + a_0 K \quad \text{(A.84)}$$

with boundary conditions like

$$B_a K \equiv \text{Cos}(\alpha)K(a) - \text{Sin}(\alpha)K'(a) = 0, \quad \text{(A.85)}$$

$$B_b K \equiv \text{Cos}(\beta)K(b) - \text{Sin}(\beta)K'(b) = 0. \quad \text{(A.86)}$$

Of course K is a green function, which means that in $t = 0$ it should reduce to a delta function. a and b are the boundaries of an interval where the generic equation is assumed to hold. We take here Robin boundary conditions which are a mix of Dirichlet and Von Neuman.

We split time and space by looking for solution of the form

$$K(x,t) = \int_\Lambda d\lambda T_\lambda(t)u_\lambda(x), \quad \text{(A.87)}$$

where $T_\lambda(t)$ is solution of $-\lambda = (1/T)(dT/dt)$ and λ is eigenvalue of L.

It is equivalent to perform a Laplace analysis of the equation. The set Λ on which the integration is performed can contain discrete values and a continuous part, depending on the operator L. To understand the relationship between eigenvalues of L and green functions of the differential equation, we write $Lu_\lambda(x) = \lambda u_\lambda(x)$ for an eigenvector of L associated with λ, we usually find a scalar product such that L is autoadjoint with this scalar product. This is done by introducing a kernel $s(x)$ such that the scalar product is expressed by:

$$\langle f, g\rangle = \int_\Omega s(x)f(x)g(x)dx. \quad \text{(A.88)}$$

Then by using this kernel, it is possible to show that the closure relationship of the problem is given by

$$\int_\Lambda u_\lambda(x)\overline{u_\lambda}(y)d\lambda = \frac{\delta(x-y)}{s(x)}, \quad \text{(A.89)}$$

where the integral over Λ may contain a discrete summation and a continuous integral.

Let us assume that we are able to find solutions of the homogeneous equation

$$L_\lambda u \equiv (L - \lambda)u = 0 \tag{A.90}$$

with only one boundary equation. Because a second-order differential equation has in general two independent solutions, we are able to find a solution $u(x)$ satisfying

$$L_\lambda u = 0, \quad B_a u = 0 \tag{A.91}$$

and another independent solution v satisfying

$$L_\lambda v = 0, \quad B_b v = 0. \tag{A.92}$$

We can then demonstrate that a solution to the green equation of L_λ with boundaries is given by:

$$g(x, y, \lambda) = \frac{-u_\lambda(x_<)v_\lambda(x_>)}{W(u_\lambda, v_\lambda, x)a_2(x)}, \tag{A.93}$$

where $x_< = \text{Min}\{x, y\}$ and $x_> = \text{Max}\{x, y\}$

$$W(u_\lambda, v_\lambda, x) = u_\lambda(x)v'_\lambda(x) - u'_\lambda(x)v_\lambda(x)$$

is the Wronskian of both solutions. Then we can show that

(1) the dependency of the Wronskian in x is given by $W(u_\lambda, v_\lambda, x) = C(\lambda)/a_2(x)$ where $C(\lambda)$ is called the Abel constant of the problem;
(2) the Eigenvalues of the operator L are given exactly by the zeros of $C(\lambda)$;
(3) if λ_n is an eigenvalue of L, an associated normalized eigenvector is given by:

$$u_n(x) = \sqrt{\frac{k_n}{C'(\lambda_n)}}v(x, \lambda_n)$$

where $k_n = u(x, \lambda_n)/v(x, \lambda_n)$ is a constant.

For more details see Bhansali (1998).

A.4.5. *Gram Charlier/Edgeworth*

A.4.5.1. *Cumulants*

An extremely useful tool to adapt a closed form formula to a slightly different environment is the Edgeworth expansion.

Everything begins with the close examination of the characteristic function associated with a random variable. Let X, Y be two independent random variables, we know that the characteristic function is defined by

$$\varphi_X(t) = E\left[e^{itX}\right] \tag{A.94}$$

verifies

$$\varphi_{X+Y}(t) = \varphi_X(t)\varphi_Y(t). \tag{A.95}$$

Therefore, it is very tempting to take the logarithm of it and define formally

$$\psi_X(t) = \mathrm{Log}(\varphi_X(t)). \tag{A.96}$$

Then, this functional cumulates over independent random variables. It could be badly defined, because we take the logarithm of something that is a complex number. Let us define now the cumulants by

$$c_n(X) = \frac{1}{i^n}\left(\frac{d}{dt}\right)^n \psi_X(t)|_{t=0} \quad \text{for } n \geq 1. \tag{A.97}$$

In other words, $c_n(X)i^n/n!$ is the nth coefficient in the power series representation of ψ_X. But it is interesting to follow the parallel with statistical physics and call ψ_X the connected generating functional. The main property of these cumulants being, of course, their accumulation over independent variables, hence their name, the first ones are:

$$c_1(X) = E(X), \tag{A.98}$$

$$c_2(X) = \mathrm{Var}(X), \tag{A.99}$$

$$c_3(X) = E\{[X - E(X)]^3\}. \tag{A.100}$$

Note that

$$c_3(X) = \nu_1 c_2(X), \tag{A.101}$$

where $\nu_1$is the skewness of X:

$$c_4(X) = E\{[X - E(X)]^4\} + 3\mathrm{Var}(X)^3 \tag{A.102}$$

We can show that for X being a standard Gaussian variable, $c_n(X) = \delta(n, 1)$. They measure how far the distribution of X is from a standard Gaussian distribution. Given a few regularity assumptions, they completely characterize the distribution of the variable X. The problem of cumulants is to recover a probability distribution from its sequence of cumulants. No solution exists in some cases, sometimes a unique solution exists, in other cases more than one solution exists.

A.4.5.2. *The cumulant modifying operator (CMO)*

If we consider the space of densities (we have to forget about the normalization condition and the positivity to get a vectorial structure). Let us take, for example, some function space defined on R, for example, with the usual Euclidian square norm. Given a sequence of real numbers $\varepsilon_n, n \in \mathbf{N}$, let us define the following operator: $\Theta : f \to \Theta(f, \varepsilon)$ defined by

$$\Theta(f, \varepsilon)(x) = \exp\left(\sum_{i=1}^{\infty} \frac{\varepsilon_i(-D)^i}{i!}\right) f(x), \tag{A.103}$$

where D is the derivative operator. It is possible to show that this operator modifies the cumulant of the density f by adding ε_n to the nth cumulant of f.

A.4.5.3. *Application to pricing*

One way to create a smile with a regular BS formula is to assume that the underlying log distribution deviates from normality, therefore the cumulants of the distributions are slightly different from what they are supposed to be on a purely lognormal market.

Let us assume that we get cumulants from historical data or from the implicit distribution associated with options markets. By injecting the definition of the CMO into the expectation computation we get:

$$E[(X-K)^+] = \int_K^\infty x \left(\sum_{i=1}^{\infty} \frac{\varepsilon_i (-D)^i}{i!} \right) f(x)dx \qquad \text{(A.104)}$$

Then by integrating by part, we can extract the deviations from the BS models. See Jarrow and Rudd (1982) for the Smile of a vanilla option, and Bhansali (1998) for application to basket options. See Vorst for application to other options and Stuart and Ord (1994) for General ideas on cumulants theory.

A.5. Summary

Itô formula	Allows obtaining the solution of an EDS of a C^2-derivable function of an Itô process from its EDS.
Girsanov theorem	Allows changing the probability measure of an expectation of an Itô process modifying its drift and keeping its "Brownian" structure.
Feynman–Kac formula	Defines a PDE problem whose solution is an option payoff value, as an alternative of computation through its expectation.
Tanaka formula	Generalizes the formula Itô for some non-differentiable functions by defining the local time notion.
Markov property	Reducing of the past time dependence of a process to its short memory or actual information.
Forward neutral probability	As a result of Girsanov formula, changes an asset dynamics such that it would be a martingale normalized respecting to a numeraire.

Copula	Describes and measures the dependence of a sample of random variables by defining the link between their joint distribution with their marginal ones.
GARCH and Nelson result	Describes the dynamics and the nonlinear relation of an asset process with its volatility process (conditional variance) in discrete time and continuous times.
PCA and Cholesky	Represents the correlation matrix of several random variables trough the same or less number of independent factors.
Newton–Raphson	Finds the extremum of a real function using the exact derivatives or a finite difference approximation in the symbolic derivative cannot be computed.
Steepest gradiant	Finds the extremum of a multi-dimensional function using the exact Hessian or a finite difference approximation in the symbolic derivative cannot be computed.
BFGS and conjugated gradient	Solves nonlinear systems using successive approximation of the Hessian matrix since its never computed explicitly.

References

[1] Bally V (2003). *An Elementary Introduction to Malliavin Calculus.* INRIA, Rapport de recherche 4718.

[2] Bender, CM and Orszag, SA (1999). *Advanced Mathematical Methods for Scientists and Engineers (Asymptotic Methods and Perturbation Theory).* Berlin: Springer.

[3] Benhamou E (2000). A Generalisation of Malliavin weighted scheme for fast computation of the Greeks. FMG Discussion Paper No. 0350. Available at SSRN: http://ssrn.com/abstract=265277

[4] Benhamou E (2004). Smart Monte Carlo: Various tricks using Malliavin calculus. CDC Ixis Capital Markets.
[5] Berestycki H, Busca J and Florent I (2004). Computing the implied volatility in stochastic volatility models, *Communications on Pure and Applied Mathematics*, LVII, 001–022.
[6] Bhansali V (1998). *Pricing and Managing Exotic Hybrid Options.* Irwin Library of Investment and Finance, New York: McGraw Hill.
[7] Dunford S and Schwartz JT (1958). *Linear Operator Part 1, General Theory.* New York: Wiley Interscience.
[8] Fournie E, Lasry JM, Lebuchoux J, Lions PL and Touzi N (1999). Application of Malliavin calculus to Monte Carlo methods in finance. *Finance and Stochastics*, 3, 391–412.
[9] Friz PK (2003). *An Introduction to Malliavin Calculus.* New York University.
[10] Hagan P, Lesniewski A and Woodward D (2004). Probability distribution in the SABR model of stochastic volatility. Working Paper Draft.
[11] Hubalek F, Teichman J and Tompkins R. Flexible complete models with stochastic volatility generalizing Hobson–Rogers. Working Paper. Vienna Institute of Technology.
[12] Jarrow R and Rudd A (1982). Approximate option valuation for arbitrary stochastic processes. *JFE*, 347–369.
[13] Kohatsu-Higa A and Montero M (2004). *Malliavin Calculus in Finance.* Handbook of Computational and Numerical Methods in Finance, pp. 111–174. Birkhauser.
[14] Malliavin P (1997). *Stochastic Analysis.* Grundlehren der Mathematischen Wissenschaften, Vol. 313. Berlin: Springer.
[15] Nualart D (1995). Probability and its applications. In *The Malliavin Calculus and Related Topics*, Berlin: Springer-Verlag.
[16] Stakgold I (1998). *Green's Function and Boundary Value Problems*, 2nd edn. New York: Wiley Interscience.
[17] Stuart A and Ord K (1994). *Kendall's Advanced Theory of Statistics*, 6th edn., Vol. 1 *Distribution Theory.* Edition Edward Arnold.
[18] Vorst T. *Analytic Boundaries and Approximations of the Price and Hedge Ratios of Average Rate Options.* Economic Institute, Erasmus University, Rotterdam.

Appendix B

MONTE CARLO

In this appendix, we are going to present the latest Monte Carlo techniques, applied to finance, which intervenes when we cannot find explicit formulae for complex derivatives. Here, after describing the Monte Carlo approach, we first explain how to generate Gaussian with random and quasi-random generators, second we show how we can improve the speed of Monte Carlo using quasi-Monte Carlo sequences and control variates, third, we detail the different methods to calculate the Greeks, and finally, we explain the methods used to price American options using Monte Carlo simulation.

B.1. Monte Carlo

B.1.1. *Theoretical background: central limit theorem*

The Monte Carlo method is derived from the fundamental central limit theorem. Roughly, the central limit theorem states that the distribution of the sum of a large number of independent, identically distributed variables will be approximately normal, regardless of the underlying distribution.

Namely, suppose that $X_1, X_2, \ldots, X_n$, are n independent identically distributed random variables with mean μ and standard deviation σ. Then the empirical average

$$\frac{1}{n}\sum_{i=1}^{n} X_i \tag{B.1}$$

is approximately normal, with mean μ and standard deviation $\sigma/\sqrt{n}$.

Moreover, let Y_n be defined as:

$$Y_n = \frac{\frac{1}{n}\sum_{i=1}^{n}(X_i - \mu)}{\sigma/\sqrt{n}}, \tag{B.2}$$

then Y_n is approximately a standard Gaussian random variable, and Y_n converges to a standard Gaussian random variable as $n \to \infty$.

B.1.2. *Pricing European securities using Monte Carlo*

A Monte Carlo simulation of a process is a way to sample random outcomes of the process. Pricing with Monte Carlo heavily relies on the central limit theorem. It consists in randomly simulating a large number of paths of a process and then in averaging payoffs on each paths.

A simple example, to understand this, is the way to price an European put on stock in the BS model. We can write the diffusion of a lognormal process as follows:

$$S_T = S_0 \exp\left(rT - \frac{1}{2}\sigma^2 T + \sigma W_T\right) \tag{B.3}$$

or equivalently

$$S_T = S_0 \exp\left(rT - \frac{1}{2}\sigma^2 T + \sigma\sqrt{T}N\right), \tag{B.4}$$

where N is a random variable that follows a standard Gaussian law (i.e. to say is with zero mean and a standard deviation of 1).

When we apply that methodology to the pricing of an European put, we get for the price:

$$P_T = \mathrm{e}^{-rT} E\big[(K - S_T)^+\big] \approx \frac{1}{N}\sum_{i=1}^{N} \mathrm{e}^{-rT}\left(K - S_T^{(i)}\right)^+, \tag{B.5}$$

where $S_T^{(i)}$ are independent identically distributed samples obtained for each of the numerical realization of the normal variable.

B.1.3. *Random numbers and quasi Monte Carlo*

As shown above, pricing with Monte Carlo relies on random number. This is why the way random numbers are generated matters. In finance, most random variables follow a Gaussian law. Therefore, we will show how to generate random numbers following that kind of law. Basically, all methods rely on first generating uniformly distributed numbers on $[0, 1]$, and transforming them into the desirable distribution, namely in our case normally distributed numbers.

Though there are many methods to generate random numbers following a given distribution or probability law from uniformly distributed numbers, we will focus on Gaussian laws only. The two well-known algorithms used by practitioners for this purpose are:

(1) the Box-Muller,
(2) the inverse repartition method.

B.1.3.1. *Box-Muller*

If u and v are two independent uniformly distributed random variables with values in $(0, 1)$, then x and y defined as below are two independent Gaussians distributed random variables:

$$\begin{aligned} x &= \sqrt{-2\ln(u)}\cos(2\pi v), \\ y &= \sqrt{-2\ln(u)}\sin(2\pi v). \end{aligned} \tag{B.6}$$

B.1.3.2. *Cumulated density inverse*

If we denote by F the cumulated density function of a Gaussian law and if u is a uniformly distributed random variable, then

$$x = F^{-1}(u) \tag{B.7}$$

is distributed according to a Gaussian variable.

As a matter of fact, computers can generate deterministic numbers only. Hence, no series of random numbers is really random. Nevertheless, it is possible to degenerate sequences of numbers which statistically behave as random numbers. Most of the time, sequences are produced within a finite set of integers $\{0, 1, \ldots, M-1\}$ and are then divided by M. Those sequences are built up recursively:

$$\begin{aligned} x_0, \ldots, x_{k-1} &\in \{0, 1, \ldots, M-1\}^k, \\ x_{n+1} &= f(x_n, \ldots, x_{n-k}) \end{aligned} \tag{B.8}$$

and then the sequence is defined as:

$$u_n = \frac{x_n}{M}. \tag{B.9}$$

Commonly, one takes a sum modulo a number:

$$f(x_1, \ldots, x_k) = (a_1 x_1 + \cdots + a_k x_k) \bmod M, \tag{B.10}$$

where $a_1, \ldots, a_k$ are integers. The sequence $x_0, \ldots, x_{k-1}$ is referred to as the seed. An important issue is to take M greater than the length of the sequences we are trying to compute. A smart choice, if this method is implemented in C, is to take the RAND_MAX constant.

Another method to compute prices using Monte Carlo method is to use sequences of not independant numbers instead of "randomly" generated ones. Proceeding this way allows us to go more quickly than $1/\sqrt{N}$ and also to have a better repartition of the numbers than with plain random number generation. In that case, the pricing method used is no longer called Monte Carlo, but quasi-Monte Carlo.

B.1.4. *Notion of discrepancy*

For a better understanding of those methods, we introduce a couple of notions:

A $[0,1]^d$-valued sequence $(u_n)_{n\geq 1}$ is said to be uniform if for any $x \in [0,1]^d$, we have

$$\lim_{N\to\infty} \frac{1}{N} \sum_{n=1}^{N} \prod_{i=1}^{d} 1_{u_n^i \leq x^i} = \prod_{i=1}^{d} x^i. \tag{B.11}$$

For a uniform $[0,1]^d$-valued sequence $(u_n)_{n\geq 1}$ we define its discrepancy as:

$$D_\infty^*(u, N) = \sup_{x\in[0,1]^d} \left| F(x) - F_N^u(x) \right| \underset{N\to\infty}{\rightarrow} 0, \tag{B.12}$$

where

$$F_N^u(x) = \frac{1}{N}\sum_{n=1}^{N}\prod_{i=1}^{d} 1_{u_n^i \le x^i} \quad \text{and} \quad F(x) = \prod_{i=1}^{d} x^i. \tag{B.13}$$

Let f be a function of finite variation, we have then the (Koshama–Hlawka) inequality:

$$\left| \int_{[0,1]^d} f(x)dx - \frac{1}{N}\sum_{i=1}^{N} f(u_i) \right| \le V(f) D_\infty^*(u, N), \tag{B.14}$$

where

$$V(f) = \sum_{k=1}^{d} \sum_{1\le i_1 \le \cdots \le i_k \le d} \int_{[0,1]^d} \left| \frac{\partial^k}{\partial_{x_{i_1}} \ldots \partial_{x_{i_k}}} f(x) \right| dx V(f) < \infty. \tag{B.15}$$

That is why the lowest our sequence's discrepancy, the better the approximation of the integral. Hence, there is a strong interest in finding algorithm to generate low-discrepancy sequence. Besides this, we observe that:

(1) f is the payoff of an option, and it is not regular in general;
(2) it becomes more and more difficult to have $V(f) < \infty$ when d increase.

B.1.5. *Examples of random numbers generation*

We describe here two random generators that have turned out to work well:

(1) the combined MRG of order 5,
(2) the Mersenne twister.

B.1.5.1. *The combined MRG of order 5*

This generator is a combination of two MRG of order 5:

$$z_n = \sum_{j=1}^{J} \delta_i x_{j,n} \bmod m_1, \quad u_0 = \frac{z_n}{m_1} \tag{B.16}$$

with the following parameters:

$$J = 2, \quad k = 5 \quad \text{and} \quad \delta_1 = -\delta_2 = 1 \tag{B.17}$$

with

$$\begin{gathered} m_1 = 2^{32} - 18269, \quad a_{1,1} = 0, \quad a_{1,2} = 1154721, \\ a_{1,3} = 0, \quad a_{1,4} = 1739991, \end{gathered} \tag{B.18}$$

$$a_{1,5} = -1108499 \tag{B.19}$$

$$\begin{gathered} m_2 = 2^{32} - 32969, \quad a_{2,1} = 1776413, \quad a_{2,2} = 0, \\ a_{2,3} = 865203, \quad a_{2,4} = 0, \end{gathered} \tag{B.20}$$

$$a_{2,5} = -1641052, \tag{B.21}$$

$$X_{1,n} = a_{1,2}X_{1,n-2} + a_{1,4}X_{1,n-4} + a_{1,5}X_{1,n-5}, \tag{B.22}$$

$$X_{2,n} = a_{2,1}X_{1,n-1} + a_{2,3}X_{1,n-3} + a_{2,5}X_{2,n-5}, \tag{B.23}$$

$$Z_n = (X_{1,n} - X_{2,n}) \bmod m_1. \tag{B.24}$$

The period is equal to $\left(m_1^5 - 1\right)\left(m_2^5 - 1\right)/2 \approx 2^{319}$. The algorithm works then as follows. First, initialize the two sequences $X_{1,j}$ and $X_{2,j}$. Then, for each call to the sequence, computation of a new point for the first and second generator and combination into the new sequence with Eq. (B.24).

B.1.6. *Examples of quasi-random numbers generation*

We show here various well-known methods used to generate such sequences, viz., Halton, Square root sequence and Sobol sequences.

B.1.6.1. *Halton quasi-random sequences*

Van Der Corput ($d = 1$) sequence: Given a prime number p, we define for any n, its p-adic decomposition as

$$n = a_0^{n,p} + a_1^{n,p} p + \cdots + a_{k_{n,p}}^{n,p} p^{k_{n,p}} \tag{B.25}$$

And then, we define the nth term of the sequence as:

$$u_n^p = \frac{a_0^{n,p}}{p} + \frac{a_1^{n,p}}{p^2} + \cdots + \frac{a_{k_{n,p}}^{n,p}}{p^{k_{n,p}+1}} \tag{B.26}$$

Sequence of Halton ($d \geq 1$) : $p_1, \ldots, p_d$ the d first prime number:

$$u_n = (u_n^{p_1}, \ldots, u_n^{p_d}). \tag{B.27}$$

For this sequence, we have the following inequality about its discrepancy:

$$D_\infty^* (u, N) \leq O\left(\frac{\ln(N)^d}{N}\right). \tag{B.28}$$

B.1.6.2. *Square quasi-random sequences*

We define

$$u_n = n\sqrt{p} - \text{Floor}\, \lfloor n\sqrt{p} \rfloor, \tag{B.29}$$

$$D_\infty^* (u, N) = O\left(\frac{1}{N^{1-\varepsilon}}\right). \tag{B.30}$$

Figure B.1 shows the difference of repartition between numbers generated with the MRG algorithm (with $k = 2$) and with the SQRT algorithm (prime numbers being 2 and 3). We can see through this that there are much more "holes" with numbers generated with the MRG algorithm than with SQRT.

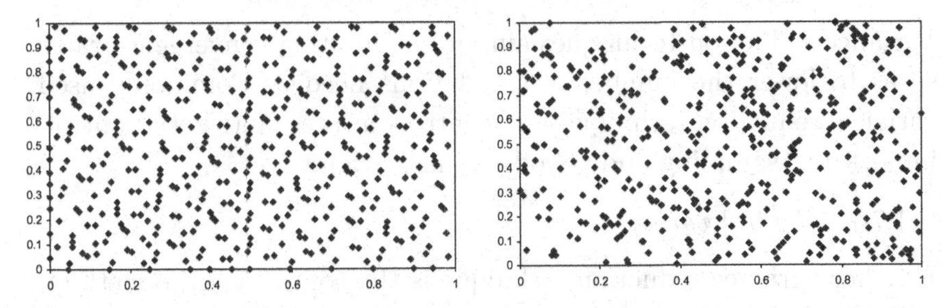

Figure B.1: Comparison quasi- and normal-random sequence. The left graphic is the Halton sequence while the right is the MRGK4 algorithm.

B.1.6.3. *Sobol sequence*

The Sobol sequence is a d-dimensional sequence in base 2. It has proved to work well in most common cases (because of additional uniform property) and have become one of the most widely used quasi-Monte Carlo sequence. First developed by Sobol, its construction is based on primitive polynomials in the field Z_2 and XOR manipulations.

Basically, each dimension is a permutation of the Halton sequence in base 2 imposing the total number to be a power of $2{:}N = 2^d$. These permutations are generated from the classical irreducible polynomials in Z_2.

The Sobol sequence is defined by

$$\xi_n = \left(a_0 V_0^{(1)} \oplus \cdots \oplus a_{R(n)} V_{R(n)}^{(1)}; \cdots ; a_0 V_0^{(d)} \oplus \cdots \oplus a_{R(n)} V_{R(n)}^{(d)}\right), \tag{B.31}$$

where the $V_i^{(j)}$ are the direction numbers (expressed as binary fraction) obtained from the d primitive polynomials and a_i denotes the coefficients of the decomposition of the number n in base 2, and $\oplus$ represents the standard bit-wise exclusive OR operator.

In order to implement this sequence, one should rather change the expression and use the formulation from Antonov and Saleev. There are some others well-known algorithm such as Faure. The Sobol and Faure sequence can reach the same speed $\ln(N)^d/N$ (for more details see Press *et al.*).

B.1.7. *Variance reduction techniques*

Variance reduction techniques aim at accelerating convergence rates so as to lower the number of simulations needed. Commonly used variance reduction techniques are control variates, antithetic control, importance sampling and payoff regularization.

B.1.7.1. *Control variate*

Another variance reduction technique is the control variate method. The idea behind is to use a known variable Y with zero mean and to correct our simulation with the bias observed on this variable.

Basically, if Y is a variable with zero mean, we have that adding this variable should still lead to the same expectation value:

$$E[F] = E[F + Y]. \tag{B.32}$$

We want to take a variable that reduces the variance so that:

$$\text{Var}[F + Y] < \text{Var}[F]. \tag{B.33}$$

In this framework, we will compute our Monte Carlo as follows:

$$E[F] \approx \sum_{i=1}^{N} F^{(i)} + Y^{(i)}. \tag{B.34}$$

In order to illustrate this, let us work out the method on a simple example and see where the variance reduction comes from. For the simplicity of the explanation, let us see on a European-style call:

$$E[g(S_T)] = E[g(S_T) + \rho^* (S_T - E(S_T))]. \tag{B.35}$$

The variance

$$\text{Var}[g(S_T) + \rho^* (S_T - E(S_T))] \tag{B.36}$$

is minimal for

$$\rho^* = \frac{\text{Cov}(g(S_T), S_T)}{\text{Var}(S_T)}. \tag{B.37}$$

B.1.7.2. *Importance sampling*

Now let us imagine that we would like to price a deeply out of the money European call on a stock using Monte Carlo. Namely,

$$E\left[(S_T - K)^+\right] \tag{B.38}$$

(where S_t follows a lognormal process) is to be computed. As our call is deeply out of the money, few simulations of

$$(S_T - K)^+ \tag{B.39}$$

will be different from 0. Hence, Monte Carlo computation will have huge variance. A way to lighten this effect is to try somehow to force simulated paths to be at the money. This is the aim of the importance sampling technique. That method consists in a probability change so that more paths of S_t are in the money, and in computing Monte Carlo using that new probability.

In order to illustrate this, let us see the case of a call options for a BS model. In this case, the asset follows a geometric Brownian motion:

$$\frac{dS_t}{S_t} = rdt + \sigma dW_t^Q. \tag{B.40}$$

We consider the following probability change:

$$\frac{dH_t}{dQ_t} = \exp\left(-hW_t - \frac{1}{2}h^2 t\right) = Y_t \tag{B.41}$$

with

$$h = \frac{1}{T}\left[\ln\left(\frac{K}{S_0}\right) - r\right] \tag{B.42}$$

In this particular measure, we have that the expectation of the asset is equal to the strike:

$$E^{H_T}[S_T] = K. \tag{B.43}$$

This change of measure basically centers the asset around the strike. The call price is then given by

$$E^Q\left[(S_T - K)^+\right] = E^H\left[\frac{1}{Y_T}(S_T - K)^+\right]. \tag{B.44}$$

Monte Carlo technique is then used to compute under probability H.

$$E^H\left[\frac{1}{Y_T}(S_T - K)^+\right]. \tag{B.45}$$

This method is better if:

$$\mathrm{Var}_H\left(\frac{1}{Y_T}(S_T - K)^+\right) \ll \mathrm{Var}_Q\left((S_T - K)^+\right). \tag{B.46}$$

No matter how miraculous this method may seem, it has a major drawback. It is very product dependant; and it is very payoff depedent. Hence one may need to think quite carefully about the product before thinking about it.

B.1.7.3. *Antithetic control*

Antithetic method is based upon the Brownian motion's symmetry, which says basically that for a Brownian motion W_t, W_t and $-W_t$ have the same law.

Hence for any monotonous function g, we can take advantage of it and compute the expectation of a function with the corresponding antithetic variables since we

$$\frac{1}{2}E[g(W_t) + g(-W_t)] = E[g(W_t)]. \tag{B.47}$$

The interest lies in the variance reduction. Indeed, it can be shown, that if W_t and W_t' are identically distributed, we have

$$\mathrm{Var}[g(W_t) + g(-W_t)] \leq \mathrm{Var}[g(W_t) + g(W_t')]. \tag{B.48}$$

Hence for any set of iid $X_1, \ldots, X_{2N}$ normally distributed variables, we have

$$E\left[\frac{1}{2N}\sum_{i=1}^{N} g(X_i) + g(-X_i)\right] = E\left[\frac{1}{2N}\sum_{i=1}^{2N} g(X_i)\right], \tag{B.49}$$

$$\mathrm{Var}\left[\frac{1}{2N}\sum_{i=1}^{N} g(X_i) + g(-X_i)\right] \leq \mathrm{Var}\left[\frac{1}{2N}\sum_{i=1}^{2N} g(X_i)\right]. \tag{B.50}$$

This inequality is an immediate consequence of the following lemma.

Let X be a random variable. For any reel decreasing transformation T that verify $T(X) \overset{\text{rule}}{=} X$ and for any monotonous function g:

$$\mathrm{cov}(g(X), \quad g(T(X)) \leq 0$$

Then, antithetic control consists in computing the standard expectation

$$E[g(W_t)]. \tag{B.51}$$

with the "symmetric" empirical average:

$$\frac{1}{2N}\sum_{i=1}^{N} g(X_i) + g(-X_i). \tag{B.52}$$

The calculations made above show that a simple computation with N simulations and antithetic control is more accurate than a computation with $2N$ simulations. Another advantage of that method is that it can be applied in a very systematic way, which will not be the case of the methods that will be shown in the following section.

B.1.8. *Computation of Greeks in Monte Carlo*

We present here three methods for Greeks computation in Monte Carlo, namely finite differences, path-wise method, likelihood ratio method, malliavin calculus and stochastic analysis.

Let us say that we have a stochastic process S that depends on a parameter θ. We know how to compute

$$C(\theta) = E\left\lfloor f(S_{T_1}(\theta), \ldots, S_{T_n}(\theta))\right\rfloor \tag{B.53}$$

with Monte Carlo, and we would like to compute its derivative function with respect to θ,

$$\nu = \frac{dC(\theta)}{d\theta}. \tag{B.54}$$

We will note that $\hat{C}(\theta)$ the value of our expectation (B.53) computed with Monte Carlo.

B.1.8.1. *Finite differences*

The finite difference method consists in approximating the derivative v with it finite difference method

$$\frac{\hat{C}(\theta + h) - \hat{C}(\theta)}{h}. \tag{B.55}$$

The computation of Eq. (B.55) is fairly easy to do for a given h. However, finding the good bump h is not an easy task, since for large bump h, the approximation becomes poor, while for small h, the difference can exhibit very bumpy result.

However, its simplicity of implementation makes it the common method for payoff not too discontinuous. In order to avoid extra noise in the difference between the two simulations, it is strongly advised to use the same random number, referred to as common random number finite difference method (Broadie, Glasserman 1996).

B.1.8.2. *Path-wise method*

For discontinuous payoff, taking the discrete derivatives function as in Eq. (B.55) can lead to very bad result as the derivative function may have strong discontinuity or kinks. The best is to smooth in a sense the payoff. This is what the path-wise or the likelihood ratio method aims at.

Let us say, for instance, that we would like to compute the derivatives function with respect to θ

$$\nu = \frac{dC(\theta)}{d\theta} \tag{B.56}$$

with

$$C(\theta) = E\left[f(S_T(\theta))\right] \tag{B.57}$$

the derivative ν, under the forward neutral world, can also be written as (under the condition that we can interchange the expectation and the derivation operator):

$$\nu = E\left[\frac{df(S_T(\theta))}{d\theta}\right] = E\left[\frac{dS_T(\theta)}{d\theta} f'\left(S_T(\theta)\right)\right]. \tag{B.58}$$

Hence, the computation of the Greek can be done through computing the price of a security with payoff

$$\frac{dS_T(\theta)}{d\theta} f'\left(S_T(\theta)\right) \tag{B.59}$$

with Monte Carlo (see Kunita 1984).

An example to understand this methodology is to consider the delta of an European call on a stock following a plain lognormal process. The payoff of a call is given by

$$f(x) = (x - K)^+, \tag{B.60}$$

while its derivatives with respect to its initial value is given by:

$$f'(x) = 1_{(x-K)\geq 0}. \tag{B.61}$$

Hence to compute the Delta of the option, we have to compute the mean of

$$\frac{S_T}{S_0} 1_{(S_T-K)\geq 0} \tag{B.62}$$

by means of simulation.

B.1.8.3. *Likelihood ratio method*

Let us assume that we know the density of the asset $S_T(\theta)$. We will denote it by $g(\theta, x)$. We have the expected value of the derivative function with respect to θ by

$$\nu = \frac{\partial}{\partial\theta} \int_{-\infty}^{+\infty} g(\theta, x) f(x) dx. \tag{B.63}$$

Interchanging the derivation and the integral, we can rewrite this as follows:

$$\nu = \int_{-\infty}^{+\infty} \frac{dg(\theta, x)}{d\theta} \frac{1}{g(\theta, x)} f(x) g(\theta, x) dx \tag{B.64}$$

or equivalently, we can express this as an expectation given by:

$$\nu = E\left[\frac{d\ln(g(\theta, S_T))}{d\theta} f(S_T)\right]. \tag{B.65}$$

This means that we only have to compute the mean of

$$\frac{d\ln(g(\theta, S_T))}{d\theta} f(S_T) \tag{B.66}$$

with Monte Carlo. Equation (B.66) can also be interpreted as computing the payoff with a weight given by the likelihood ratio,

$$\frac{d\ln(g(\theta, S_T))}{d\theta} f(S_T), \tag{B.67}$$

hence the name of the method. We can remark that this method can be used if and only if we know the density of the underlying $S_T(\theta)$.

As title of example, we can exhibit explicit Greek's formulas for vanilla options in one-dimensional BS model, under the forward neutral world:

$$\begin{aligned}\Delta_0 &= E\left(f(S_T)\frac{W_T}{S_0\sigma T}\right),\\ \Gamma_0 &= E\left(f(S_T)\frac{((W_T^2/\sigma T) - W_T - 1/\sigma)}{S_0^2\sigma T}\right).\end{aligned} \tag{B.68}$$

We can extend these results in multi-dimensional BS model. Besides this, this method could not be applied to all models, like, for example, BGM. If one is interested in computing the weights without knowing the density, one can use sophisticated tools such as Malliavin calculus.

B.1.8.4. *Malliavin calculus*

In a general context where the joint density of $(S_T(\theta), \partial_\theta S_T(\theta))$ is unknown or undefined, but $S_T(\theta)$ is defined on the Wiener space, one can write:

$$\partial_\theta E(f(S_T(\theta))) = E(f'(S_T(\theta))\partial_\theta S_T(\theta)) = E(f(S_T(\theta))H_\theta). \tag{B.69}$$

For some (explicit) random variable H_θ. This formula is due to integration by parts on the infinite Wiener space which is called Malliavin calculus. No uniqueness in the representation: $E(H_\theta\,|S_T(\theta)) = E(H'_\theta\,|S_T(\theta))$ (see Fournie, 1999).

B.1.8.5. *Stochastic analysis*

Delta of vanilla options: Suppose that the dynamic of $S_t = \begin{pmatrix} S_t^1 \\ \vdots \\ S_t^d \end{pmatrix}$

under the forward neutral probability:

$$S_t = \underbrace{S_0 + \int_0^t B(s, S_s)ds}_{\neq 0, \text{ if dividends}} + \underbrace{\sum_{i=1}^{d} \int_{\acute{a}}^t \sigma_i(s, S_s)dW_s^i}_{=\int_{\acute{a}}^t \sigma(s,S_s)dW_s}. \tag{B.70}$$

Let $Y_t := \partial_{S_0} S_t$ the matrix valued process. Set $u(t, y) = E[f(S_T) | S_t = y]$: u solves the PDE evaluation.

Under the assumption that σ is an invertible volatility, we can show that the process $M = (M_t)_{0 \leq t < T}$ defined by $M_t = \partial_S u(t, S_t) Y_t$ for $t < T$ is a martingale, and knowing that the delta is equal to: $\Delta_0 = E\left(\frac{1}{T}\int_0^T M_s ds\right)$, we can exhibit an explicit formula:

$$\Delta_0 = E\left(\frac{f(S_T)}{T}\left[\int_0^T \left[\sigma(s, S)^{-1} Y_s\right]^* dW_s\right]^*\right). \tag{B.71}$$

Remark: Coincides with the previous formula in the BS model.

We can extend this stochastic analysis to calculate delta for European barrier option, European lookback options (see Bernis and Kohatsu-Higa, 2003), and European Asian options (Fournié *et al.* 1999; Benhamou, 2000).

B.1.9. *Extension of Monte Carlo methods for American-style securities pricing*

Monte Carlo methods shown above can be used for European-style securities pricing only: it works through simulating pay-offs at a certain date and doing a mean of those simulated prices.

Securities with early exercise features are not so straightforward to compute with Monte Carlo. Solving that problem has involved much work amongst practitioners during the last years, which gave birth to many algorithms. We present here methods that work in almost every case: The least square method (also referred to as Longstaff–Schwartz algorithm), and Andersen's algorithm. Those pricing algorithms work in two steps. Both methods are based upon a common framework: First, an exercise boundary for the option is computed; second, it is used for forward pricing with Monte Carlo (once known, the exercise boundary can be used to price the option like a trigger option).

This methodology can be explained easily: as shown in Section B.1.2, the price of a Bermudan option is given by its expected value provided its best exercise strategy (or stopping time). Hence, a Bermudan option can be priced by first approximating its best exercise strategy, and then using it to price the option as a trigger.

The difference between these two numerical methods (Longstaff–Schwartz and Andersen) lies in the way exercise boundaries approximations are computed.

B.1.9.1. *Andersen method*

We will first have a look at Andersen's algorithm. As explained earlier, it works with two steps.

Step 1. Computation of the exercise boundary: In the case of Andersen, the options' best exercise strategy is approximated by stopping times of the form: $\tau = T_i$ if for the first time $I_i(x_i) \geq a_i$. We will refer to the set of those stopping times as Γ_N, and we will note $\tau_{a_1,\ldots,a_N}$ an element of Γ_N (because a stopping time $\tau \in \Gamma_N$ can be identified with an N-uple $(a_1,\ldots,a_N)$). Hence, the best exercise strategy in the Andersen algorithm is given by:

$$(a_1^*,\ldots,a_N^*) \equiv \tau_{a_1^*,\ldots,a_N^*} \equiv B(T_0,T_N)\underset{\tau\in\Gamma_N}{\text{ArgMax}}\left[E^{Q_N}\left(\frac{I_\tau(x_\tau)}{B(\tau,T_N)}\right)\right]. \tag{B.72}$$

The value of a Bermudan option is then approximated by:

$$\begin{aligned} V(0) &\approx B(T_0,T_N)\text{Sup}\left[\underset{\tau\in\Gamma_N}{E}\left[E\left(\frac{I_\tau(x_\tau)}{B(\tau,T_N)}\right)\right]\right] \\ &= B(T_0,T_N)\underset{\tau=\tau_{a_1^*,\ldots,a_N^*}}{E}\left[E\left[\frac{I_\tau(x_\tau)}{B(\tau,T_N)}\right]\right]. \end{aligned} \tag{B.73}$$

Now, the point is about the way to find out the optimal boundary

$$(a_1^*,\ldots,a_N^*). \tag{B.74}$$

A first method would be to use the expression:

$$\tau_{a_1^*,\ldots,a_N^*} \equiv B(T_0, T_N) \operatorname*{ArgMax}_{\tau \in \Gamma_N} \left[E\left(\frac{I_\tau(x_\tau)}{B(\tau, T_N)} \right) \right]. \tag{B.75}$$

This problem can be solved by approximating

$$E\left(\frac{I_\tau(x_\tau)}{B(\tau, T_N)} \right) \tag{B.76}$$

with

$$\frac{1}{M} \sum_{i=1}^{M} \left(\sum_{n=1}^{N} \frac{I_n(x_n^{(i)})}{B(T_n, T_N)} 1_{\tau(x^{(i)})=n} \right), \tag{B.77}$$

where

$$(x^{(1)}, \ldots, x^{(M)}) \tag{B.78}$$

are M simulated paths and by solving:

$$\tau_{a_1^*,\ldots,a_N^*} = \operatorname*{ArgMax}_{\tau \in \Gamma_N} \sum_{i=1}^{M} \left(\sum_{n=1}^{N} \frac{I_n(x_n^{(i)})}{B(T_n, T_N)} 1_{\tau(x^{(i)})=n} \right). \tag{B.79}$$

This expression involves an optimization over N variables, which is quite complicated to achieve. That is why Andersen proposed a solution involving N one-variable optimizations.

This solution lies in a recursive backward algorithm. It works as follows. First, build M trajectories

$$(x^{(1)}, \ldots, x^{(M)}). \tag{B.80}$$

Second, we set the optimal boundary to zero $a_N^* = 0$. This is due to the fact that on last exercise date, the option is exercised if and only if its exercise value is positive.

Third, assume that $\left(a_{n+1}^*, \ldots, a_N^*\right)$ are computed. We would like to compute the optimal boundary a_n^*. We define the function

$f_n(a_n)$ as

$$f_n(a_n) = \sum_{i=1}^{M} \left(\frac{I_n(x_n^{(i)})}{B(T_n, T_N)}^* 1_{I_n(x_n^{(i)}) \geq a_n} + C_n(x_n^{(i)})^* 1_{I_n(x_n^{(i)}) < a_n} \right), \tag{B.81}$$

where

$$C_n(x_n^{(i)}) \tag{B.82}$$

is defined as

$$C_n(x_n^{(i)}) = I_q(x_q^{(i)}) \tag{B.83}$$

with

$$q = \inf(p > n / I(x_p^{(i)}) \geq a_p^*). \tag{B.84}$$

Obviously, the optimal strategy is given by the best

$$a_n^* = \arg\max(f_n(a_n)). \tag{B.85}$$

Common practice is to use a Brent minimization to find the optimal.

For a better understanding of the method, $C_n(x_n^{(i)})$ can be interpreted as the continuation value of the option. In other words, it is the value of the option at date t_n in the state of the world $x_n^{(i)}$ in case it is not exercised at time t_n. Hence, $f_n(a_n)$ is the value of the option at date T_n provided that its holder exercise if

$$I_n(x_n) \geq a_n. \tag{B.86}$$

This method is continued until the last step $n = 0$.

Step 2. Computation of the price provided the exercise boundary: Once the optimal exercise boundary has been determined $(a_1^*, \ldots, a_N^*)$, the pricing leads to the one of a trigger options. It works as follows.

Draw M other paths $\left(x^{(1)}, \ldots, x^{(M)}\right)$ for the underlying process. Compute the price of the security as its empirical average like in a standard Monte Carlo given by

$$P = B(T_0, T_N)\frac{1}{M}\sum_{i=1}^{M}\frac{I_{n(i)}(x_{n(i)}^{(i)})}{B(T_{n(i)}, T_N)} \tag{B.87}$$

with

$$n(i) = \inf\left(p \geq 0 / \frac{I_p(x_p^{(i)})}{B(T_p, T_N)} \geq a_p^*\right). \tag{B.88}$$

B.1.9.2. *Longstaff–Schwartz method*

The Longstaff–Schwartz algorithm works almost the same way as Andersen, namely estimate an optimal exercise boundary and compute the option knowing it. The only difference lies in the estimation of the optimal exercise boundary given as the result of a regression of the continuation value on a basis of representative polynomial functions. Let us give a more detail presentation of this algorithm.

Step 1. Computation of the exercise boundary (i.e. of the continuation values). First, draw M trajectories $\left(x^{(1)}, \ldots, x^{(M)}\right)$. Second, initialize the optimal exercise boundary $C_N^*(x_N)$ at the last step as $C_N^*(x_N) = 0$. This is because the exercise decision at the last exercise date is taken if $I_N^*(x_N) \geq 0$. Third, let us now assume that $C_{n+1}^*, \cdots, C_N^*$ are computed. We want to compute C_n^*. Let $\Phi = (\phi_1, \ldots, \phi_k)$ be a set of R^d-valued functions. We write for all $j \in [1, M]$

$$C_n^j = \frac{I_q(x_q^{(i)})}{B(T_q, T_N)}, \tag{B.89}$$

where

$$q = \inf\left(p \geq n \Bigg/ \frac{I_p(x_p^{(i)})}{B(T_p, T_N)} \geq C_p^*(x_p^{(i)})\right) \tag{B.90}$$

and we define the optimal exerciser boundary C_n^* as

$$C_n^* = \alpha_1^*\phi_1 + \cdots + \alpha_k^*\phi_k \tag{B.91}$$

with

$$(\alpha_1^*, \ldots, \alpha_k^*) = \text{ArgMin}\left(\sum_{i=1}^{M}(\alpha_1\phi_1(x_n^{(j)}) + \cdots + \alpha_k\phi_k(x_n^{(j)}) - C_n^j)^2\right) \tag{B.92}$$

Step 2. Given the optimal exercise boundary $C_0^*, \ldots, C_N^*$, the price of the option is trivially given by:

$$P = B(T_0, T_N)\frac{1}{M}\sum_{i=1}^{M}\frac{I_{n(i)}(x_{n(i)}^{(i)})}{B(T_{n(i)}, T_N)}, \tag{B.93}$$

where

$$n(i) = \inf\left(p \geq 0 \Big/ \frac{I_p(x_p^{(i)})}{B(T_p, T_N)} \geq C_p^*(x_p^{(i)})\right). \tag{B.94}$$

B.2. Summary

In this appendix, we have seen how to improve the speed of Monte Carlo simulation, using control variates like antithetic control or sampling importance, we have also seen the different techniques employed to calculate the Greeks, among its methods, we have Malliavin calculus which is a strong method that can be applied for general models (see for more details Bally), and the new approach stochastic analysis which is a method that use the martingale methods to exhibit formula for the Greeks, in the last way, we have detailed two known methods to calculate American Monte Carlo price , using backward algorithm for optimal time exercise (Anderson and Longstaff).

References

[1] Bernis G, Gobet E and Kohatsu-Higa A (2003). Monte Carlo evaluation of Greeks for multidimensional barrier and lookback options. *Mathematical Finance*, **13**, 99–113.

[2] Duffy DJ (2004). *Financial Instrument Pricing Using C++*. New York: Wiley (Book and CD edition).

[3] Fournie E, Lasry J-M, Lebuchoux J, Lions PL and Touzi N (1999). Applications of Malliavin calculus to Monte Carlo methods in finance. *Finance and Stochastics*, **3**(4), 201–236. Available at http://springerlink.metapress.com/content/xodulvwve7vxgjlh/

[4] Glasserman P (2003). *Monte Carlo Methods in Financial Engineering* (*Stochastic Modeling and Applied Probability*), 1st edn. Berlin: Springer.

[5] Jaeckel P (2002). *Monte Carlo Methods in Finance*. New York: Wiley (Book and CD edition).

[6] London J (2004). *Modeling Derivatives in C++*. New York: Wiley (Book and CD-Rom edition).

[7] Pardoux E, Lapeyre B and Sentis R (1998). Méthodes de Monte Carlo pour les equations de transport et de diffusion. *Mathematiques Applications*, 29, 2.

[8] Press WH, Teukolsky SA, Vetterling WT and Flannery BP. *Numerical Recipes in C: The Art of Scientific Computing*, 2nd edn. Cambridge: Cambridge University Press.

[9] Shaw WT. *Modeling Financial Derivatives with Mathematica*. Cambridge: Cambridge University Press.

[10] Tavella D (2002). *Quantitative Methods in Derivatives Pricing: An Introduction to Computational Finance*, 1st edn. New York: Wiley.

Appendix C

TREE AND PDE METHODS

Across this appendix, we will show some numerical solutions for partial derivative equations: the tree approach, the finite difference schemes and the finite elements method.

C.1. Trees

C.1.1. *Arrow Debreu prices*

An arrow Debreu price is the price of a security that pays 1 at a given date and in a certain state of the world. Hence, once Arrow Debreu prices are known, the computation of the price of any European style security is straight forward: it is given by the sum of all payoffs balanced with arrow Debreu prices.

The way to find arrow Debreu prices in a tree is quite straightforward: it is given by the product of the transitional probabilities and the discount factors from the first node to the last node.

C.1.2. *Computing probabilities in a generated trees*

A tree is a representation of the evolution of the value of market variables for the purpose of valuing an option or another derivative. A tree can be defined by three things:

(1) a set of dates,
(2) a set of states for each date,

(3) transition probabilities (to move from one state at a certain date to another at another date).

When a tree is built, those three issues need to be solved. We will illustrate this through the building of the HW trinomial. As seen in Chapter 8, this model can be written as:

$$r_t = \varphi(t) + X_t^{(1)} + \cdots + X_t^{(n)}, \tag{C.1}$$

$$dX_t^{(k)} = -\lambda_k X_t^{(k)} dt + \sigma_k dW^{(k)}. \tag{C.2}$$

We will first describe how to build that tree for a 1-factor HW tree (this is the case of a one-dimensional tree n=1). We will write $(t_1, \ldots, t_p)$ the set of dates used to build the tree with Δt the time discretization, and Δx the space discretization of variable X for each date t_j. The process X_t will be approximated at date t_j by a discretized process x_j with $x_{kj} = k\Delta_1$. We will take three standard deviation as this is an optimal choice (it fulfills the third and fourth conditional moment when residual expectation is equal to 0).

$$\begin{aligned} \Delta_1 &= \sqrt{3^*\text{Var}(X_{t_j})}, \\ \Delta_1 &= \sigma\sqrt{3\Delta t}. \end{aligned} \tag{C.3}$$

As the tree is trinomial, at date t_i each node x_{ki} can branch to three consecutive nodes at date t_{i+1}. We will write h the index of the middle target node. See Figure C.1.

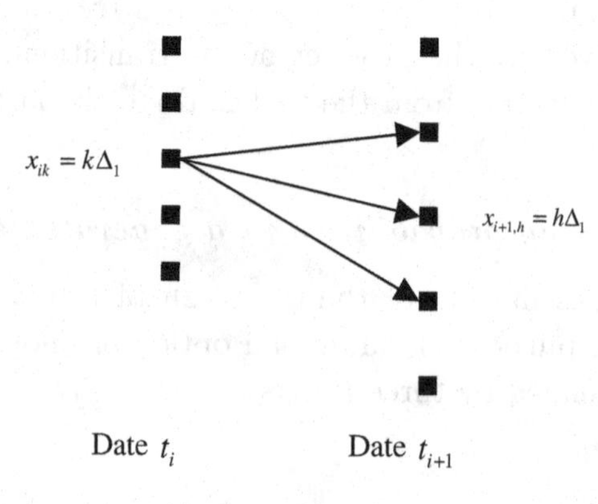

Figure C.1: Tree branching.

Moment matching between the two date imposes that we have:

$$E(X_{t_{i+1}}|X_{t_i} = x_{kj}) = E(x_{j+1}|x_j = x_{kj}), \tag{C.4}$$
$$\mathrm{Var}(X_{t_{i+1}}|X_{t_i} = x_{kj}) = \mathrm{Var}(x_{j+1}|x_j = x_{kj}). \tag{C.5}$$

In the present case, this means that we want:

$$\begin{aligned} x_{kj}\mathrm{e}^{-\lambda\Delta t} &= p_u x_{h+1,j+1} + p_m x_{h,j+1} \\ &\quad + p_d x_{h-1,j+1}, \end{aligned} \tag{C.6}$$

$$\begin{aligned} \int_{t_j}^{t_{j+1}} \sigma^2 \mathrm{e}^{-2\lambda(t_{j+1}-t)} dt &= p_u \left(x_{h+1,j+1} - x_{kj}\mathrm{e}^{-\lambda\Delta t}\right)^2 \\ &\quad + p_m \left(x_{h,j+1} - x_{kj}\mathrm{e}^{-\lambda\Delta t}\right)^2 \\ &\quad + p_d \left(x_{h-1,j+1} - x_{kj}\mathrm{e}^{-\lambda\Delta t}\right)^2. \end{aligned} \tag{C.7}$$

Moreover, we need to add that the sum of the probability is equal to one.

$$p_u + p_m + p_d = 1. \tag{C.8}$$

Here, there are three possible alternative branching processes at each node:

(1) Normal branching: we can move up by Δ_1, stay the same and move down by Δ_1.
(2) Up branching: we can move up by $2\Delta_1$, move up by Δ_1 and stay the same.
(3) Down branching: we can saty the same, move down by Δ_1 and move down by $2\Delta_1$.

The system of Eqs. (C.6) and (C.7) can then be solved. This leads to for normal branching

$$p_u = \frac{1}{6} + \frac{\eta^2}{2} + \frac{\eta}{2},$$

$$p_m = \frac{2}{3} - \eta^2,$$

$$p_d = \frac{1}{6} + \frac{\eta^2}{2} - \frac{\eta}{2},$$

for up branching

$$p_u = \frac{1}{6} + \frac{\eta^2}{2} - \frac{\eta}{2},$$

$$p_m = -\frac{1}{3} - \eta^2 + 2\eta,$$

$$p_d = \frac{7}{6} + \frac{\eta^2}{2} - \frac{3\eta}{2},$$

for down branching

$$p_u = \frac{7}{6} + \frac{\eta^2}{2} + \frac{3\eta}{2},$$

$$p_m = -\frac{1}{3} - \eta^2 - 2\eta, \tag{C.9}$$

$$p_d = \frac{1}{6} + \frac{\eta^2}{2} + \frac{\eta}{2},$$

where η is defined as the difference in space step for normal branching, for up branching, for down branching between the new and old node:

$$\eta = \frac{(x_{h,j+1} - x_{kj}\mathrm{e}^{-\lambda\Delta t})}{\Delta x}. \tag{C.10}$$

It is worth noticing that having probabilities between 0 and 1 imposes at least to have

$$-\sqrt{\frac{2}{3}} \leq \eta \leq \sqrt{\frac{2}{3}}. \tag{C.11}$$

At this stage, we can notice that we have built a tree for a zero-centered Ornstein–Uhlenbeck process, but not for the short rate.

In order to change this tree into a short rate tree, a simple solution is to apply a shift α_i at each slice of the tree so as to fit the zero-coupon curve. To do this, let us introduce a few notations: we will write Q_{ik} the arrow Debreu price of a security that pays 1 at date t_i in the state of the world corresponding to node k. We will also write $q_i(k, h)$ the probability to move from node k at date t_i to node h at date t_{i+1}. We have the following recursive relation between Q_{ik}:

$$Q_{i+1,p} = \sum_k Q_{ik} q(k,p) \mathrm{e}^{-\Delta t(x_{ik}+\alpha_i)}. \tag{C.12}$$

Moreover Q_{ik} can be used to compute zero coupons:

$$\begin{aligned} B(0,t_{i+1}) &= \sum_h 1 \times Q_{i+1,h} \\ &= \sum_h \sum_k Q_{ik} q(k,h) \mathrm{e}^{-\Delta t(x_{ik}+\alpha_i)} \\ &= \sum_k Q_{ik} \left(\sum_h q(k,h) \right) \mathrm{e}^{-\Delta t(x_{ik}+\alpha_i)} \\ &= \sum_k Q_{ik} \times 1 \times \mathrm{e}^{-\Delta t(x_{ik}+\alpha_i)}. \end{aligned} \tag{C.13}$$

These two relations can be used as follows:

$$\alpha_i = \frac{1}{\Delta t} \ln \frac{\sum_k Q_{ik} \mathrm{e}^{-x_{ik}\Delta t}}{B(0,t_{i+1})}, \tag{C.14}$$

which is used to find the α_is once the Q_{ik}s are known. The recursive relationship over Q_{ik} is used to find them out with already computed α_is. The tree changes according to the following way:

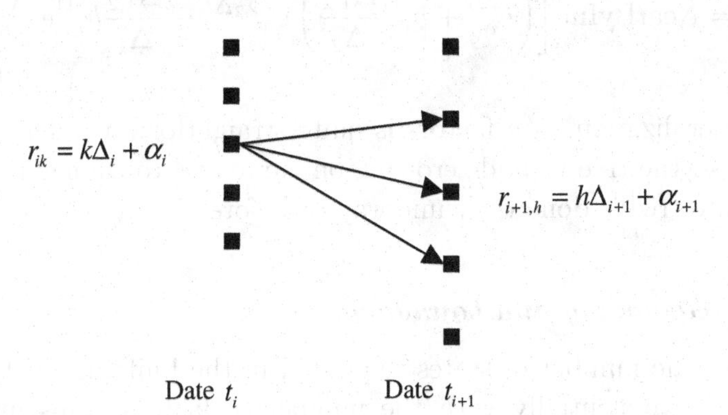

Let us now see how to build a tree in the two-factors case. First, we introduce some notations:

$$\Delta_{1,2} = \sqrt{3^*\mathrm{Cov}(X^{(1)}, X^{(2)})} \tag{C.15}$$

and we will introduce the processes

$$x_i = (x_i^{(1)}, x_i^{(2)}), \tag{C.16}$$

where $x_i^{(j)}$ represents the value at date t_i of the process, $X_t^{(j)}$. We will write

$$x_i^{(1)} = h_i^{(1)} \Delta_1 \tag{C.17}$$

and

$$x_i^{(2)} = h_i^{(1)} \Delta_{1,2} + h_i^{(2)} \Delta_2 \tag{C.18}$$

Let us see how to move from a state x_i to a state x_{i+1}. To do so, we have to find out all the coordinates of state x_{i+1}. The same kind of reasoning as before shows us that

$$h_{i+1}^{(1)} = \text{NearByInt}\left[h_i^{(1)} \mathrm{e}^{-\lambda^{(1)} \Delta t}\right] \tag{C.19}$$

and

$$h_{i+1}^{(2)} = \text{NearByInt}\left[\left(h_i^{(2)} + h_i^{(1)} \frac{\Delta_{1,2}}{\Delta_2}\right) \mathrm{e}^{-\lambda_2 \Delta t} - \frac{\Delta_{1,2}}{\Delta_1} h_i^{(1)} \mathrm{e}^{-\lambda_1 \Delta t}\right]. \tag{C.20}$$

The generalization to n factors is quite straightforward. The recalibration of the tree to find zero-coupon curve and to change it into a short rate tree is done the same way as before.

C.1.3. *Geometry and boundaries*

Roughly, the number of states requested in the building of the tree increases exponentially with the number of factors. This involves high computation time. To avoid this, tree-truncation methods can be used. It consists in arbitrarily preventing the tree from increasing. Most of the time, this is done by setting a limit to the number of plausible states that can be reached. We can see this in the HW tree example we are studying in this appendix. We wrote

$$x_i^{(j)} = \sum_{k=1}^{j} A_{j,k} N_k, \tag{C.21}$$

where N_k are independent standard Gaussian laws. We can also write

$$N_i = \sum_{k=1}^{j} B_{i,k} X_k. \tag{C.22}$$

As X_k are approximated by $x_k = \sum_{i=1}^{k} \Delta_{k,i} j_i$, we can then write

$$n_i = \sum_{k=1}^{i} C_{i,k} j_k. \tag{C.23}$$

It is often considered that standard Gaussian laws rarely move beyond n times its standard deviation, and hence limitation of the tree growth is given by

$$|n_i| \leq n_\sigma. \tag{C.24}$$

Hence, we get

$$|j_i| \leq \frac{n_\sigma + \sum_{k=1}^{i-1} C_{i,k} j_k}{C_{i,i}} \leq \frac{n_\sigma + \sum_{k=1}^{i-1} C_{i,k} j_k^{\max}}{C_{i,i}}. \tag{C.25}$$

The tree is then built as usual for states that involve

$$|j_i| \leq j_i^{\max}. \tag{C.26}$$

In fact, two cases may happen:

(1) *Central target node is on the edge of the tree*: The variance condition can no longer be matched. The current node connected with two destination nodes only, and we have (in case, for instance, current node is on the top of the tree)

$$p_m = 1 - \eta, \quad p_d = \eta. \tag{C.27}$$

with the notations defined above.

(2) *Central target node is out of bound*: Then connection is straight (with probability one) to the closest node that is in bound.

C.1.4. *Smoothing*

Smoothing of a tree is a way to deal with the problem of irregularities of a payoff. To understand how this methodology works, let us address the following problem: the pricing of a call in a tree.

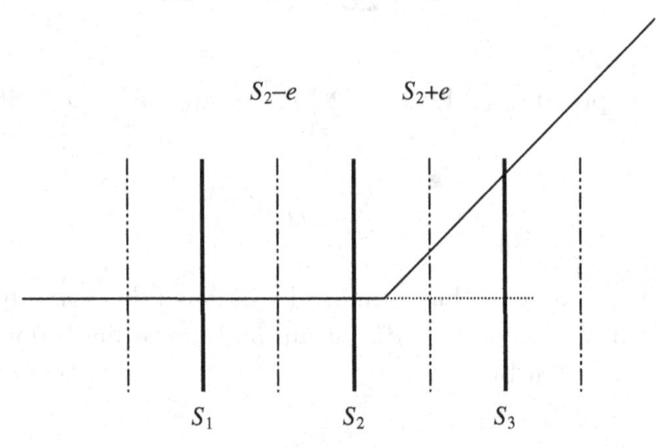

S_1, S_2, S_3 stands for values of the nodes of the tree at current date. The value of the option is given by the mean of its values at each node weighted by the probability to access those nodes. In the present case, this leads to an underestimation of the price as proceeding this way contains the implicit assumption that payoff is constant and equal to 0 on interval $[S_2 - e, S_2 + e]$.

A way to solve this is to consider that instead of a pay-off that equals to 0 the pay-off equals the surface of the small forgotten triangle on the right divided by 2e.

C.2. Partial Differential Equations

C.2.1. *Motivation*

Let S_t be a stock following a lognormal process:

$$\frac{dS}{S} = rdt + \sigma dW_t. \tag{C.28}$$

The value at time t of a European call maturing at time T is given by

$$C_t = E[e^{-r(T-t)}(S_T - K)^+|S_t]. \tag{C.29}$$

Let us consider a portfolio π made up with a call C_t and with Δ stocks such that it is perfectly hedged against spot price movements. The value of this portfolio is given by

$$\pi_t = C_t + \Delta S_t. \tag{C.30}$$

As it is perfectly hedged, we have that

$$\begin{aligned} d\pi_t &= \left(\frac{\partial C}{\partial t} + \frac{1}{2}\frac{\partial^2 C}{\partial S^2}\sigma^2 S^2\right) dt \\ &\quad + \frac{\partial C}{\partial S}S(rdt + \sigma dW) + \Delta S(rdt + \sigma dW) \\ &= \left(\frac{\partial C}{\partial t} + \frac{1}{2}\frac{\partial^2 C}{\partial S^2}\sigma^2 S^2\right) dt. \end{aligned} \tag{C.31}$$

as we must take

$$\Delta = -\frac{\partial C}{\partial S} \tag{C.32}$$

so as to cancel the source incertitude. Due to non-arbitrage reasons, the income of this portfolio must be the same as any risk-free asset, namely we have:

$$\frac{d\pi}{\pi} = rdt, \tag{C.33}$$

where r is the risk free interest rate. This gives birth to the following equation:

$$\frac{\partial C}{\partial t} + \frac{1}{2}S^2\sigma^2\frac{\partial^2 C}{\partial S^2} + rS\frac{\partial C}{\partial S} - rC = 0 \tag{C.34}$$

with limit condition:

$$C_T(S) = (S - K)^+. \tag{C.35}$$

Remark C.1: This equation, known as the BS PDE can also be obtained through direct application of the Feynman–Kac theorem (see Appendix A).

Remark C.2: If we make a variable change t into $T - t$, the problem changes into a problem with limit condition at $t = 0$.

So, any issue of derivatives pricing in finance can be changed into a PDE solving issue. During the rest of this appendix, we will concentrate on a general PDE of the form:

$$a(C,x,t)\frac{\partial^2 C}{\partial x^2} + b(C,x,t)\frac{\partial C}{\partial x} + c(C,x,t) = \frac{\partial C}{\partial t}, \quad C(x,0) = f(x). \tag{C.36}$$

C.2.2. *Finite difference*

So, let us consider the following PDE:

$$\frac{\partial C}{\partial t} = a(C,x,t)\frac{\partial^2 C}{\partial x^2} + b(C,x,t)\frac{\partial C}{\partial x} + c(C,x,t) \tag{C.37}$$

with initial conditions:

$$C(x,0) = f(x), \tag{C.38}$$

which we would like to solve for $t \in [0,T]$.

There is no known general closed form for that problem. That is why it has to be solved numerically. A well-known numerical method to solve that problem is the finite difference method. It consists in finding approximate values of $C(x,t)$ for $(x,t) \in [0, K\Delta x] \times [0, N\Delta T]$.

Approximated values of $C(x_i, t_n)$ is written C_i^n. The approximation consists in writing:

$$\frac{\partial^2 C}{\partial x^2}(x_i,t_n) \approx \frac{C(x_{i+1},t_n) + C(x_{i-1},t_n) - 2C(x_i,t_n)}{\Delta x^2} \approx \frac{C_{i+1}^n + C_{i-1}^n - 2C_i^n}{\Delta x^2}, \tag{C.39}$$

$$\frac{\partial C}{\partial x}(x_i,t_n) \approx \frac{C(x_{i+1},t_n) - C(x_i,t_n)}{\Delta x} \approx \frac{C_{i+1}^n - C_i^n}{\Delta x}, \tag{C.40}$$

$$\frac{\partial C}{\partial t}(x_i,t_n) \approx \frac{C(x_i,t_{n+1}) - C(x_i,t_n)}{\Delta t} \approx \frac{C_i^{n+1} - C_i^n}{\Delta t}. \tag{C.41}$$

Hence, the PDE changes into a plain relationship linking C_i^n together:

$$\frac{C_i^{n+1} - C_i^n}{\Delta t} = a(C_i^n, x_i)\frac{C_{i+1}^n + C_{i-1}^n - 2C_i^n}{\Delta x^2} + b(C_i^n, x_i)\frac{C_{i+1}^n - C_{i-1}^n}{\Delta x} + c(C_i^n, x_i)C_i^n, \quad (C.42)$$

with

$$C_i^0 = f(x_i). \quad (C.43)$$

Those relationships can also be written in summarized way as:

$$C^{n+1} = (I + \Delta t M^n)C^n. \quad (C.44)$$

This way, the computation is quite straight forward: C^0 is given by initial conditions (payoff of the option), and C^{n+1} is computed from C^n through matrix multiplication. The solution of the problem is then given by the element of C^Ncorresponding the state of the world at time $t = 0$. This finite difference scheme is known as "explicit finite difference scheme".

There exists namely two other schemes, depending on the way $\partial^2 C/\partial x^2$ and $\partial C/\partial x$ are approximated. One of those schemes, known as "implicit finite difference scheme", is obtained for

$$\frac{\partial C}{\partial x} \approx \frac{C_i^{n+1} - C_i^n}{\Delta x} \quad (C.45)$$

And we approximate the second-order derivative by the self-centered scheme given by:

$$\frac{\partial^2 C}{\partial x^2}(x_i, t_n) \approx \frac{C_{i+1}^{n+1} + C_{i-1}^{n+1} - 2C_i^{n+1}}{\Delta x^2}. \quad (C.46)$$

Summarizing this leads to

$$C^{n+1} - C^n = \Delta t M^{n+1} C^{n+1} \quad (C.47)$$

or equivalently

$$C^{n+1} = (I - \Delta t M^{n+1})^{-1} C^n. \quad (C.48)$$

The latter numerical scheme is named "implicit" due to the fact that in Eq. (C.48), the right term depends on the terms we are trying to compute.

C.2.3. *Crank Nicholson*

If we make a linear combination of the two preceding relationships with a certain weight $\theta \in [0,1]$, we get for the first direction:

$$C^{n+1} - C^n = \Delta t(((1-\theta)M^{n+1}C^{n+1} + \theta M^n C^n) \tag{C.49}$$

for $\theta \in [0,1]$. And we have also:

$$(I - (1-\theta)\Delta t M^{n+1})C^{n+1} = (I + \theta \Delta t M^n)C^n. \tag{C.50}$$

This numerical scheme is named θ scheme. We can notice that for $\theta = 0$, we get the implicit scheme while $\theta = 1$ changes it into an explicit scheme.

If we take $\theta = 1/2$, we get a numerical scheme named the Crank–Nicholson scheme. The systems to solve then becomes:

$$\left(I - \frac{1}{2}\Delta t M^{n+1}\right)C^{n+1} = \left(I + \frac{1}{2}\Delta t M^n\right)C^n. \tag{C.51}$$

We can notice that for $\theta \neq 0$, solving the theta scheme involves linear system solving. We will see later the differences between the three numerical schemes described above.

In order to solve this problem many well-known methods, such as SSOR, can be used. However, these methods are beyond the scope of this book.

C.2.4. *Stability and robustness*

The purpose of this section is to give criteria that shows how well-numerical schemes converge. To realize this aim, we need to introduce three concepts, namely consistency, stability and convergence.

First introduce some notations: We will write H the finite difference operator, and L the original PDEs operator.

Such that the solution of the problem is given by $H(u) = 0$ and $L(u) = 0$. In the example of the Heat equation, we have:

$$H(u) = \frac{u(x, t+\Delta t) - u(x,t)}{\Delta t} - c\frac{u(x+\Delta x, t) + u(x-\Delta x, t) - 2u(x,t)}{\Delta x^2}. \quad \text{(C.52)}$$

This leads to

$$L(u) = \frac{\partial u}{\partial t} - c\frac{\partial^2 u}{\partial x^2}. \quad \text{(C.53)}$$

A finite difference representation of a PDE is said to be consistent if the difference between the PDE and the numerical scheme (ie the truncation error) tends to zero as grid interval (discretization in space and time) approach zero.

Consistency deals with how well the finite difference scheme approximates the underlying PDE. With the previous notations, consistency means that

$$H(u) - L(u) \to 0 \quad \text{(C.54)}$$

as Δt and Δx tends by upper value to 0.

We say that a numerical scheme has order of accuracy

$$O(\Delta t^p + \Delta x^q) \quad \text{(C.55)}$$

if there exists

$$(C_1, C_2) \in \Re^2_{+*} \quad \text{(C.56)}$$

such that

$$|H(u) - L(u)| \leq C_1 \Delta t^p + C_2 \Delta x^q. \quad \text{(C.57)}$$

A numerical scheme is said to be stable if an error from one source will not grow unboundedly with time. A numerical scheme is said to be convergent if it approaches the true solution of the PDE as both time and space discretization tends to zero.

Lax's Equivalence theorem states that a finite difference scheme is convergent if and only if it is both stable and consistent.

There are two main differences between the numerical schemes exposed above (implicit, explicit and Crank–Nicolson):

The first one lies in their order of accuracy: implicit and explicit schemes have order $O(\Delta t + \Delta x^2)$ whereas Crank–Nicolson has order $O(\Delta t^2 + \Delta x^2)$. Since the space approximation has same order for all these schemes, it makes sense to say Crank–Nicolson is second-order accurate whereas explicit and implicit schemes are first-order accurate.

Another difference between all those methods is that the explicit scheme is stable at certain conditions only given by:

$$\frac{2c\Delta t}{\Delta x^2} < 1, \tag{C.58}$$

which is usually referred to as the Courant–Friedrich–Levy condition. The implicit and the Crank–Nicolson methods are unconditionally stable and convergent.

C.2.5. *Finite differences schemes for multi-dimensional PDE: alternating direction iterative (ADI)*

In this section, we will treat the case of two-dimensional PDEs; such a case is useful for pricing with multifactor models.

Any second-order diffusion PDE can be set back to a PDE of the form using simple variable changes:

$$\frac{\partial C}{\partial t} = a(C, x, y)\frac{\partial^2 C}{\partial x^2} + b(C, x, y)\frac{\partial^2 C}{\partial y^2}. \tag{C.59}$$

For simplicity reasons we will show how to solve the following PDE (generalization to PDE of the previous form is quite straight forward):

$$\frac{\partial C}{\partial t} = \frac{\partial^2 C}{\partial x^2} + \frac{\partial^2 C}{\partial y^2}. \tag{C.60}$$

In the same way as before, this PDE will be solved on $[0, K\Delta x] \times [0, L\Delta y] \times [0, N\Delta t]$ with:

$$C(x_i, y_j, t_n) = C^n_{i,j}. \tag{C.61}$$

As in the one-dimensional case, we discretize the derivatives with finite difference:

$$\begin{aligned}\frac{\partial^2 C}{\partial x^2} &\approx \frac{C^n_{i+1,j} + C^n_{i-1,j} - 2C^n_{i,j}}{\Delta x^2},\\ \frac{\partial^2 C}{\partial y^2} &\approx \frac{C^n_{i,j+1} + C^n_{i,j-1} - 2C^n_{i,j}}{\Delta y^2},\end{aligned} \tag{C.62}$$

with

$$C^n = (C^n_{1,1}, C^n_{2,1}, \ldots, C^n_{k,1}, C^n_{1,2}, \ldots, C^n_{k,l})^T \tag{C.63}$$

So, we have

$$C^n_{i,j} = C_{j\cdot k+i}. \tag{C.64}$$

Hence, our finite difference scheme changes our PDE to:

$$\begin{aligned}\frac{C^{n+1} - C^n}{\Delta t} = \big(\theta\big(A^n + B^n\big)C^n \\ + (1-\theta)\big(A^{n+1} + B^{n+1}\big)C^{n+1}\big)\end{aligned} \tag{C.65}$$

or

$$\big(I - (1-\theta)\Delta t\big(A^{n+1} + B^{n+1}\big)\big)C^{n+1} = \big(I + \theta\Delta t\big(A^n + B^n\big)\big)C^n, \tag{C.66}$$

where

$$A^n = \frac{1}{\Delta y^2}\begin{pmatrix} -2 & 0 & \cdots & 0 & 1 & 0 & \\ 0 & \ddots & & & & \ddots & 0 \\ \vdots & & & & & & 1 \\ 0 & & & \ddots & & & 0 \\ 1 & & & & & & \vdots \\ 0 & \ddots & & & & \ddots & 0 \\ & 0 & 1 & 0 & \cdots & 0 & -2 \end{pmatrix} \tag{C.67}$$

and

$$B^n = \frac{1}{\Delta x^2} \begin{pmatrix} -2 & 1 & 0 & \cdots & & 0 \\ 1 & -2 & 1 & & & \\ 0 & 1 & \ddots & & & \vdots \\ & & & & & \\ \vdots & & & & 1 & 0 \\ & & & 1 & -2 & 1 \\ 0 & & \cdots & 0 & 1 & -2 \end{pmatrix}. \quad \text{(C.68)}$$

In the mono-dimensional case, the matrix that needed to be inverted was tri-diagonal (like B^n), which is quite fast to compute, whereas in the present case, $A^n + B^n$ has 5-diagonals filled. This case is not as fast to compute. That is why there is the ADI algorithm.

We can notice that A^n and B^n do not depend on n in the present case (through they do in the general case of equation). Hence we can write

$$A = \frac{\Delta t}{2} A^n \quad \text{(C.69)}$$

and

$$B = \frac{\Delta t}{2} B^n. \quad \text{(C.70)}$$

This gives for $\theta = 1/2$ the following equation

$$(I - (A + B))C^{n+1} = (I + (A + B))C^n. \quad \text{(C.71)}$$

This can be rewritten as:

$$(I - A)(I - B)C^{n+1} = (I + A)(I + B)C^n + AB(C^{n+1} - C^n) \quad \text{(C.72)}$$

with

$$AB(C^{n+1} - C^n) = O(\Delta t^3). \quad \text{(C.73)}$$

As solving of the problem is done with $O(\Delta t^2)$, we can consider this term negligible and come up with the following expression:

$$(I - B)(I - A)C^{n+1} = (I + B)(I + A)C^n. \quad \text{(C.74)}$$

That last problem is equivalent to:

$$(I - B)C^{n+1/2} = (I + A)C^n \tag{C.75}$$

and

$$(I - A)C^{n+1} = (I + B)C^{n+1/2}. \tag{C.76}$$

This way, there are two linear systems that need to be solved. The first one involves a tri-diagonal matrix $(I - B)$, which is quite straight forward.

The second one involves manipulation of a much more complicated matrix. However, there is a method to handle the latter problem in a quite easy way: This method involves vector basis changes. It consists in finding a vector basis such that A can be expressed as a tri-diagonal matrix.

More precisely A^n would be expressed as

$$A = \frac{1}{\Delta y^2}\begin{pmatrix} -2 & 1 & 0 & \cdots & & & 0 \\ 1 & -2 & 1 & & & & \\ 0 & 1 & \ddots & & & & \vdots \\ & & & & & & \\ \vdots & & & & & 1 & 0 \\ & & & & 1 & -2 & 1 \\ 0 & & & \cdots & 0 & 1 & -2 \end{pmatrix}. \tag{C.77}$$

It can be easily shown that if C^n was defined as:

$$C^n = (C^n_{1,1}, C^n_{1,2}, \ldots, C^n_{1,l}, C^n_{2,1}, \ldots, C^n_{k,l})^T \tag{C.78}$$

instead of the way it is defined above, A would be tri-diagonal and B difficult to inverse.

Hence, to simplify our problem, we can permute lines of vector $(I + B)C^{n+1/2}$, which will give birth to a new vector that we will write $P(I + B)C^{n+1/2}$, and the problem we have to solve is

$$(I - A't)C'^{n+1} = P(I + B)C^{n+1/2}, \quad \text{(C.79)}$$

where the variable to find is C'^{n+1} and A' is the tri-diagonal (and hence easy to inverse) matrix shown above. And then we only have to permute lines of C'^{n+1} the other way round to find C^{n+1}.

More rigorously, if we write P the permutation matrix that changes

$$C = (C_{1,1}, C_{2,1}, \dots, C_{k,1}, C_{1,2}, \dots, C_{k,l}^{n})^T \quad \text{(C.80)}$$

into

$$C' = (C_{1,1}, C_{1,2}, \dots, C_{1,l}, C_{2,1}, \dots, C_{k,l}^{n})^T \quad \text{(C.81)}$$

for every $K \times L$-uplets

The equation

$$(I - A)C^{n+1} = (I + B)C^{n+1/2} \quad \text{(C.82)}$$

can then be rewritten as

$$(P(I - A)P^{-1})(PC^{n+1}) = P(I + B)C^{n+1/2}, \quad \text{(C.83)}$$

which can be reformulated as:

$$(I - A')(PC^{n+1}) = P(I + B)C^{n+1/2}. \quad \text{(C.84)}$$

This last equation can be solved in a straight-forward manner; since this is once again a linear system.

C.2.6. *Finite elements*

Another way to solve a PDE is to use finite elements numerical schemes. For each $t \in [0, T]$, $x \mapsto C(x, t)$ is approximated by

$$\tilde{C}(x, t) = \sum_{i=0}^{n+1} \alpha_i(t)\beta_i(x). \quad \text{(C.85)}$$

To do this, we have to reformulate the problem in its weak formulation. That is, solving the PDE described above is equivalent to finding $C(x,t)$ such as for any function $f(x)$:

$$-\int \frac{\partial f}{\partial x}\frac{\partial C}{\partial x}dx + \int f(x)b(x,t)\frac{\partial C}{\partial x}dx + \int f(x)C(x,t)dx = \int f(x)\frac{\partial C}{\partial t}dx. \quad \text{(C.86)}$$

As we are trying to find $\tilde{C}(x,t)$, this equation leads to:

$$\sum_{i=0}^{i=n+1} -\int \frac{\partial \beta_j}{\partial x}(x)\alpha_i(t)\frac{\partial \beta_i}{\partial x}(x)dx + \int \beta_j(x)b(x,t)\alpha_i(t)\frac{\partial \beta_i}{\partial x}(x)dx + \int \beta_j(x)\beta_i(x)\alpha_i(t)dx = \sum_{i=0}^{i=n+1}\frac{\partial \alpha_i}{\partial t}\int \beta_j(x)\beta_i(x)dx. \quad \text{(C.87)}$$

So that we have to solve:

$$A\dot{\alpha} = B\alpha, \quad \text{(C.88)}$$

where A is a matrix defined by

$$A_{ij} = \int \beta_j(x)\beta_i(x)d \quad \text{(C.89)}$$

and

$$B_{ij} = -\int \frac{\partial \beta_j}{\partial x}(x)\frac{\partial \beta_i}{\partial x}(x)dx + \int \beta_j(x)b(x,t)\frac{\partial \beta_i}{\partial x}(x)dx + \int \beta_j(x)\beta_i(x)dx \quad \text{(C.90)}$$

We can see that the PDE to solve becomes an n-dimensional linear ODE which is quite straight forward to compute.

The β_i are chosen to make the matrix A a tridiagonal matrix, and are given by:

$$\beta_j(x) = \begin{cases} \dfrac{x - x_{j-1}}{x_j - x_{j-1}}, & \text{if } x_{j-1} \leq x < x_j, \\ \dfrac{x_{j+1} - x}{x_{j+1} - x_j}, & \text{else if } x_j \leq x < x_{j+1}, \\ 0, & \text{else.} \end{cases} \quad \text{(C.91)}$$

This method has the advantage to be dimension-free, in the sense that to solve a PDE on $\Re^n$, the only thing to do is to change the basis of functions used for solving into such a set defined on $\Re^n$. The methodology remains the same.

C.3. Summary

In this chapter, we had seen how to build a tree for a model like the HW, using this building, we have then shown how to price some interest rate derivatives. Besides this, we detail another numerical method for PDE, which include the trees approach: the finite difference method and we describe in dimension one three known schemes: the implicit and the Crank–Nicolson ones which are unconditionally stable and convergent, but the explicit scheme is stable under some conditions. After this, we explain the ADI method in dimension two: it is an extension of Crank–Nicolson scheme in higher dimensions. Finally, we describe the finite elements method which have the advantage to be a dimension free one.

References

[1] Clewlow L and Strickland C (1998). *Implementing Derivative Models.* New York: Wiley.
[2] Duffy DJ (2004). *Financial Instrument Pricing Using C++*. New York: Wiley (Book and CD edition).
[3] Kloeden PE and Platen E (1995). *Numerical Solution of Stochastic Differential Equations*, Applications of Mathematics, Vol 23, Springer.
[4] London J (2004). *Modeling Derivatives in C++*. New York: Wiley (Book and CD-Rom edition).
[5] Morton KW and Mayers DF (1994). *Numerical Solution of Partial Differential Equations.* Cambridge: Cambridge University Press.
[6] Tavella D (2002). *Quantitative Methods in Derivatives Pricing: An Introduction to Computational Finance*, 1st edn. New York: Wiley.

Index